INSTRUCTION & ANSWER GUIDE

BUILDING THINKING SKILLS®

Book 1

SERIES TITLES:

BUILDING THINKING SKILLS® HANDS-ON PRIMARY
BUILDING THINKING SKILLS® BOOK 1
BUILDING THINKING SKILLS® BOOK 2
BUILDING THINKING SKILLS® BOOK 3 FIGURAL
BUILDING THINKING SKILLS® BOOK 3 VERBAL

SANDRA PARKS AND HOWARD BLACK

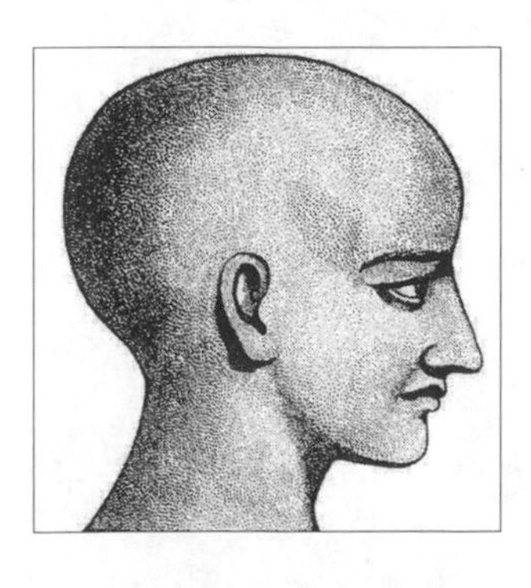

THE CRITICAL THINKING CO.
(BRIGHT MINDS™)
www.CriticalThinking.com
P.O. Box 1610 • Seaside • CA 93955-1610
Phone 800-458-4849 • FAX 831-393-3277
ISBN 0-89455-320-8

Table of Contents

INTRODUCTION

Program Design vii
Types of Instructional Methods ix
Rationale and Description of Skills xii
Evaluating Building Thinking Skills Instruction xv
Evaluation Recommendations xvii
Instructional Recommendations xviii
Vocabulary and Synonyms xix
Guide to Using the Lesson Plans xxii

LESSON PLANS

CHAPTER ONE—DESCRIBING SHAPES

Describing Shapes—Select 1
Describing Shapes—Explain 3
Following Directions 5
Writing Directions 7
Describing Position—A 9
Describing Position—B 11
Characteristics of a Shape 13

CHAPTER TWO—FIGURAL SIMILARITIES AND DIFFERENCES

Matching Shapes 15
Which Shape Does Not Match?/Matching Shapes That Have Been Turned 17
Finding Shapes 20
Combining Interlocking Cubes 22
Combining Shapes 24
Finding and Tracing Patterns 27
Dividing Shapes into Equal Parts—A 29
Dividing Shapes into Equal Parts—B 31
Dividing Shapes into Equal Parts—C 33
Which Shape Completes the Square? 35
Which Shapes Make Squares? 37
Copying a Figure/Drawing Identical Shapes 39
Comparing Shapes—Select 41
Comparing Shapes—Explain 42

CHAPTER THREE—FIGURAL SEQUENCES

Copying a Pattern 45
Which Color Comes Next?—Select 46
Which Shape Comes Next?—Select 47
Tumbling 50
Which Figure Comes Next?—Select (Flips) 52
Which Figure Comes Next?—Select (Changes in Detail) 56
Which Figure Comes Next?—Draw It! 59
Describe a Sequence/Sequence of Polygons and Angles 62
Paper Folding—Select 64
Paper Folding—Draw 67

CHAPTER FOUR—FIGURAL CLASSIFICATIONS

Match a Shape to a Group 71
Match a Pattern to a Group 73
Select a Shape that Belongs to a Group 74
Describing Classes 77
Matching Classes by Shape/Pattern 80
Classifying by Shape—Find the Exception 82
Classifying by Pattern—Find the Exception 85
Classifying More than One Way 87
Classifying by Color/Shape—Sorting 89
Complete the Class 91
Form a Class 93
Draw Another 95
Classifying by Shape/Color/Size—Sorting 97
Overlapping Classes—Intersection 99
Overlapping Classes—Matrix 103
Writing Descriptions of Classes 105

CHAPTER FIVE—FIGURAL ANALOGIES

Analogies with Shapes 107
Analogies with Shapes—Complete 110

CHAPTER SIX—DESCRIBING THINGS

Describing Things—Select 115
Describing Things—Explain 118
Describing Words—Select 120
Describing Words—Supply/Identifying Characteristics 122

CHAPTER SEVEN—VERBAL SIMILARITIES AND DIFFERENCES

Opposites—Select (pictures and words) 125
Opposites—Select (words) 130
Opposites—Supply 135
Similarities—Select (pictures and words) 139
Similarities—Select (words) 142
Similarities—Supply 147
How Alike?—Select 149
How Alike and How Different? 151
Compare and Contrast—Graphic Organizer 154

CHAPTER EIGHT—VERBAL SEQUENCES

Following Yes-No Rules—A 159
Writing Yes-No Rules 162
Following Yes-No Rules—B 164
Completing True-False Tables 166
Finding Locations on Maps 169
Describing Locations—A 171
Describing Directions—A/Locations—B 174
Describing Directions—B 176
Select the Word that Continues the Sequence 178
Ranking 185
Supply a Word that Continues a Sequence 191
Warm-up Deductive Reasoning 197
Deductive Reasoning 199
Ranking Time/Length Measures 203
Ranking in Geography 203
Flowchart—Arithmetic 205

CHAPTER NINE—VERBAL CLASSIFICATIONS

Parts of a Whole—Select 209
Parts of a Whole—Graphic Organizer 213
Class and Members—Select 215
What Is True of Both Words?—Select 219
How Are These Words Alike?—Select 221
How Are These Words Alike?—Explain 223
Explain the Exception 225
Picture Dictionary—Sorting into Classes 228
Word Classes—Select/Sorting by Class and Size 232
Recognizing Classes—Graphic Organizer 234

CHAPTER TEN—VERBAL ANALOGIES

Picture Analogies—Select 237
Picture Analogies—Name the Relationship 239
Analogies—Select 241
Analogies—Supply 244
Analogies—Select the Right Pair 246

TRANSPARENCY MASTERS

Transparency Masters 249

INTRODUCTION

PROGRAM DESIGN

Skills in *Building Thinking Skills*

- **Figural Skills**: describing shapes, figural similarities and differences, figural sequences, figural classifications, figural analogies
- **Verbal Skills**: describing things, verbal similarities and differences, verbal sequences, verbal classifications, verbal analogies

The five cognitive skills developed in this series (describing, finding similarities and differences, sequencing, classifying, and forming analogies) were selected because of their prevalence and relevance in academic disciplines. These analysis skills are required in all content areas, including the arts. Since improved school performance is an important goal of thinking skills instruction, many variations of each of these thinking skills are demonstrated in *Building Thinking Skills* exercises.

These five analysis skills are sequenced in the same order that a child develops intellectually. A child first learns to observe and describe objects, recognizing the characteristics of an object and then distinguishing similarities and differences between the objects. These skills are integral to a learner's ability to put things in order, to group items by class, and to think analogically. Each analysis skill is presented first in the concrete figural form and then in the abstract verbal reasoning form.

Teaching Options

The teacher may select one of three alternatives for teaching the activities: (1) teaching skills in the order in which they are presented in the student book, i.e., completing the figural exercises first and then the verbal exercises; (2) alternating between figural and verbal forms of each skill; or (3) scheduling the thinking skills exercises as they occur in content objectives. Using any of these alternatives, the exercises dealing with *description* and *similarities and differences* in figural and verbal forms should be offered before more complex ones.

The following flowchart illustrates the first two teaching options:

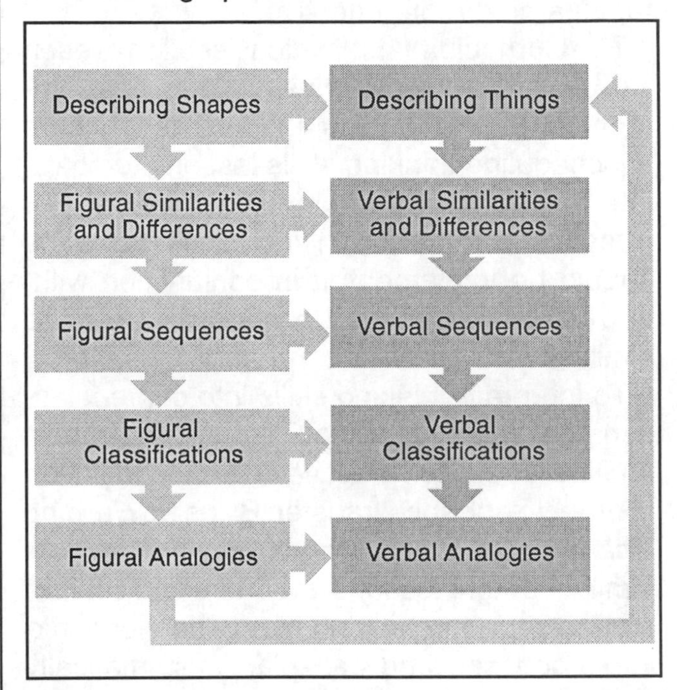

In the first two options, *Building Thinking Skills* is used as a structured program. Students are offered considerable explanation of and practice in skillful thinking and are able to recognize their growing competence in carrying out the thinking tasks required in school. The careful sequencing of the complexity of the exercises and the repeated practice of consecutive lessons produces greater cognitive development gains than spreading the skill exercises over the academic year to correlate with similar content lessons. For special education instruction, compensatory education classes, or bilingual programs, teaching thinking skills sequentially produces better results on tests of cognitive ability, vocabulary, and achievement than correlating *Building Thinking Skills* activities to content objectives does.

The third teaching option involves linking thinking skills instruction to content objectives. By identifying thinking skills in the curriculum, one can offer a structured thinking skills program tailored for teaching content objectives to district needs rather than offering a "packaged program." This articulated program allows supervisors to identify and evaluate instruction, but is accepted by teachers because thinking skills instruction makes content learning more effective.

The Curriculum Applications section of each *Building Thinking Skills* lesson identifies common content objectives and assists teachers in scheduling thinking skills lessons to correlate with appropriate curriculum material. For effective transfer, thinking skills activities should be implemented in conjunction with, and just prior to, content lessons which feature similar processes.

To integrate thinking skills into content lessons, teachers must provide sufficient opportunities for explicit practice and metacognition in order to enhance transfer. By using *Building Thinking Skills* activities in conjunction with similar content lessons, teachers can confirm students' understanding of a variety of thinking processes and can plan systematically through the school year to develop students' competence and confidence in their thinking and learning.

The decision whether to use thinking skills instruction as a lesson supplement or as a sequential series to improve learning skills depends on several factors:

(1) how thinking skills instruction can be effectively scheduled within the school program

(2) how thinking skills instruction can be most easily managed and evaluated

(3) whether teachers are more receptive to using a structured program with a carefully developed sequence of lessons or to correlating thinking skills lessons to existing instructional objectives

(4) the extent to which student achievement on objective tests and in classroom performance is expected to improve

Item Design

In both the figural and verbal strands, exercises have been designed in the manner that the developing child learns: *cognition, evaluation,* and *convergent production.* The simplest form of a task is recognizing the correct answer among several choices. These cognition exercises have the direction *select.*

Next in difficulty is the ability to explain or rank items. This evaluation step helps the learner clarify relationships between objects or concepts. The evaluation exercises contain the direction *rank* or *explain.*

When the learner must supply a single correct answer from his own background and memory, the task becomes more difficult. This convergent production step is designated by the heading *supply.* Teachers may find it helpful to explain concepts in any discipline by remembering the simple "select, explain, then supply" process. The increasing difficulty of cognition, evaluation, and convergent production processes follows J. P. Guilford's *Structure of Intellect* model.

Vocabulary Level

The vocabulary level of *Building Thinking Skills Book 1* utilizes the first thousand words that a child learns to read. Vocabulary designation is based on the *New Horizon Ladder Dictionary of the English Language* by John and Janet Shaw (New American Library, Inc., 1970). Lesson directions and a few items contain words from the second thousand words. *Building Thinking Skills Book 1* contains pictures to allow learners with limited reading abilities to develop more complex thinking skills.

Building Thinking Skills Book 2 includes the second thousand words, with occasional word choices in the third thousand words. This level of difficulty corresponds to the reading vocabulary typically used in grades 4–6. Because the vocabulary level is compounded by the thinking skills component, the resulting exercises in both books may be more difficult than the vocabulary level suggests.

TYPES OF INSTRUCTIONAL METHODS

Piagetian learning theory indicates that the learner proceeds from the concrete manipulative form of a task to the semiconcrete paper-and-pencil form of the task and, finally, to the abstract verbal form. The *Building Thinking Skills* program is based on that progression. Ideally, students should practice each cognitive task in concrete form. Manipulatives, such as attribute blocks, pattern blocks, and interlocking cubes, are commonly available or easily made from inexpensive materials.

The student book provides the paper-and-pencil form of the thinking task. Doing paper-and-pencil tasks alone does not offer the same cognitive benefit as combining thinking skills tasks in all three forms—manipulative, paper-and-pencil, and discussion.

The third step in this process—abstract, verbal expression of the task—involves class discussion of the exercises. Discussion reinforces and confirms the thinking processes which the learner used to carry out the task. The discussion process clarifies the information the learner considered in formulating an answer and differentiates that thinking process from similar ones.

Discussion demonstrates differences in learning styles, allowing students to recognize and understand other ways of arriving at an answer and to value other people's processes for solving problems. For gifted students, discussion provides insight regarding how other equally bright learners can produce correct answers by different analyses.

Discussion reinforces the learner's memory of the thinking process, increasing transfer to similar tasks in content applications. Carrying out a thinking task in a nonthreatening learning situation enhances the learner's confidence in his or her ability to solve similar problems in a different context.

Figural and Verbal Development

Class discussion also provides verbal stimulation for figural learners and is particularly helpful for the student whose language skills are underdeveloped. Since the intellectual development of the learner continues in spite of the lack of language stimulation, figural tasks are likely to be cognitive strengths of students with a limited language background or auditory impairment. Implementation of *Building Thinking Skills* with limited English, learning disabled, or hearing impaired students has indicated that developing one's thinking through figural tasks is also an effective strategy for language acquisition.

Figural observation skills are integral to scientific observation. Undeveloped figural skills may explain why a student who can make good grades on textbook tests may sometimes perform less satisfactorily in a laboratory exercise.

Figural exercises offer figural stimulation for verbal learners. The verbal proficiency of academically achieving learners may mask limited conceptualization. For example, students may memorize the patterns of words for solving various types of mathematical problems, rather than conceptualizing the mathematical principles and visualizing the process. When the learner forgets the formula or algorithm, he or she no longer has a basis for knowing how to do the problem. As verbal learners practice the figural exercises, such students may need to use verbal skills to reinforce and clarify figural perceptions until they become skillful enough to perceive relationships without verbalizing the task.

Discussion Principles

Discussion allows learners to clarify subtle aspects of their mental processing of the exercises. This clarification distinguishes a thinking task from other kinds of instructional tasks and provides alternative and creative ways of getting an answer. Through discussion, the learner ties a thinking skill to other tasks in his or her experience and anticipates situations when using that thinking skill will be helpful.

For effective explanation and transfer of the skill, the lesson proceeds from previous experience to new experience. When introducing a skill, the teacher should identify a real world or academic experience in which the learner has used that skill, cueing the learner that he or she has already had some experience and competence doing that thinking task.

Tying the activity into previous experience signals the learner that the activity is useful and reduces his or her anxiety about being able to accomplish it skillfully. While improved thinking skills often enhance school performance, low-functioning students seldom expect such improvement. For learners who have not been successful in school, the relevance and usefulness of thinking skills may influence how thoroughly the learner will attend to the lesson.

After explanation and guided practice of the thinking skill, the learner should identify other contexts in which he or she has used this skill. This association with past personal experience increases the learner's confidence in reasoning and encourages transfer of the skill.

Because an important goal of thinking skills instruction is improved school performance, both teachers and students should identify thinking skills as they experience them in the content curriculum . Skillful thinking enhances newly mastered skills, improves student confidence, and facilitates new content learning.

Those benefits are best realized by frequent identification of the five thinking skills (describing, finding similarities and differences, sequencing, classifying, and forming analogies) whenever teachers and students encounter similar examples or applications.

The Thinking About Thinking section of each lesson reminds students of the thinking process practiced in the lesson and prompts students to express their thinking clearly. Research on thinking process instruction indicates that unless students can express the thinking they practice in such instruction, subsequent transfer and demonstrated competence in improved thinking are greatly reduced. Metacognition is fostered by peer discussion, by class discussion as outlined in the lesson plan, by creating posters to describe the thinking process, or by student journaling.

Types of Graphic Organizers

The following graphic organizers are used in *Building Thinking Skills* activities. Each diagram cues a different thinking process.

Central idea graphs are used to aid in writing descriptions; to depict a main idea and supporting details; to depict parts of a given object, system, or concept; to depict general classes and subclasses of a system; to depict factors leading to or resulting from a given action; to narrow or broaden proposed topics for a paper or speech; to organize thoughts in writing essay questions or in preparing a speech; to depict alternatives or creative connections in decision making and creative thinking.

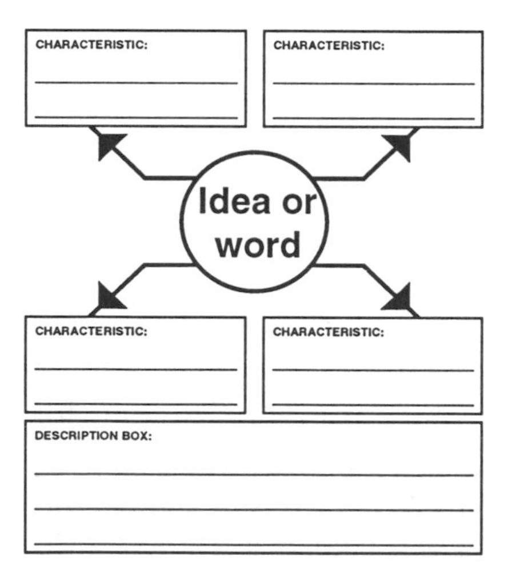

Transitive Order graphs are used to record the inferred order or sequence of information from written materials.

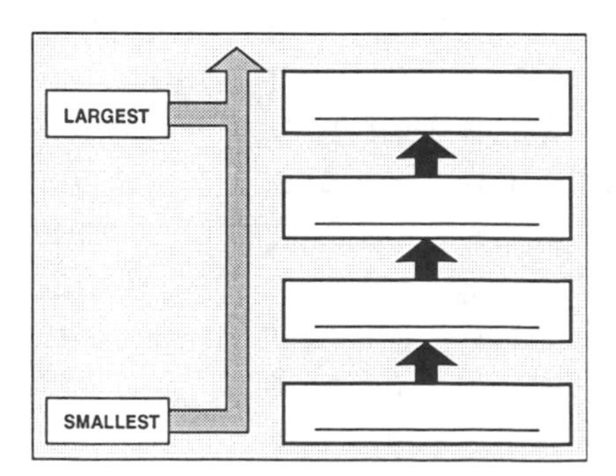

Compare and contrast diagrams are used to compare and contrast two terms or ideas, to organize thinking to respond to essay questions, to clarify the meaning of terms in reviewing for a test.

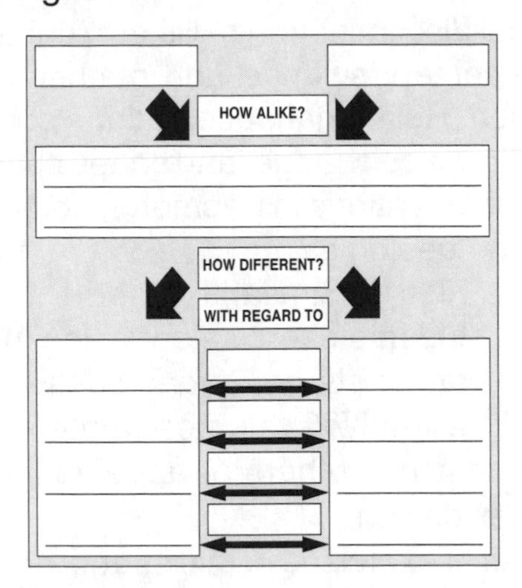

Flowcharts are used to sequence steps in an arithmetic problem; to write instructions; to depict the consequences of decisions; to plan a course of action; to sequence events in plots, historical eras, or laboratory instructions; to picture stages in the development of organisms, social trends, or legislative bills.

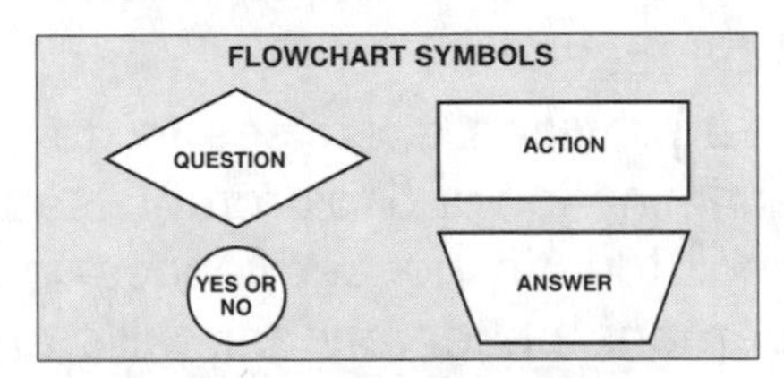

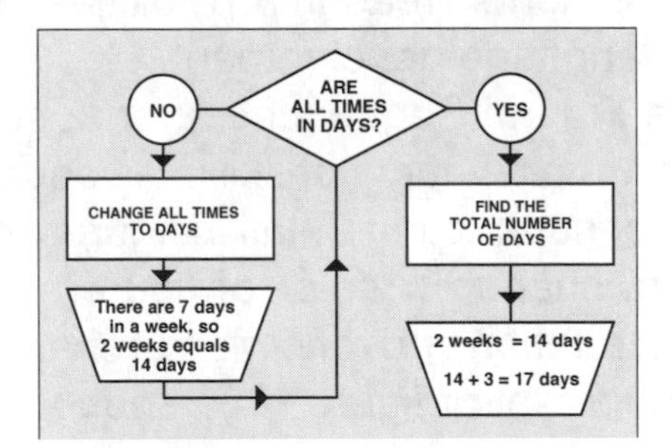

Class relationship diagrams are used to depict class membership and to depict part/whole relationship.

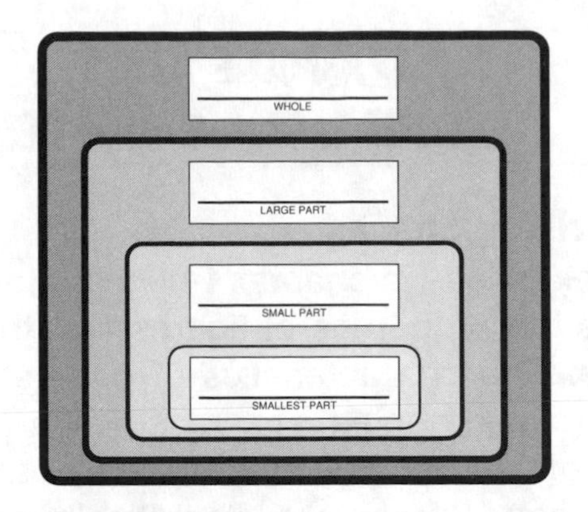

Supplemental Materials

The following books and software published by CTB&S may be used to extend the lessons in *Building Thinking Skills Book 1:*

- *Building Thinking Skills Book 2*—parallels the lessons in Book 1 and can be used to further extend or advance students' skills in all areas
- *Visual Perceptual Skill Buildng Book 2*—can be used in conjunction with the Figural Similarities and Differences and the Figural Sequences sections
- *Language Smarts Book A-1* and *B-1*—offer supplemental activities in Verbal Similarities and Differences, Verbal Sequences, and Verbal Classification
- *Mind Benders®A-1* and *A-2*—provide further practice in the deductive reasoning exercises presented in Verbal Sequences
- *A Case of Red Herrings*—practice discerning multiple word meanings (Verbal Classification)
- *ThinkAnalogy* books and software—can be used in conjunction with Verbal Analogies
- *Organizing Thinking Book 1*—is an excellent resource in using the graphic organizers presented in Verbal Sequences

RATIONALE AND DESCRIPTION OF SKILLS

Describing Shapes

The exercises in chapter 1 help students express the properties of figures that they observe. The characteristics of geometric figures (sides, angles, area), as well as definitions of basic polygons, are featured in these exercises. Students systematically observe basic features of figures, discuss the features, and write about the properties of the figures.

Types of exercises include the following:

1. Selecting a description of a shape
2. Writing the description of a shape
3. Following directions
4. Writing directions
5. Describing position
6. Describing characteristics of a shape

Figural reasoning instruction allows primary gifted students or mature students with low reading ability to carry out complex analysis and evaluation tasks. Since these activities do not require an extensive English vocabulary or well-developed reading skills, these exercises can be used successfully by non-English speaking students.

Figural Similarities and Differences

Chapter 2 features activities to develop visual discrimination skills and to improve students' perception of congruence and similarity. The ability to discern similarities and differences is necessary before the learner can place objects in order, classify them, or make analogous comparisons.

Visual discrimination in its simplest forms involves recognizing geometric shapes and the appearance of letters. In elementary school mathematics programs, the concepts of congruence and similarity are involved in establishing geometric definitions and developing perceptions of area and volume.

Visual discrimination activities may be helpful in promoting word decoding for upper elementary students having reading difficulty. Reading development requires the learner to recognize subtle differences in the shapes and sequences of letters or the appearance of a whole word.

Visual discrimination skills are fundamental to elementary science and mathematics instruction. Relational observations of rotation, reflection, size change, and shape change are basic observations in geometry, botany, zoology, and geology.

In the Figural Similarities and Differences strand, students exercise cognition in selecting the correct shape among subtly different ones. The learner evaluates whether or not a shape matches others or appears in a more complex design.

Types of exercises in this strand include the following:

1. Analyzing and matching shapes
2. Finding and combining shapes
3. Evaluating and producing equal shapes
4. Recognizing shapes necessary to complete a whole figure

Figural Sequences

Chapter 3 provides exercises to develop visual discrimination and to promote sequential reasoning. Identifying sequences in figural form sharpens observational skills and promotes students' reasoning abilities regardless of language development.

In the Figural Sequences strand, students demonstrate a variety of skills in sequencing: adding or subtracting detail in figures; changing size, shape, or color of figures in a sequence; rotation and reflection of shapes; and rearrangement of figures in a sequence.

Types of exercises in this strand include the following:

1. Recognizing the next figure in a sequence
2. Producing the next figure in a sequence
3. Recognizing rotation and reflection
4. Folding paper

Figural Classification

Figural Classification exercises in chapter 4 develop the ability to group or organize objects by similar characteristics. Classification is a significant concept-building process, associated with science concepts. However, classification is a helpful study skill, promoting visual discrimination, memory, observation, and organizing skills.

The student uses classification as an observational tool. By identifying similar or different characteristics, the learner systematically examines and understands new material. Classification is involved in assimilating new information and accommodating old categories to include new information or experiences.

Classification assists the student in visualizing relationships and provides a practical problem-solving technique. Venn diagrams and matrices are useful for showing class relationships by showing that items have some or all variables in common. Classification techniques are integral to computer logic and set theory.

The Figural Classification exercises increase in complexity and difficulty throughout the chapter. The exercises proceed through the following learning sequence:

1. Classifying by shape
2. Classifying by pattern
3. Classifying by shape and pattern
4. Describing characteristics of a class
5. Matching classes
6. Completing or forming a class
7. Producing another member of a class
8. Using diagrams to depict overlapping classes

Figural Analogies

Chapter 5 provides exercises to develop visual discrimination skills and to promote analogical reasoning. Teaching students to identify analogous relationships in figural form sharpens observational skills and promotes students' reasoning abilities.

Analogous relationships are basic to all fields of study. Analogies are expressed as imagery in literature, ratios in mathematics, and analysis techniques in geometry, natural science, and social sciences. While teachers use analogies in explaining concepts, students seldom practice this useful relational technique before they encounter analogies on objective tests.

In the Figural Analogies strand, students are introduced to analogous relationships ("A" is to "B" as "C" is to "D"). Students analyze the components, recognize the relationships, and complete the analogies by selecting or drawing the missing figure.

The exercises in this strand are nongraded and require little reading skill.

Types of Figural Analogies include the following:

1. Color or size change
2. Rotations and reflections
3. Change in detail

Describing Things

In chapter 6, students discuss and define the characteristics of key concepts in the elementary school curriculum such as occupations, food, animals, vehicles, buildings, common objects, etc. Students practice using precise language to describe or define things they study.

Types of exercises include the following:

1. Matching a description to a drawing
2. Writing descriptions of objects
3. Matching a description to a word
4. Supplying a word that fits a description

Verbal Similarities and Differences

Chapter 7 introduces synonyms and antonyms. The ability to discern similarity and difference in meaning is integral to reading comprehension, vocabulary development, and writing skills.

Synonym and antonym exercises are presented in picture form, using line drawings suitable for primary children or nonreading adults.

Types of exercises in this chapter include the following:

1. Selecting similar and opposite words
2. Selecting how words are alike and how they are different
3. Explaining how words are alike and how they are different
4. Supplying similar and opposite words

Verbal Sequences

Chapter 8 introduces verbal sequences. Students must recognize word relationships in order to understand subtle differences in meaning, to recognize chronological order, and to organize and retrieve information. Language arts research indicates that students learn vocabulary effectively through context. Context may be paragraphs or clusters of words that give meaning to the word or words being learned.

In some of the Verbal Sequence exercises, two of the three words suggest a progression in size, rank, or order that the next word should continue. If the student knows the meaning of some of the words in the sequence, he or she can infer the meaning of the missing word.

Verbal sequence exercises include distinguishing transitive order. From a written passage, students rank objects or people being compared according to some characteristic (weight, age, height, score, etc.). Transitive order is applied in solving deductive reasoning puzzles.

Some verbal sequences involve degree. Recognizing degree of meaning fosters correct inference in reading and listening and promotes clarity in writing and speech. This clarification alerts students to differences in meaning and reinforces the meaning of new words. Vocabulary and reading comprehension test items frequently involve slight differences in meaning.

Achievement tests contain items which require the student to number sentences in chronological order. The student recognizes and organizes a commonly known sequence or relies on the context of the passage to determine order.

Verbal sequence is also involved in following simple directions and distinguishing Yes-No rules and True-False values. More complex directions include describing direction and location, as commonly found in map reading.

Verbal Classifications

Chapter 9 features activities to improve students' conceptualization of class relationships and formation of clear definitions. Classification promotes the understanding and recall of the meaning of words. *Classes* are the categories in any definition of nouns. In the definition of *bicycle* as a "vehicle having two wheels," "vehicle" is the class and "two wheels" are the descriptors.

Classification provides a basis for storing and retrieving information. Just as a computer stores, organizes, and recalls bits of information, our human memory uses categories or classes to organize and retain otherwise unconnected items and ideas. The learner remembers categories or associations as an aid to recalling details.

Classification of collections allows learners to find items or information easily. Commonly, students are introduced to conventional classification systems (library classification systems, biological phyla, the periodic chart, etc.). Less often, however, do students learn how to classify. Classification brings order to everyday tasks, such as arranging items on storage shelves, managing a family budget, keeping records, sorting collections, or scheduling family activities.

Verbal classification exercises are presented in picture form, using pictures sufficiently sophisticated to allow use with either primary students or nonreading adults.

Types of exercises include the following:

1. Distinguishing parts of a whole
2. Distinguishing between class and members of that class
3. Selecting and explaining common characteristics of a class
4. Explaining the exception to a class
5. Sorting words into classes

Verbal Analogies

The Verbal Analogies strand in Chapter 10 features activities to sharpen students' perceptions of analogical relationships and to sharpen vocabulary. Analogies are basic to all fields of study: expressed as imagery in literature, ratios in mathematics, and analysis techniques in geometry and the natural and social sciences. Practicing verbal analogies prepares students for similar exercises on achievement tests.

The types of analogies include synonyms, antonyms, part of, kind of (classification), something used to, and association.

Analogy exercises include the following:

1. Selecting the word to complete an analogy
2. Naming the kind of analogy
3. Supplying the word to complete an analogy
4. Selecting analogous pairs of words to complete an analogy

EVALUATING BUILDING THINKING SKILLS INSTRUCTION

School district implementation of the *Building Thinking Skills* program has been evaluated using many assessment procedures:

- Student performance on cognitive abilities tests
- Student performance on normed-referenced achievement tests
- Student performance on teacher-designed, criterion-referenced tests
- Student performance on language proficiency tests
- Student performance on writing assessments
- Number of students placed in heterogeneous grouped classes, or advanced academic programs, as well as students' subsequent successful performance in gifted or academic excellence classes

Cognitive Abilities Tests

Cognitive abilities tests indicate students' capabilities to perform thinking tasks that are related to school performance but do not require content knowledge. Such tasks include figural and verbal forms of tasks that require certain analysis skills. Because *Building Thinking Skills* is a cognitive development program that emphasizes language development, the figural and verbal subtests of cognitive abilities tests are closely correlated to *Building Thinking Skills* goals and activities.

Pretesting with cognitive abilities tests offers baseline data on student performance on various thinking skills and guides teachers' planning regarding suitability of the level of vocabulary and difficulty of the exercises. Pretest information also can identify cognitive strengths that are not easily uncovered in daily classwork.

Post-testing gives information to teachers, students, parents, administrators, and the community about the effectiveness of thinking instruction in improving students' thinking and learning. Correlated with achievement information (tests, products, performance), post-instruction performance on cognitive abilities tests offers an indicator of how well a student is performing relative to his or her ability.

A list of cognitive abilities tests that have been used in program effectiveness evaluation of thinking instruction using the *Building Thinking Skills* series is provided on page xi.

Norm-referenced Achievement Tests

Composite scores on norm-referenced achievement tests are generally poor indicators of improved thinking skills. Total language scores include grammar, spelling, and literacy items that do not reflect analysis meaningfully. Total mathematics scores include a significant number of items to measure computation, rather than improved analysis in mathematics.

However, some subtests do reflect the analysis skills addressed in *Building Thinking Skills* instruction. Program evaluation using this series has indicated substantial gains in subtests which measure reading comprehension, mathematics concepts, and mathematics problem solving. If achievement test information is needed to report the effectiveness of *Building Thinking Skills* instruction, only those subtests should be monitored.

Teacher-designed, Criterion-referenced Tests

To address improved content learning as a result of thinking skills instruction, criterion-referenced tests (tests that measure students' understanding of what was taught) should also show improvement in students' comprehension of content. Teacher-designed tests, given at the end of an instructional unit, should show gain in several kinds of items:

- Open-response writing involving definition, description, chronological order or prioritizing, or analogy
- Multiple choice, short response, or matching items involving definition, analogy, transitive order, rank, chronological order, or prioritizing
- Interpreting graphs or diagrams or completing graphic organizers to show learning

Language Proficiency Tests

Because one of the goals of the *Building Thinking Skills* program is language development, increased vocabulary can be shown on a variety of language tests. Tests commonly used to evaluate the effect of thinking instruction on language development include the following:

- *Peabody Picture Vocabulary Test*
- ESOL proficiency tests

Writing Assessments

Writing assessment can involve many forms of evaluation:

- Students' evaluation of their own writing portfolios, showing vocabulary and writing skills early in the school year and after participation in thinking instruction
- Pre- and post-instruction on the types of writing used on state writing assessment tests: descriptive, expository, and narrative writing are the types of prompts most related to the *Building Thinking Skills* program
- End-of-unit essay questions requiring description, compare/contrast, definition, sequential order, or analogy

Inclusion and Performance in Mainstream or Advanced Academic Programs

Building Thinking Skills is commonly used to promote access to academic excellence programs or to prepare students to be successful in mainstream classes from special services programs (Chapter 1 classes, bilingual programs, ESOL classes, special education classes, or remedial programs). The key statistics for evaluating this goal are the number of students who gain access to programs, the speed with which the transition is accomplished, and the students' level of achievement when included in general classes.

EVALUATION RECOMMENDATIONS

1. Any standardized content test currently utilized by your district will reflect increases in students' academic performance that result from better thinking skills and can be used to measure the effectiveness of the *Building Thinking Skills®* series.
2. Tests are also available to measure growth in cognitive skills specifically. These tests include the following:

- *Cognitive Abilities Test* (*Woodcock-Johnson*)
  Riverside Publishing Company
  425 Spring Lake Dr.
  Itasa, IL 60143
  800-323-9540 • 312-693-0325 (fax)
- *Developing Cognitive Abilities Test*
  American College Testronics
  (formerly American Testronics)
  P.O. Box 2270
  Iowa City, IA 52244
  800-533-0030 • 319-337-1578 (fax)
- *Differential Aptitude Tests*
  Psychological Corporation
  Order Service Center
  P.O. Box 839954
  San Antonio, TX 78283-3954
  800-228-0752 • 800-232-1223 (fax)
- *Otis-Lennon School Ability Test*
  (OLSAT-7)
  Harcourt Brace Educational Measurement
  555 Academic Court
  San Antonio, TX 78204
  800-228-0752, 210-299-1061
  (fax) 800-232-1222
- *Structure of Intellect Learning Abilities Test*
  S.O.I. Institute
  P.O. Box D
  Vida, OR 97488
  503-896-3936 • 503-896-3983 (fax)
- *Test of Cognitive Skills*
  CTB-McGraw Hill
  P.O. Box 150
  Monterey, CA 93942-0150
  800-538-9547 • 800-282-0266 (fax)
- *WISC-III*
  Psychological Corporation
  Order Service Center
  P.O. Box 839954
  San Antonio, TX 78283-3954
  800-228-0752 • 800-232-1223 (fax)

INSTRUCTIONAL RECOMMENDATIONS

- **Use real objects or detailed pictures whenever possible.** Observation and vocabulary acquisition are enhanced by carefully examining pictures. Supplement the pictures, whenever possible, with real objects.
- **Do description exercises first.** Description requires that students observe significant characteristics of food, plants, animals, people, buildings, and vehicles.
- **Do similarities and differences exercises next.** Sequencing, classification, and analogy require that students observe significant characteristics and distinguish important similarities and differences.
- **Encourage peer discussion.** Program evaluation suggests that the quality of student responses, students' willingness to participate, and the attentiveness of easily distracted students significantly improve when peers discuss their answers prior to full class discussion.
- **Conduct short exercises.** Do only a few activities in each session—lasting not more than twenty minutes. Thoroughly discuss a few items with ample time for students to explain their thinking, rather than attempting to conduct additional exercises.
- **Identify and use students' background knowledge.** Teachers who have used *Building Thinking Skills* comment that the richness of the group's background exceeds the limited experience of individual marginal students in the class. Use these sessions as diagnostic indicators of students' prior knowledge. Remember the language that students use in their descriptions. Use the same words to remind students of the thinking processes in subsequent content lessons.
- **Use *Building Thinking Skills* lessons to introduce an activity in content areas.** The curriculum applications supplied in the lesson plans offer a wide variety of ways to integrate thinking skills into content areas.
- **Examine concepts in other content areas using the same processes.** Use correct terms to cue students to use a thinking process in other lessons. Use the same methods (peer and class discussion, observation of pictures or objects) in other contexts.
- ***Building Thinking Skills*** lessons are not to be given as strictly independent activities, such as homework assignments. The lessons in *Building Thinking Skills* are designed to enhance cognitive development through discussion and observation. Independent practice exercises may be used to reinforce mastery or build confidence but should never be used as a substitute for class discussions.
- **Use the "language of thinking" in thinking skills activities and in other lessons.** Encourage students to express their thinking with the terms listed in the Vocabulary section (p. xix). Your use of the vocabulary of thinking helps students transfer the thinking skills they practice in this program and other contexts. Encourage students to use these analysis terms when they discuss their thinking in *Building Thinking Skills* exercises, in content lessons, and in personal applications. Frequent, natural use of the language of thinking promotes precision in using these terms and confidence in expressing one's thoughts.
- **Use graphic organizers to cue various thinking processes.** *Building Thinking Skills* features the following graphic organizers: central idea graph, compare and contrast diagram, transitive order graph, flowchart, and class relationship diagram. Each diagram cues a different thinking process.

VOCABULARY AND SYNONYMS

Reinforce the language of thinking in thinking skills activities and in other lessons. The following list includes terms that teachers and students can use in *Building Thinking Skills* lessons, content lessons, and personal applications. Students may create a "thinking thesaurus" of the words and idioms they use to describe their thinking. Encourage them to express their thinking using the terms below.

WORD	SYNONYMS
Adjust	modify, adapt, fit
Alternate	replace, substitute, vary, use instead of
Analogies	comparisons, metaphors, similarities
Antonyms	opposites, least like
Appropriate	fitting, correct, proper, suitable
Arrange	place, order, rank, organize
Assemble	put together, gather, build, organize
Associate	correlate, link, connect, is similar to
Attribute	feature, property, characteristic
Category	classification, class, kind, type, form
Characteristic	attribute, quality, style, traits
Chart	map, outline, graph, diagram, matrix
Chronological	time-order, consecutive, sequential
Clarity	being clear, direct, well-defined, understanding
Class	group, set, category, kind, type, sort
Classification	arrangement, assortment, grouping
Clockwise	rotating to the right like the hands of a clock
Combine	connect, link, join, assemble, put together
Common	familiar, ordinary, similar, have the same characteristic, frequently occurring
Compare	relate, match, find similarities
Conclude	decide, complete, end, determine, finish, realize, comprehend
Conclusion	understanding, result, finish, outcome, completion, decision, determination, realization
Confirm	prove, explain, agree, make sure, show, determine, check
Consecutive	next, sequential, following, continuing
Construct	build, create, compose, invent, put together, assemble
Contrasts	differs, is unlike
Corresponds to	is the same as, compares with, parallels, is like, is similar, has the same qualities as
Counterclockwise	turning in a left direction, opposite movement from the hands of a clock
Decreasing	lessening, diminishing, shrinking, becoming smaller
Definition	meaning, explanation, description
Demonstrate	show, illustrate, enact, make clear
Describe	explain, clarify, give details of
Detail	part, piece, feature, component
Determine	decide, find out, learn, arrange, conclude, figure out, show

Diagram .... graph, layout, outline, design
Differences .... dissimilarity, unlike qualities, contrasts
Disassemble..... separate, divide, take apart, dismantle, disconnect
Discuss .... talk about, describe, explain, explore, express, exchange ideas
Distinguish .... clarify, differentiate, analyze, show how it is different
Elaborate .... expand, explain, add details, tell more
Elements .... parts, features, portions, details, components, particles
Eliminate .... remove, take out, erase, end
Enable .... allow, approve, permit, make possible
Equal .... same, matching, evenly divided, same size, congruent
Examine .... find the details, analyze, inspect, explore, investigate, look at, observe closely
Express .... declare, say, tell, signify, verbalize
False .... not true, not real, untrue, unreal, not valid
Figural .... pictured, drawn, illustrated, geometric
Figure .... picture, shape, structure, illustration, diagram, outline, drawing
Frequency .... occuring often, recurrence, regularity
Geometric......... Figural, having shape, many-sided
Graph .... diagram, chart, outline
Hidden .... unclear, not distinct, camouflaged
Identify.............. find, recognize, pick out, know, indicate, show, specify
Illustrate............ draw, depict, show, represent, picture, portray, model
Indicate............. show, identify, point out, specify
Inequality.......... difference, unequal, not matching, dissimilar
Interpret .... convey, explain, make clear, define, show, mode, paraphrase, tell another way
Intersect .... cross, divide, meet, converge, pass through
Justify .... explain, offer reasons for, defend, show how, evaluate
Locate .... identify, place, find
Location............ place, position, point
Matching........... equal, making an equal pair, congruent
Matrix .... chart, graph with rows and columns, grid
Member .... belongs to a group or class
Observe............ pay attention to, examine, look at carefully, perceive
Order .... rank, sequence, organization
Overlap............. intersection, comes together, overlay, belongs to both
Part................... piece, fragment, segment, section, detail
Pattern .... arrangement, design, repetition
Point/position.... location, spot
Precise .... clear, accurate, definite, correct
Prepare .... produce, create, arrange, ready, plan
Produce .... generate, make, create, assemble
Quality .... characteristic, attribute, trait, standard
Rank .... order, class, grade, type
Recognize .... identify, be familiar with, verify, indicate
Reinforce .... strengthen, make sure
Relationship ..... tie, connection, how related or similar
Represent......... show, depict, model, typify
Select .... pick out, identify, locate, decide
Sequence .... steps, order, rank, consecutive arrangement, change or shift
Series .... group, succession, sequence, order
Set.................... group, category, collection, class, type

Shape .... figure, form, pattern, outline, drawing
Significant .... important, key, basic, meaningful
Similarity .... likeness, sameness, resemblance
Solution .... answer, result, resolution, clarification, outcome
Solve .... figure out, think, interpret, resolve, find answer, find out why
Sort .... group, classify, file, organize
Specify .... identify, show, point out, detail, stipulate, clarify
Sufficient .... acceptable, adequate, enough, satisfactory
Supply .... provide, furnish, produce
Support .... assist, defend, reason, detail, give evidence for, explain
Transparent ...... revealing, clear, see-through
True .... real, accurate, precise
Unique .... rare, one of a kind, distinctive
Verbal .... spoken, expressed in words, oral
Visualize .... imagine, picture mentally, envision
Whole .... entire, complete, total

Chapter No.

GUIDE TO USING THE LESSON PLANS

EXERCISES

LESSON TITLE

ANSWERS
Lists exercises and pages covered in student book.
Guided Practice: Provides answers to guided practice exercises.
Independent Practice: Lists independent practice answers.

LESSON PREPARATION

OBJECTIVE AND MATERIALS

OBJECTIVE: Explains for the teacher the thinking objective of the lesson.
MATERIALS: Lists materials or supplies for modeling the lesson

CURRICULUM APPLICATIONS

Lists content objectives which feature the skill or require it as a prerequisite.

TEACHING SUGGESTIONS

Alerts the teacher to special vocabulary or concepts in the lesson. Guided practice should be followed by class discussion of the exercises, with the teacher clarifying significant terms and concepts. Discussion should also include student explanations, reasons for rejecting incorrect answers, and confirmation of correct responses.

MODEL LESSON

LESSON

Introduction

- Indicates to the learner when he or she has seen or used a similar kind of learning.

Explaining the Objective to the Students

Q: Explains to the student what he or she will learn in the lesson.

Class Activity

- Offers a concrete form of the thinking task (and/or illustrates the task by modeling). Students may also prepare models or materials similar to those suggested. When conducting the lesson, teachers should verbalize their own thinking process. This modeling provides cues to students for executing the thinking task skillfully.

GUIDED PRACTICE

Controlled practice allows the teacher to identify errors or omissions in students' processing. Guided practice should be followed by class discussion.

INDEPENDENT PRACTICE

- Provides practice exercises for promoting skill mastery.

THINKING ABOUT THINKING

Helps the student clarify and verbalize the thinking process: metacognition.

PERSONAL APPLICATION

Relates the skill to the learner's experience and cues the learner regarding possible future uses of the skill.

Chapter 1

DESCRIBING SHAPES
(Student book pages 1–21)

EXERCISES A-1 to A-27

DESCRIBING SHAPES—SELECT

ANSWERS A-1 through A-27 — Student book pages 2–7
NOTE: In **A-1** through **A-4,** the answers may be given in either order.
Guided Practice: A-1 short, wide; **A-2** short, wide
Independent Practice: A-3 short, narrow; **A-4** tall, wide; **A-5** four; **A-6** one; **A-7** no; **A-8** two; **A-9** three; **A-10** five, two; **A-11** four, all OR four OR all four; **A-12** five, three; **A-13** six, four; **A-14** three, two; **A-15** four, all OR four OR all four; **A-16** three, all OR three OR all three; **A-17** four, two; **A-18** three, none; **A-19** three, none; **A-20** five, none; **A-21** four, two; **A-22** six, all OR six OR all six; **A-23** five, none; **A-24** four, four, rectangle; **A-25** five, five, pentagon; **A-26** three, three, triangle; **A-27** six, six, hexagon

LESSON PREPARATION

OBJECTIVE AND MATERIALS
OBJECTIVE: Students will choose the appropriate words to describe shapes.
MATERIALS: Large rectangle • transparency of Transparency Master 1 (TM 1) on page 249 in this book

CURRICULUM APPLICATIONS
Language Arts: Visual discrimination needed for reading readiness
Mathematics: Naming geometric shapes
Science: Recognizing shapes of leaves, insects, or shells
Social Studies: Recognizing geographic features on map puzzles
Enrichment Areas: Recognizing shapes of road signs, discerning different patterns in art

TEACHING SUGGESTIONS
Reinforce the following vocabulary emphasized in this lesson: *hexagon*, *pentagon*, *rectangle*, *triangle*, *square*, *sides*, *square corner*, and *equal*. (A square corner is a right angle.) Encourage students to use these terms in their discussion. If students are reluctant to use new terms, ask them to discuss their answers with a partner before participating in class discussion. If students do not know or use *angle*, introduce the concept. Young children may not commonly call the top and the bottom of a shape a side. If you hear this confusion emerging, explain that, in the language of mathematics, all the lines are called sides.

You may explain that triangles with two equal sides (isosceles) also have two equal angles. Exercises **A-19** through **A-23** show the relationship between number of sides and number of angles (corners), i.e., three-sided shapes have three corners, etc. Use attribute or pattern blocks to teach shape names. If you have transparent attribute or pattern blocks, use them with the overhead projector.

MODEL LESSON

LESSON

Introduction

Q: What is this shape called?

- Hold up a large rectangle.

  Q: Find some rectangles in this room (tables, desktops, door, windows, ceiling tiles, etc.). Pick one rectangle and describe it to your partner.

- Give students time to discuss the object.

Explaining the Objective to the Students

Q: In this lesson, you are going to select some words that describe shapes.

Class Activity

- Project the transparency of TM 1.

  Q: Which shapes are rectangles?

  A: Shapes 1 and 2

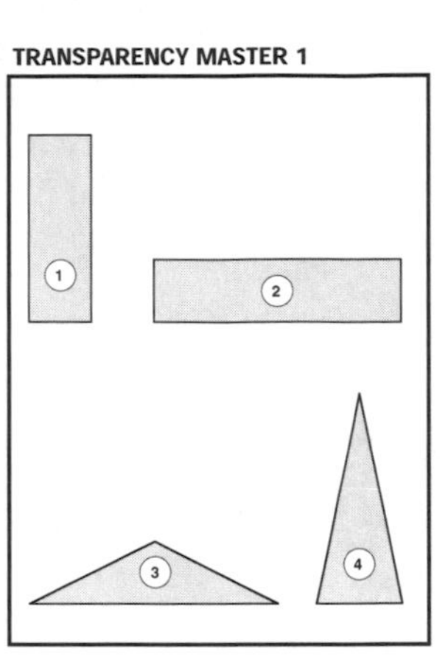

  Q: How many square corners does each rectangle have?

  A: Four

  Q: How many equal sides does each rectangle have?

  A: Two pairs; the top and bottom sides are equal and the opposite ends are equal.

  Q: What words from the choice box would you use to describe how rectangle 1 is different from rectangle 2?

  A: *Tall* and *narrow*

  Q: Which shapes are triangles?

  A: Shapes 3 and 4

- Point to figures 3 and 4.

  Q: How many corners does each triangle have?

  A: Three

  Q: How many equal sides does each triangle have?

  A: Two (The slanted sides are equal.)

  Q: What words from the choice box would you use to describe how triangle 3 is different from triangle 4?

  A: *Short* and *wide*

GUIDED PRACTICE

EXERCISES: **A-1**, **A-2**

- When students have had sufficient time to complete the exercises, check by discussion that they have answered correctly. Discuss *why* the incorrect answers were eliminated.
- Note that the answers for **A-1** through **A-4** may be given in either order. You may want to explain to students that in this example, as in others, there may be more than one way to answer correctly.

INDEPENDENT PRACTICE

- Assign exercises **A-3** through **A-27**. Note: This lesson may require two sessions (**A-3** through **A-13** and **A-24** through **A-27**).

THINKING ABOUT THINKING

Q: What did you pay attention to when you picked a word that described a shape?

1. I read the words (characteristics) in the choice box.
2. I looked for those characteristics in each shape.
3. I picked the word that best described the shape.
4. I checked to see why the other words didn't describe the shape.

PERSONAL APPLICATION

Q: When might you have to describe shapes at home?

A: Examples include describing objects or places (e.g., "Our table has two long sides and two short sides").

EXERCISES A-28 to A-33

DESCRIBING SHAPES—EXPLAIN

ANSWERS A-28 through A-33 — Student book pages 8–10
NOTE: Answers may vary.
Guided Practice: A-28 This rectangle is two inches wide (across) and three inches tall (high).
Independent Practice: A-29 This triangle with a square corner is two inches wide and three inches tall. OR This right triangle has a two-inch base and a three-inch height. **A-30** This pentagon is two inches wide and three inches tall. The vertical lines of the pentagon are equal. The slanted parts of the pentagon are equal. OR This pentagon has a two-inch base and a three-inch height. **A-31** This hexagon is one inch on each side. It is made up of two trapezoids. **A-32** This hexagon is one inch on each side. It is made up of three diamonds (rhombuses). **A-33** This hexagon is one inch on each side. It is made up of six triangles.

LESSON PREPARATION

OBJECTIVE AND MATERIALS

OBJECTIVE: Students will practice describing many shapes.
MATERIALS: Photocopies and a transparency of Transparency Master 2 (TM 2) on page 250 of this book (Students may make flash cards from photocopies of TM 2 to practice learning shape names.) • transparency of page 8 of the student workbook • washable transparency marker • pattern blocks are desirable to demonstrate exercises **A-28** through **A-30**

CURRICULUM APPLICATIONS

Language Arts: Visual discrimination for reading readiness
Mathematics: Naming geometric shapes
Science: Recognizing shapes of leaves, insects, or shells
Social Studies: Recognizing geographic features on map puzzles
Enrichment Areas: Recognizing shapes of road signs, discerning different patterns in art

TEACHING SUGGESTIONS

Reinforce the following vocabulary emphasized in this lesson: *diamond, hexagon, pentagon, rectangle, right triangle, rhombus, square, square corner, trapezoid,* and *triangle*. Encourage students to use these terms in their discussion and respond in complete sentences. To demonstrate *perimeter,* ask students to calculate the perimeter of the hexagons. (Each hexagon has six one-inch sides for a perimeter of six inches.)

MODEL LESSON

LESSON

Introduction

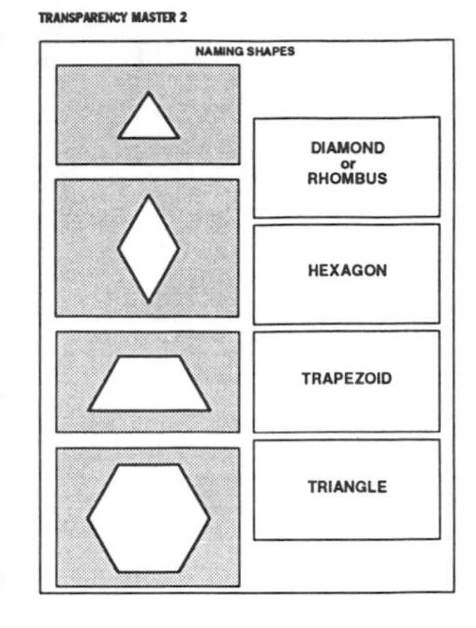

- Project transparency of TM 2. Point to the triangle.

  Q: What is this shape called?

  A: Triangle

  Q: Describe the shape to your partner.

Explaining the Objective to the Students

Q: In this activity, you are going to describe shapes in your own words.

Class Activity

- Distribute photocopies of TM 2 and project the transparency.

  Q: Match the top shape to its name. (Ask a student to draw a line from the picture to the name.)

  A: Triangle

  Q: Match the next shape to its name. (Ask a student to draw a line from the picture to the name.)

  A: Diamond or rhombus

  Q: Try to remember this new name. Match the next shape to its name. (Ask a student to draw a line from the picture to the name.)

  A: Trapezoid

  Q: Try to remember this new name. Match the next shape to its name. (Ask a student to draw a line from the picture to the name.)

  A: Hexagon

- Students may cut apart the shapes and the shape names to make their own set of "flash cards" for practice in learning these shape names. Students may paste the shape names on the backs of the shapes. Leave the projected correct answers in view to ensure that the students make correct flash cards.

DESCRIBING SHAPES—EXPLAIN

DIRECTIONS: In each description box, describe the shape in the picture at the left. Use complete sentences in your descriptions.

EXAMPLE — 2 in. — 2 in.

DESCRIPTION: *This triangle is two inches wide and two inches high.*

A-28 — 3 in. — 2 in.

DESCRIPTION

- Project transparency of student book page 8.

  Q: Notice the arrows on the triangle. These arrows tell the size of the shape. There are numbers on the arrows. The letters *in* after each number are the symbol for inches. How tall is the triangle?

A: Two inches high

Q: How wide is the triangle?

A: Two inches wide

Q: How can you describe this triangle?

A: The triangle is two inches wide and two inches tall.

GUIDED PRACTICE

EXERCISES: **A-28**

- Check by discussion that students have answered correctly.

INDEPENDENT PRACTICE

- Assign exercises **A-29** through **A-33**.

THINKING ABOUT THINKING

Q: What did you pay attention to when you picked a word that described a shape?

1. I recalled the name of the shape.
2. I looked at special features of the shape (square corners, equal sides, length of sides).
3. I named the shape and described its special features.

PERSONAL APPLICATION

Q: When might you have to describe shapes at home?

A: Examples include making requests (e.g., "Hand me that triangular pillow, not the square one"), giving directions, following directions to assemble toys, etc.

EXERCISES A-34 to A-43

FOLLOWING DIRECTIONS

ANSWERS A-34 through A-43 — Student book pages 11–12
Guided Practice: A-34 to **A-36** See below.
Independent Practice: A-37 to **A-43** See below.

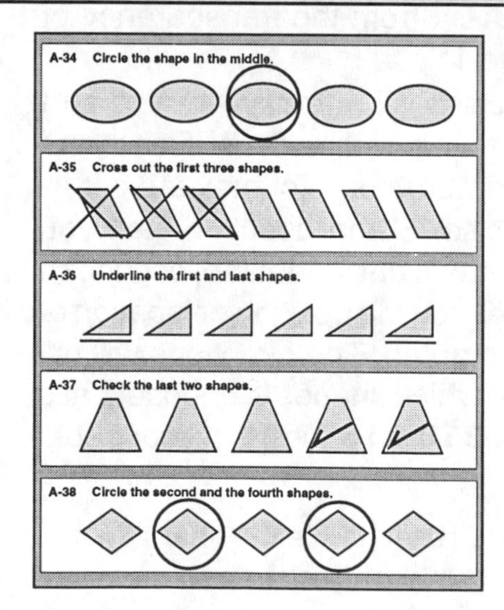

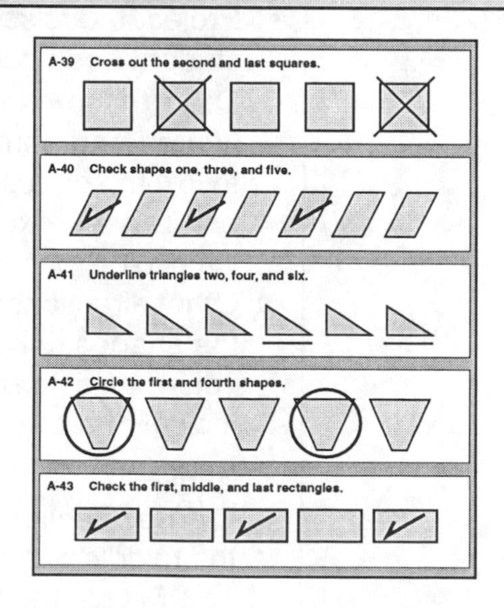

LESSON PREPARATION

OBJECTIVE AND MATERIALS

OBJECTIVE: Students will follow directions to carry out tasks similar to those they do in school.

MATERIALS: Transparency of student book page 11 • washable transparency marker

CURRICULUM APPLICATIONS

Language Arts: Reading and following directions in any situation (text, worksheet, test) involving order, number, or position

Mathematics: Following the correct sequence in solving word problems

Science: Following sequence of instructions in laboratory experiments

Social Studies: Answering questions regarding charts, graphs, or schedules; reading and constructing time lines

Enrichment Areas: Art, music, or physical education activities involving order, number, or position; following dance, drill, or sports instructions

TEACHING SUGGESTIONS

Encourage students to discuss reasons for their choices. Notice the language that the students use to describe their own thinking and use this same language as you remind students of the key characteristics in this exercise and subsequent exercises. Students' explanations of their thinking are often more meaningful to them than an adult's explanation.

MODEL LESSON

LESSON

Introduction

Q: Think about a school activity when you were asked to follow directions. What did you think about in order to follow those directions?

A: I had to remember what to do, what order to do it in, and where to do it.

Explaining the Objective to the Students

Q: In these exercises, you will follow directions to carry out tasks similar to those you do in school.

Class Activity

- Project exercise **A-34** from the transparency of student book page 11.

  Q: In these exercises, you are given a row of figures and some written directions. Read and follow the directions carefully, for they are different in each exercise. Some directions tell you what to do (circle, cross out, underline, or check). Other directions tell you the position or location of the shapes to be marked. The directions will tell you what to do and which shapes to choose. Let's look at exercise **A-34** from page 11 in your book.

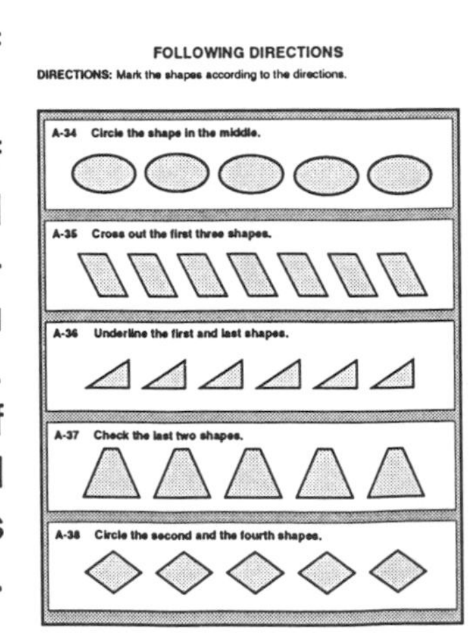

- Pause.

  Q: The directions tell us to circle the shape in the middle. How can we find the middle of this group of five shapes?

  A: Middle means halfway from one end to the other.

- Draw a dotted line down the middle of the box.

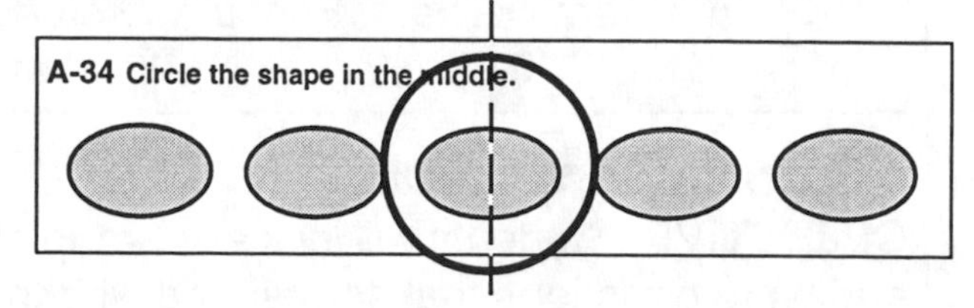

Q: Now that we have found the middle, how can we identify the shape that is in the middle?
  A: The shape that the line passes through is in the middle.

Q: What do the directions tell us to do when we find the middle shape?
  A: Circle the shape in the middle.

- Circle oval three on the transparency.

GUIDED PRACTICE

EXERCISES: **A-34** through **A-36**

- Make sure students understand what to do and which shapes to mark. When students have had sufficient time to complete these two exercises, check answers by discussion to determine whether they have answered correctly.

INDEPENDENT PRACTICE

- Assign exercises **A-37** through **A-43**.

THINKING ABOUT THINKING

Q: What did you pay attention to when you followed directions?

1. I carefully read the directions.
2. If the instructions involved number or position, I counted the shapes.
3. I found the right shape(s).
4. I marked the shape(s) the correct way.

PERSONAL APPLICATION

Q: At what other times is it necessary to follow written instructions?
  A: Examples include crafts or sports activities involving order, number, or position; assembling games or models; completing questionnaires or answer sheets; taking tests; following a recipe; etc.

EXERCISES A-44 to A-53

WRITING DIRECTIONS

ANSWERS A-44 through A-53 — Student book pages 13–15
Guided Practice: A-44 Check the second and fourth circles.
Independent Practice: A-45 Cross out the third and fourth triangles. **A-46** Underline the first two rectangles. OR Underline the first and second rectangles. **A-47** Cross out the third and fifth circles. **A-48** Circle the first two rectangles. **A-49** Check the first, third, and fifth squares. **A-50** Underline the second, fourth, and sixth triangles. **A-51** through

A-53, answers will vary. Encourage students to name the shape being marked, i.e., **A-51** pentagon, **A-52** trapezoid, **A-53** square.

LESSON PREPARATION

OBJECTIVE AND MATERIALS

OBJECTIVE: Students will look at a group of shapes that have been marked and write directions that correspond with the markings.
MATERIALS: Transparency of student workbook page 13 • washable transparency marker

CURRICULUM APPLICATIONS

Language Arts: Writing a how-to paragraph or series of steps; planning and remembering events in a story; following directions in any situation (text, worksheet, test) involving order, number, or position
Mathematics: Explaining how to solve a problem, writing and following the correct sequence in solving word problems
Science: Recording a sequence of steps in a science activity
Social Studies: Writing directions using a map; completing charts, graphs, or schedules; reading and constructing time lines
Enrichment Areas: Writing and following instructions for arts and crafts projects; art, music, or physical education activities involving order, number, or position; giving dance, drill, or sports instructions

TEACHING SUGGESTIONS

Encourage students to answer in complete sentences, to name the shape being marked, and to describe the marking.

MODEL LESSON

LESSON

Introduction

Q: Think about a time when you were asked to give directions about how to do something. What did you think about?

Explaining the Objective to the Students

Q: In these exercises, you will look at a group of shapes that have been marked and then write directions describing how to mark the shapes this way.

Class Activity

- Project the example from the transparency of student book page 13.

  Q: In these exercises, you see a row of shapes with some of the shapes marked. Ask yourself, "How would I tell someone to mark the shapes this way?"

  A: In the first example, I would tell a person to "Circle the first and last squares." I would then write the directions above the row of shapes as shown in the example.

WRITING DIRECTIONS
DIRECTIONS: Look at the shapes that are marked. Ask yourself, "How would I tell someone to mark the shapes this way?" Write the directions you would give.
EXAMPLE *Circle the first and last squares.*
A-44
A-45
A-46

GUIDED PRACTICE

EXERCISE: **A-44**

INDEPENDENT PRACTICE

- Assign exercises **A-45** through **A-50**.
- Assign exercises **A-51** through **A-53**, and explain to students that they will be doing something a little different with this page.
  Q: Exercises **A-51** through **A-53** ask you to write directions for marking each row of shapes. Exchange papers with a classmate and see if you can follow each other's directions.

THINKING ABOUT THINKING

Q: What did you pay attention to when you described how to mark shapes?

1. I looked at which shapes were marked. Sometimes I needed to count the shapes.
2. When the shapes were marked (circled, checked, crossed out, or underlined), I looked at how they were marked.
3. I planned what to say to tell which shapes to mark, which marks to use, and how to name the shapes in the row.

PERSONAL APPLICATION

Q: When might you have to write directions at home?

A: Leaving notes for someone, explaining chores or rules (e.g., Mom said to hang our coats on the second and third hooks because the first one is broken), explaining how to play a game, etc.

EXERCISES A-54 to A-57

DESCRIBING POSITION—A

ANSWERS A-54 through A-57 — Student book pages 16–17
Guided Practice: A-54 left, square, add △ near the upper right corner; **A-55** circle, right, add △ near the upper right corner
Independent Practice: A-56 left, right, triangle, add ● near the lower left corner; **A-57** right, circle, left, add ▲ near the upper left corner

LESSON PREPARATION

OBJECTIVE AND MATERIALS

OBJECTIVE: Students will complete sentences describing locations of shapes and follow directions for adding other shapes, helping them learn to write clear directions and accurately describe locations.

MATERIALS: Transparency of student workbook page 16 • washable transparency marker

CURRICULUM APPLICATIONS

Language Arts: Diagramming sentences, placing words in sentences according to function (subject, direct object, etc.), writing descriptive or instructive paragraphs

Mathematics: Plotting graph coordinates, constructing geometric shapes from written or oral directions

Science: Plotting an area or reading a map of a nature study area, observing and describing natural formations

Social Studies: Interpreting and constructing maps, graphs, or diagrams

Enrichment Areas: Art, music, physical education, or dance activities involving position or location; following square dance, drill, or sports instructions involving location, direction, or movement

TEACHING SUGGESTIONS

Encourage students to discuss reasons for their choices and why the other items don't fit. Students' explanations of their thinking are often more meaningful than an adult's explanation.

MODEL LESSON

LESSON

Introduction

Q: You have followed written directions which ask you to do something to certain figures or letters within a given sequence.

Explaining the Objective to the Students

Q: In this activity, you will complete sentences describing the locations of shapes and you will follow directions for adding other shapes. This will help you to write clear directions and accurate descriptions.

Class Activity

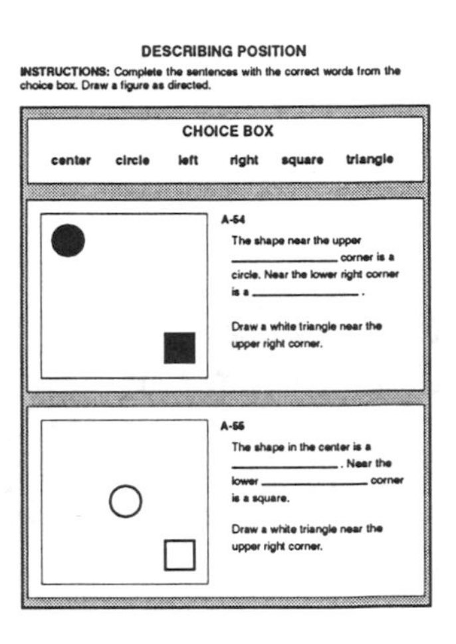

DESCRIBING POSITION

INSTRUCTIONS: Complete the sentences with the correct words from the choice box. Draw a figure as directed.

CHOICE BOX

center circle left right square triangle

A-54

The shape near the upper __________ corner is a circle. Near the lower right corner is a __________ .

Draw a white triangle near the upper right corner.

A-55

The shape in the center is a __________ . Near the lower __________ corner is a square.

Draw a white triangle near the upper right corner.

- Project exercise **A-54** from the transparency of student book page 16.
  Q: Let's look at exercise **A-54** on page 16. In this exercise, you will finish a sentence that describes the drawings at the left. Choose your answers from the group of words listed in the choice box at the top of the page.
- Pause
  Q: Look at the first sentence. "The shape near the upper __________ corner is a circle." Find the circle. Which word in the choice box makes the sentence complete and accurate? Is the circle on the left or right side of the large box?
  A: Left
- Write "left" on the line.
  Q: Now look at the second sentence. "Near the lower right corner is a __________." Find the lower right part of the large box.
- Point to the bottom right corner.
  Q: Which word in the choice box describes the shape in this corner?
  A: Square
- Write "square" on the blank.
  Q: Read the last sentence. "Draw a white triangle near the upper right corner." Find the upper part of the large box.
- Point to the top half.
  Q: Now find the right side of the upper part of the large box.
- Point to the top right corner.

Q: What do the directions tell us to do after we have found the location?
A: "Draw a white triangle."

- Draw the triangle on the transparency.

GUIDED PRACTICE

EXERCISES: **A-54**, **A-55**

- Remind students to read carefully. When they have had sufficient time to complete this exercise, check by discussion to determine whether students have answered correctly.

INDEPENDENT PRACTICE

- Assign exercises **A-56** and **A-57**.

THINKING ABOUT THINKING

Q: What did you pay attention to when you described the location of something?

1. I read the clue.
2. I looked for the shape mentioned in the clue.
3. I decided what word from the "choice box" completes the sentence.
4. I wrote the answer on the blank.

PERSONAL APPLICATION

Q: At what other times might you need to find or place something in a given location or describe the location of something?

A: Examples include craft or sports activities involving position, assembling games or models, using computer-tabulated questionnaires or answer sheets, playing games involving location or direction, giving directions to someone else, reading directions regarding placement or location of an answer.

EXERCISES A-58 to A-60

DESCRIBING POSITION—B

ANSWERS A-58 through A-60 — Student book pages 18–19
Guided Practice: A-58 The red block is above the green block. The blue block is to the right of the green block. OR The green block is left of the blue block and below the red block. OR The green block is below the red block and left of the blue block. OR Any acceptable variation.
Independent Practice: A-59 The blue cube is above the red cube. The green cube is to the right of the red cube. The black cube is below the red cube. OR Any acceptable variation. **A-60** The blue cube is left of the black cube. The green cube is below the black cube, and the red cube is below the green one. OR Any acceptable variation.

LESSON PREPARATION

OBJECTIVE AND MATERIALS

OBJECTIVE: In order to help them write clear directions and accurately describe positions, students will build a pattern of cubes and describe the patterns they build.

MATERIALS: Transparency of student workbook page 18 • cubes (Multilinks or Unifix) one each of four different colors • crayons • washable transparency marker

CURRICULUM APPLICATIONS

Language Arts: Placing words in sentences according to function (subject, direct object, etc.), writing descriptive or instructive paragraphs

Mathematics: Plotting graph coordinates, constructing geometric shapes from written or oral directions

Science: Plotting an area or reading a map of a nature study area, observing and describing natural formations

Social Studies: Interpreting and constructing maps, graphs, or diagrams

Enrichment Areas: Art, music, physical education, or dance activities involving position or location; following square dance, drill, or sports instructions involving location, direction, or movement

TEACHING SUGGESTIONS

Encourage students to discuss reasons for their word choices. Students' explanations of their thinking are often more meaningful than an adult's explanation.

MODEL LESSON

LESSON

Introduction

Q: You have completed sentences describing locations of shapes.

Explaining the Objective to the Students

Q: In this activity, you will build a pattern of interlocking cubes and describe the pattern. This will help you learn to write clear directions and accurate descriptions.

Class Activity

- Project the example from the transparency of page 18.
  Q: Let's look at the example on page 18.
- Ask students to build a similar pattern. (Note: It is not necessary that all students use red, green, and blue cubes; they may use one cube each of any three different colors.)
  Q: Build a train of 3 cubes of 3 different colors. I will match the colors used in the example.
- Allow time for building block trains.
  Q: Describe the "train" I built.
  A: The green block is above the red block. The blue block is below the red block. OR The green block is on top, the red block is in the middle, and the blue block is on the bottom. OR Any acceptable variation.

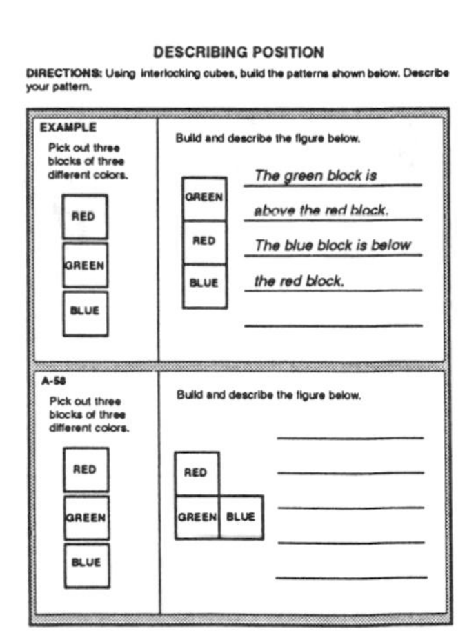
DESCRIBING POSITION

DIRECTIONS: Using interlocking cubes, build the patterns shown below. Describe your pattern.

EXAMPLE

Pick out three blocks of three different colors.

RED
GREEN
BLUE

Build and describe the figure below.

GREEN
RED
BLUE

The green block is above the red block. The blue block is below the red block.

A-58

Pick out three blocks of three different colors.

RED
GREEN
BLUE

Build and describe the figure below.

RED
GREEN BLUE

- Ask a few students to describe the trains they built with other colors.

GUIDED PRACTICE

EXERCISE: **A-58**

- If you use colors other than those in the exercise, color your pattern to match your cubes. Allow time for building and class discussion.

INDEPENDENT PRACTICE

- Assign exercises **A-59** and **A-60**.

THINKING ABOUT THINKING

Q: What did you pay attention to in describing the position of the cubes?

1. I picked a cube in the pattern.
2. I found the positions of the other cubes compared to the cube I picked.
3. I described the positions of the other cubes as above, below, left, or right of the cube I picked.

PERSONAL APPLICATION

Q: When might you have to describe the positions of things?

A: Examples include making requests or giving directions (e.g., "The key is in the red box on top of the blue book"), explaining how to play a game or build a model.

EXERCISES A-61 to A-62

CHARACTERISTICS OF A SHAPE

ANSWERS A-61 through A-62 — Student book pages 20–21
Guided Practice: A-61 See below.
Independent Practice: A-62 See below.

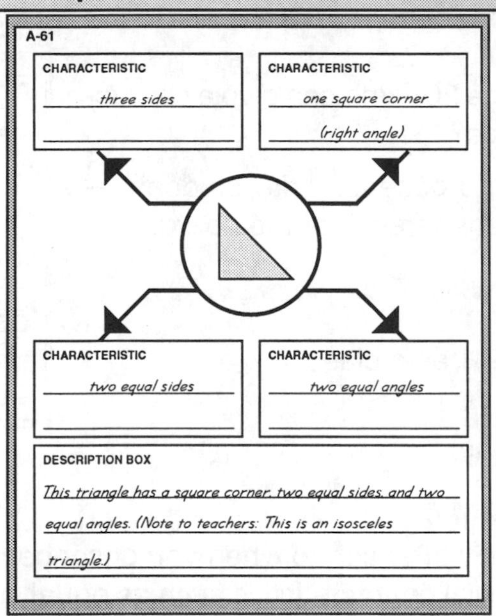

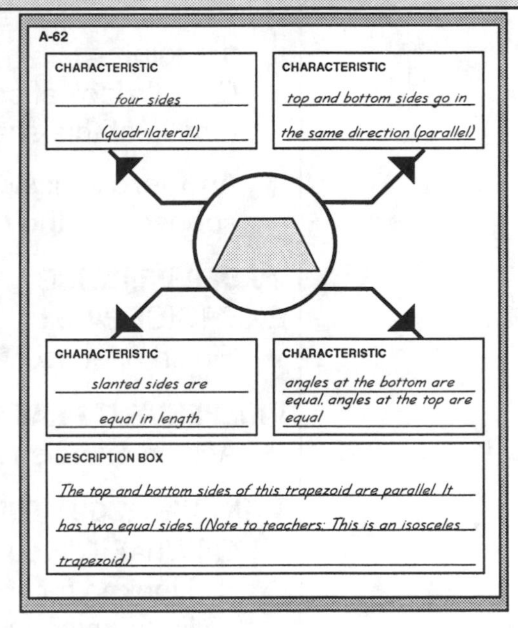

LESSON PREPARATION

OBJECTIVE AND MATERIALS

OBJECTIVE: Students will look carefully at a shape, write down the characteristics that they see, then combine these characteristics to describe the shape.

MATERIALS: Transparency of student workbook page 20 • washable transparency marker

CURRICULUM APPLICATIONS

Language Arts: Visual discrimination exercises for reading readiness

Mathematics: Naming geometric shapes
Science: Recognizing shapes of leaves, insects, or shells
Social Studies: Recognizing geographic features on map puzzles
Enrichment Areas: Recognizing road signs, discerning patterns in art

TEACHING SUGGESTIONS

Reinforce the following vocabulary emphasized in this lesson: *characteristic, equal, parallel, triangle, trapezoid.* Encourage students to use these terms in their discussion. If students are familiar with the term *angle*, encourage them to discuss angle when describing the shapes.

Encourage students to discuss reasons for their word choices. As you remind students of the key characteristics in this exercise and subsequent exercises, remember the language that the students use to describe their own thinking. Students' explanations of their thinking are often more meaningful to them than an adult's explanation.

MODEL LESSON

LESSON

Introduction

Q: In previous lessons, you have built and described a pattern of shapes.

Explaining the Objective to the Students

Q: In this activity, you will combine the characteristics of a shape to write a description of the shape.

Class Activity

- Project transparency of page 20 and point to the triangle.

  Q: Let's look at page 20. Who can give a characteristic of this shape?

- As the discussion proceeds, write student responses on the transparency or the board.

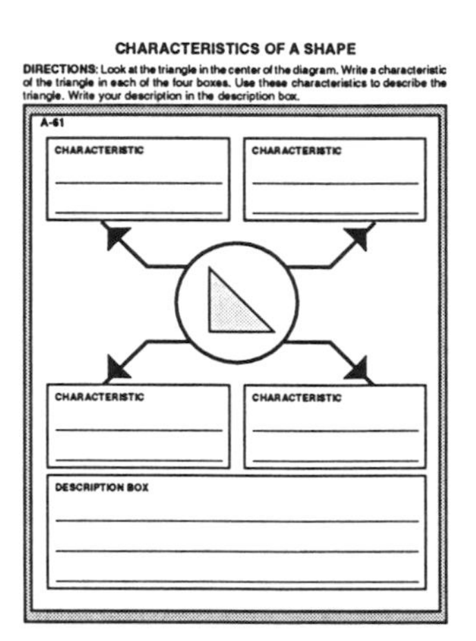

GUIDED PRACTICE

EXERCISE: **A-61**

- Finish exercise **A-61** as a class.

INDEPENDENT PRACTICE

- Assign exercise **A-62**.

THINKING ABOUT THINKING

Q: What did you pay attention to when you described the shape?

1. I looked for special characteristics such as equal sides, equal angles, lines going in the same direction (parallel lines).
2. I described each characteristic.
3. I combined the characteristics to make a complete description.

PERSONAL APPLICATION

Q: When might you have to describe shapes?

A: Examples include giving directions, following the directions to assemble toys or models, explaining how to play a game.

Chapter 2

FIGURAL SIMILARITIES AND DIFFERENCES

(Student book pages 23–70)

EXERCISES B-1 to B-17

MATCHING SHAPES

ANSWERS B-1 through B-17 — Student book pages 24–30
Guided Practice: B-1 a; **B-2** c; **B-3** b
Independent Practice: B-4 c; **B-5** b; **B-6** a; **B-7** c; **B-8** b, d; **B-9** a, e; **B-10** b, d; **B-11** c, e; **B-12** b, c; **B-13** a, d; **B-14** a, d; **B-15** to **B-17** See below.

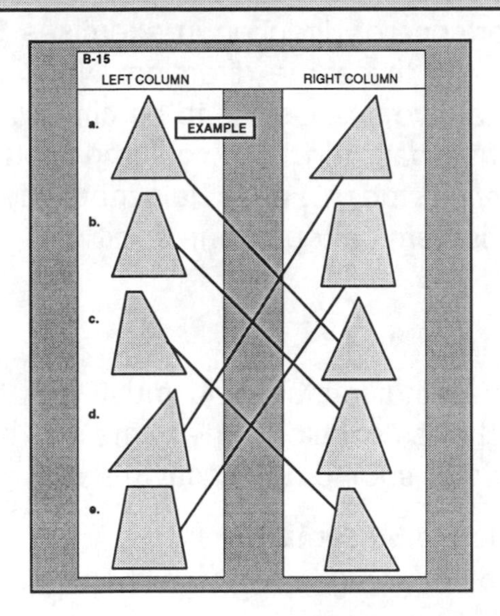

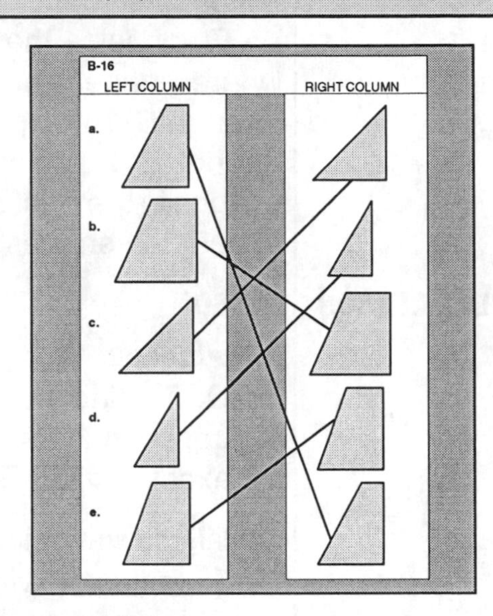

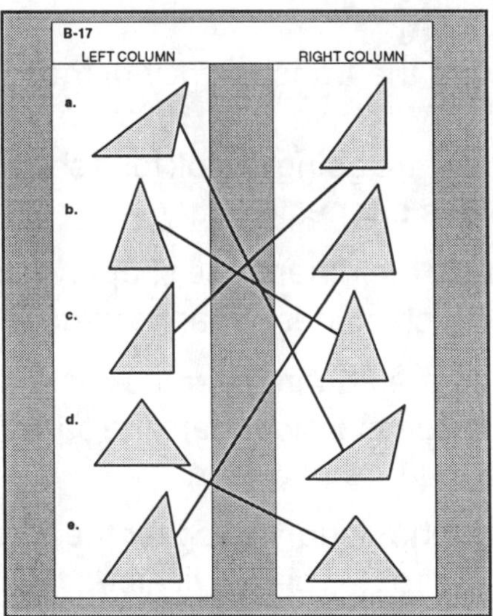

LESSON PREPARATION

OBJECTIVE AND MATERIALS

OBJECTIVE: Students will match a shape with one that is like it.
MATERIALS: TM 3 (p. 251, top section, cut out movable square)

CURRICULUM APPLICATIONS

Language Arts: Visual discrimination exercises for reading readiness
Mathematics: Similarity and congruence exercises, writing numerals correctly
Science: Recognition of similarly shaped leaves, insects, or shells
Social Studies: Matching puzzle sections to geographic features on map puzzles
Enrichment Areas: Recognizing shapes of road signs, discerning different patterns in art

TEACHING SUGGESTIONS

Reinforce the following vocabulary emphasized in this lesson: *square*, *circle*, *rectangle*, *parallelogram*, *triangle*, *right triangle*, and *trapezoid*. Encourage students to use these terms in their discussion. Students should describe shapes as pointed, shorter, taller, wider, etc.

Emphasize the importance of direction in exercises **B-4** through **B-10**, where shapes must face in the same direction. Note: In exercises **B-11** through **B-14**, explain that two shapes are to be circled. In exercises **B-15** through **B-17**, point out that shapes in the left column should be matched with shapes in the right column. If students need help matching shapes, they may cut out the shapes and fit them on top of one another.

MODEL LESSON

LESSON

Introduction

Q: To find two shapes that match, you must look for the similarities between them. The shapes should be the same size and have the same exact outline. Each detail should be exactly the same.

Explaining the Objective to the Students

Q: In this activity, you are going to match a shape with one that is like it.

Class Activity

- Project the transparency of the movable square from TM 3.
  Q: You are going to look for a shape that is exactly like this shape.
- Project the other three shapes at the top of TM 3.
  Q: Which of these shapes matches the first one?
- Place the square over the rectangle.
  Q: These two pieces are the same height, but this one (square) is wider.
- Move the square to cover the triangle.
  Q: These two pieces fit along the bottom, but the triangle has a pointed top, so the sides don't match.
- Move the square to cover the other square.
  Q: The square is an exact match. It is just as tall and just as wide and all the edges touch.

GUIDED PRACTICE

EXERCISES: **B-1** through **B-3**

- When students have had sufficient time to complete the exercises, check by discussion that they have answered correctly. Discuss why the incorrect answers were eliminated.

INDEPENDENT PRACTICE

- Assign exercises **B-4** through **B-17**. This lesson may require two sessions.

THINKING ABOUT THINKING

Q: What did you pay attention to when you matched these figures?

1. I looked carefully at the details (size of the figure, length of the sides, size of the angle, pattern or color, similarity to some common object, etc.).
2. I matched equal parts (side, angle, etc.).
3. I checked that all the sides and angles were the same.
4. I checked how those figures that didn't match are really different.

PERSONAL APPLICATION

Q: When might you have to match shapes at home?

A: Examples include putting away toys or tools, matching building blocks, matching parts or sections from construction toys, putting away dishes or silverware.

EXERCISES B-18 to B-26

WHICH SHAPE DOES NOT MATCH?/MATCHING SHAPES THAT HAVE BEEN TURNED

ANSWERS B-18 through B-26 — Student book pages 31–34
Guided Practice: B-18 b (opposite) **B-19** c (opposite) **B-20** a (opposite)
Independent Practice: B-21 c (upside down) **B-22** b (opposite) **B-23** a (flipped and turned) **B-24** c (opposite) **B-25** to **B-26** See below.

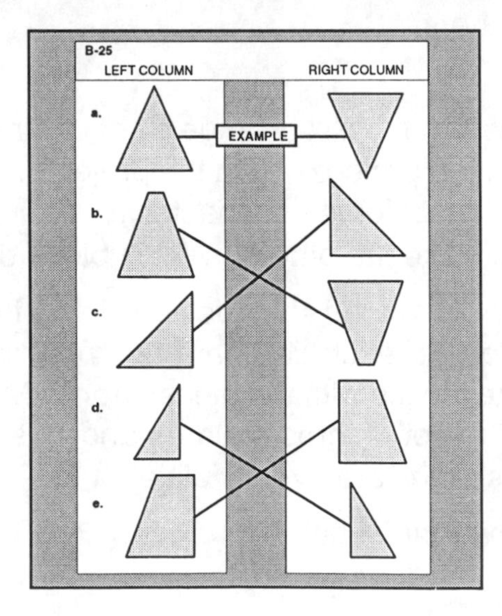

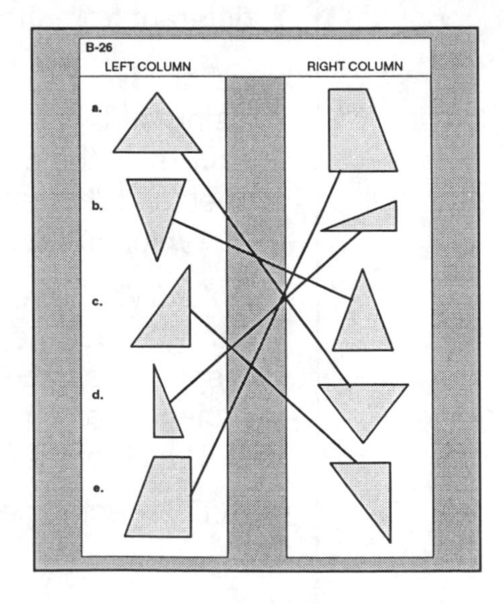

LESSON PREPARATION

OBJECTIVE AND MATERIALS

OBJECTIVE: Students will practice finding the shape that is different from the others.

MATERIALS: Transparency of TM 3 (p. 251, bottom section, cut apart as indicated) • attribute blocks or teacher-made manipulatives • cards with letters *b*, *d*, *p*, *q*, and *a* printed in manuscript letters

CURRICULUM APPLICATIONS

Language Arts: Visual discrimination exercises for reading readiness, discriminating between punctuation marks

Mathematics: Distinguishing geometrically different shapes

Science: Sorting leaf, shell, or insect collections by shape

Social Studies: Distinguishing between landforms according to size and shape

Enrichment Areas: Distinguishing between purposes of road signs by shape, distinguishing different note values in music

TEACHING SUGGESTIONS

Reinforce the following vocabulary emphasized in this lesson: *square*, *circle*, *rectangle*, *parallelogram*, *triangle*, *pentagon*, and *right triangle*. Remind students of the words they used for shapes and sizes in the previous lesson. The difference between this exercise and the previous one is the emphasis on the word *not*.

MODEL LESSON

LESSON

Introduction

Q: Do you remember learning to tell the difference between *b* and *d?* (Hold up letters.) Between *p* and *q?* (Hold up letters.) Between *a* and *d?* (Hold up letters.) Small lines (hold up *a* and *d)* or the direction the letter faces (hold up *p* and *q)* made a big difference. You were learning to see small differences in the shapes of letters.

Explaining the Objective to the Students

Q: In this exercise, you are going to practice finding the shape that is different from the others.

Class Activity

- Using attribute or design blocks, arrange a set of three triangles and one square on the overhead projector. (On the projector, these blocks will be seen as silhouettes, so color is not a factor.)
  Q: Which block is not like the others? Which block doesn't belong?
- Note: If you have a prearranged set of design blocks for each student, they can pick out the block in each set that does not match. After students respond, replace the blocks with the pieces from TM 3. On the overhead projector, arrange the set of three squares and a rectangle.
  Q: Which of these shapes are exactly alike?
- Put the rectangle over the square.

Q: The rectangle is not wide enough.

- Remove the rectangle. Stack the squares to confirm that all are the same size.
- Hold up the *p* and *q* cards.
  Q: The direction the shape faces is important, just like the difference between *p* and *q*.
- Hold up the *p* and *d* cards. Slowly turn the *p* card upside down.
  Q: If I turn the *p* upside down, it begins to look like a *d* instead. So direction makes a difference, too.
- Turn to page 31 in the student book.
  Q: Look at the example at the top of page 31. The shape that is different from the others has been crossed out. How is this shape different from the others?
  A: It faces in the opposite direction. (It has been turned.)

GUIDED PRACTICE

EXERCISES: **B-18** through **B-20**

- When students have had sufficient time to complete the exercises, check by discussion that they have answered correctly. Discuss why each nonmatching figure was eliminated.

INDEPENDENT PRACTICE

- Assign exercises **B-21** through **B-26**. Note that **B-25** and **B-26** are matching exercises involving similar skills. When students have had sufficient time to complete the exercises, check by discussion that they have answered correctly. Discuss why each nonmatching figure was eliminated. If students have trouble matching shapes, they may cut out the shapes and fit them on top of one another.

THINKING ABOUT THINKING

Q: What did you pay attention to when you matched these figures?

1. I looked carefully at the details (size of the figure, length of the sides, size of the angle, pattern or color, similarity to a common object, etc.).
2. I matched equal parts (side, angle, etc.).
3. I checked that all the sides and angles were the same.
4. I checked how those figures that didn't match are really different.

PERSONAL APPLICATION

Q: When might you need to find something that is not shaped like the others?

A: Examples include identifying the wrong block or the wrong eating utensil stored with others, sorting edges from interior puzzle pieces.

EXERCISES B-27 to B-34

FINDING SHAPES

ANSWERS B-27 through B-34 — Student book pages 35–37
Guided Practice: B-27 b, c; **B-28** a, d
Independent Practice: B-29 b, c, d; **B-30** a, b, c; **B-31** b, c, d; **B-32** c; **B-33** a, b, c; **B-34** b, c
See below.

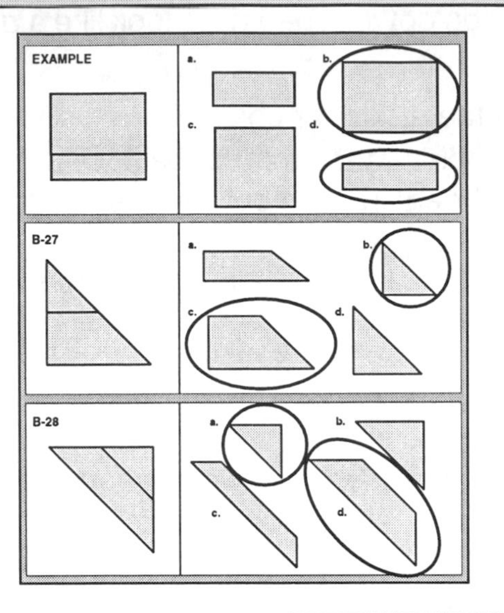

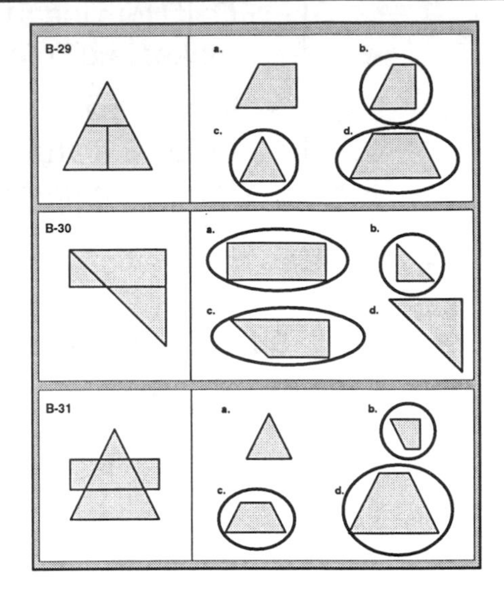

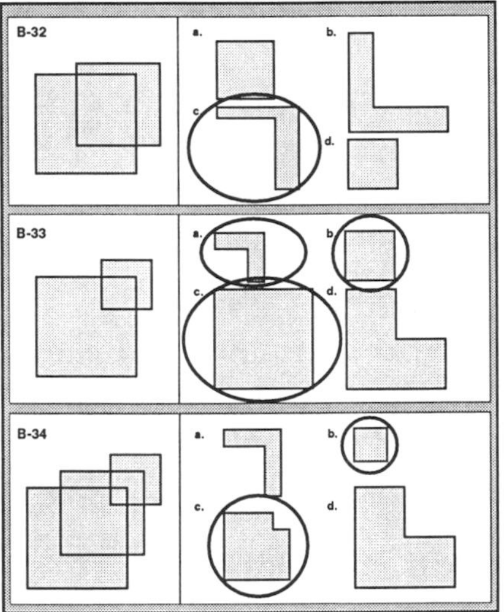

LESSON PREPARATION

OBJECTIVE AND MATERIALS

OBJECTIVE: Students will decide which shapes make up a whole figure.

MATERIALS: Transparency of TM 4 (p. 252, cut apart as indicated) • piece of clothing with obvious seams • attribute blocks, pattern blocks, or tangrams (optional)

CURRICULUM APPLICATIONS

Language Arts: Recognizing syllables of words
Mathematics: Identifying names and functions of numbers in arithmetic problems
Science: Naming component parts of equipment or compound organisms,

such as compound flowers that are made up of multiple flowerets; differentiating among skeletal parts
Social Studies: Recognizing map divisions, finding and tracing routes on road maps
Enrichment Areas: Duplicating sections of an art project, naming parts of a stanza of music (clef sign, bar, time signature, etc.), recognizing different steps in a dance

TEACHING SUGGESTIONS

Reinforce the following vocabulary emphasized in the lesson: *taller*, *shorter*, *wider*, *narrower*, or *more pointed*. Students should know the words which describe the shapes and sizes used in the exercises since these words have been practiced in the previous lessons. It may be difficult for students to describe these figures, so encourage students to use words with qualifiers, such as slightly smaller, much larger, etc. These exercises are valuable language training only if the terminology is used and reinforced so it becomes functional in the learner's vocabulary.

MODEL LESSON

LESSON

Introduction

- Show a piece of clothing that is seamed, and indicate the parts of the whole.
  Q: Have you ever watched anyone put together a piece of clothing? He or she seems to "see" just how things fit. The pieces become the whole. Sometimes, if you look closely, you can see where the pieces have been put together.

Explaining the Objective to the Students

Q: You are going to decide which shapes make up a whole figure.

Class Activity

- Choose any two shapes (use tangram or any design-type blocks) and join them so the outlined figure shows on the overhead projector. Select four other shapes, two of which are the same as the first two, and place them to the right of the figure.
  Q: Which two pieces, if put together, would make the figure on the left?
- Try different combinations as the children say them. When they have arrived at a choice, you can prove correct choices by stacking the figures.
- Project the transparency of TM 4 and have the students follow in the book.
  Q: This is a picture of the example on page 35. Which of these shapes can you find in the figure at the top?
- Point to figures *a*, *b*, *c*, and *d*. Place figure *a* over each part of the pattern.
  Q: This piece is wide enough, but it is too short for the top part of the pattern and too tall for the bottom part.

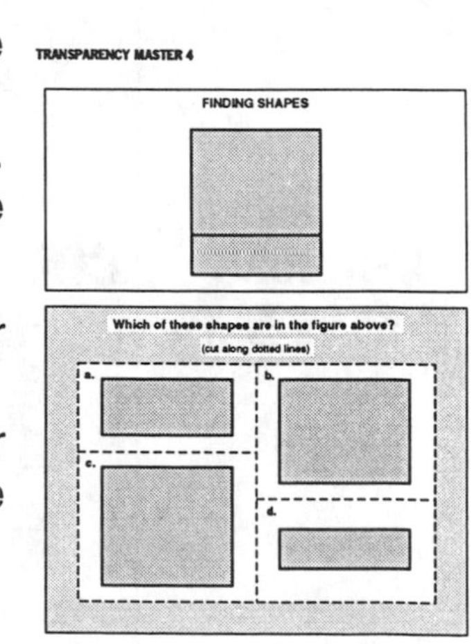

- Remove figure *a*; place figure *b* over the pattern parts.
  Q: This piece is the same width and the same height as the top part. It matches exactly.
- Replace this piece with figure *c*.
  Q: This piece is too tall for either part of the pattern.
- Replace this piece with figure *d*.
  Q: This piece is the same width and the same height as the bottom part of the pattern. It fits exactly.
- Put pieces *b* and *d* together to show that they form a pattern like the one on the left.

GUIDED PRACTICE

EXERCISES: **B-27, B-28**

- If students have trouble matching shapes, they may cut out the shapes and fit them on top of one another. (Note that in the following exercises, the shapes may face in any direction.) When students have had sufficient time to complete the exercises, check by discussion that they have answered correctly. Encourage explanations of why the uncircled figures were not chosen.

INDEPENDENT PRACTICE

- Assign exercises **B-29** through **B-34**.

THINKING ABOUT THINKING

Q: What did you pay attention to when you found parts of these figures?

1. I looked carefully at the details of the shapes in the choice box (size of the figure, length of the sides, size of the angle, similarity to some common object, etc.).
2. I looked at the figure box to see which shapes I might find there.
3. I matched equal parts (side, angle, etc.).
4. I checked that all the sides and angles were the same.
5. I checked how those shapes that didn't match were really different.

PERSONAL APPLICATION

Q: When might you be asked to recognize the parts of something after it has been put together?

A: Examples include art activities, parts of furniture, parts of tools or engines, parts of picture puzzles before you take them apart.

EXERCISES B-35 to B-39

COMBINING INTERLOCKING CUBES

ANSWERS B-35 through B-39 — Student book pages 38–40
Guided Practice: B-35 a, c
Independent Practice: B-36 b, c; **B-37** a, d; **B-38** b; **B-39** a, d
See next page.

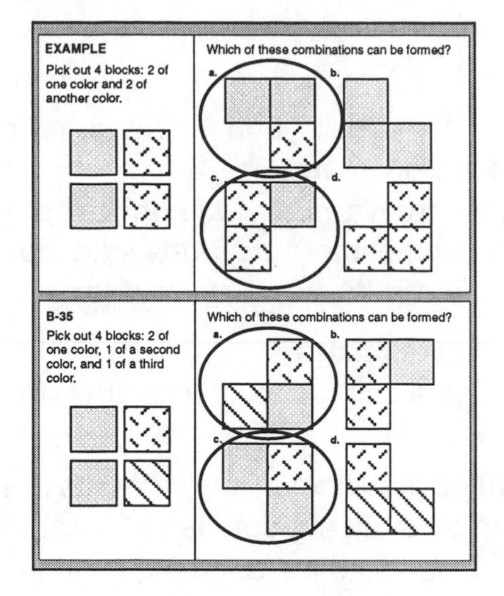

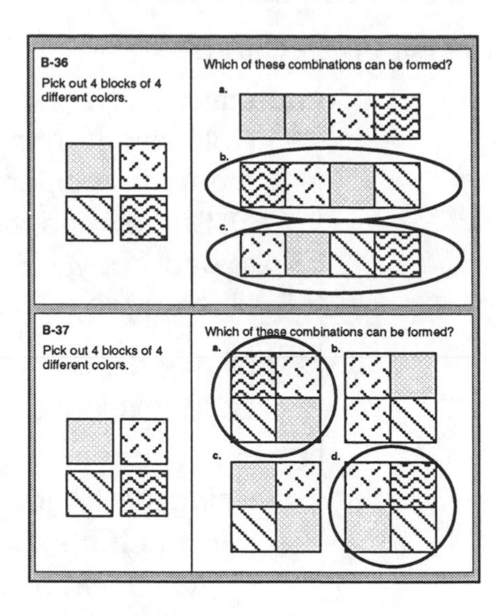

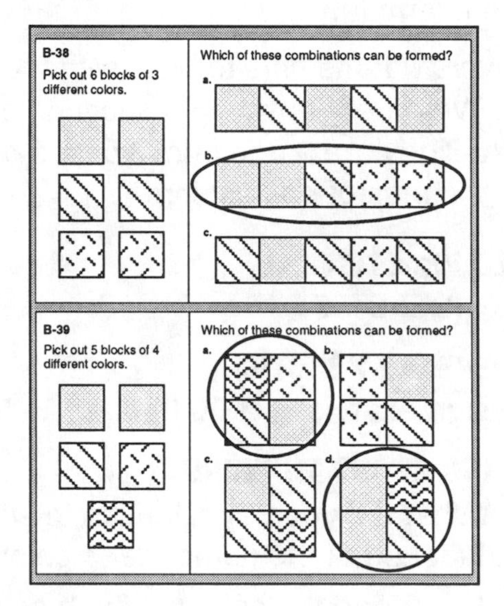

LESSON PREPARATION

OBJECTIVE AND MATERIALS

OBJECTIVE: Students will practice building combinations of cubes.
MATERIALS: Interlocking cubes (Unifix or Multilink cubes)

CURRICULUM APPLICATIONS

Language Arts: Recognizing parts that make up words
Mathematics: Recognizing fractional parts
Science: Recognizing parts of skeletal structure, parts of a body
Social Studies: Finding subdivisions of political divisions (wards, cities, counties, states, regions, countries, continents) on maps
Enrichment Areas: Sections of orchestras or bands, movements in classical music, art collages, divisions of a music or drama production from a printed program

TEACHING SUGGESTIONS

Ask the students to identify objects in the classroom which are wholes made up of geometric shapes (examples include ceiling tiles, floor tiles, brick or concrete block walls, window panes, etc.). Note how the part appears within the whole.

MODEL LESSON

LESSON

Introduction

- Hold out 4 cubes: 2 of one color and 2 of another color. Fix the cubes together in different arrangements.

  Q: Sometimes, when you put a model together, the parts don't look like you expect they should. Sometimes the parts can be put together many different ways. Notice that there are many ways to put these cubes together.

Explaining the Objective to the Students

Q: In this lesson, you are going to build patterns with cubes.

Class Activity

- Using the 4 cubes mentioned above, ask students to show that the circled items in the example on student book page 38 can be built.

  Q: As shown on page 38, build the cube combinations that are circled in the example.

- Give students time to build with the cubes.

  Q: Why are the two combinations on the right not circled?

  A: They are made from 3 cubes of the same color, and we only have 2 cubes of each color.

GUIDED PRACTICE

EXERCISE: **B-35**

INDEPENDENT PRACTICE

- Assign exercises **B-36** through **B-39**.

THINKING ABOUT THINKING

Q: What did you pay attention to when you matched these figures?

1. I looked at the number and color of the blocks.
2. I matched the colors of the blocks to the combinations to be formed.
3. I checked that the combination I made matched the combination to be formed.
4. If I could not build a combination, I checked the number and color of the blocks to see why they didn't match.

PERSONAL APPLICATION

Q: In what objects can you see that parts have been combined to make a whole?

A: Examples include parts of a sandwich or pizza; sections of oranges or grapefruit; parts of a clock, gear, or motor; puzzles, tangrams, or other construction toys; pieces of furniture.

EXERCISES B-40 to B-42

COMBINING SHAPES

ANSWERS B-40 through B-42 — Student book pages 41–42
Guided Practice: B-40 b, d
Independent Practice: B-41 a, d; **B-42** a, d See next page.

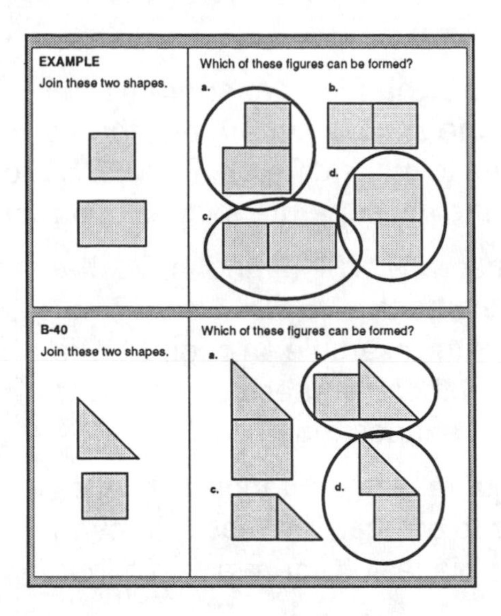

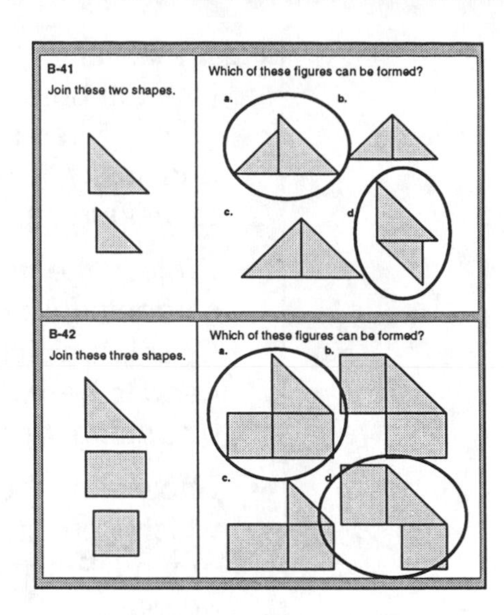

LESSON PREPARATION

OBJECTIVES AND MATERIALS

OBJECTIVE: Students will recognize what the whole looks like when the parts are put together.

MATERIALS: Transparency of TM 5 (p. 253, cut apart as indicated) • dress or shirt pattern • attribute blocks, pattern blocks, or tangrams (optional)

CURRICULUM APPLICATIONS

Language Arts: Recognizing syllables
Mathematics: Recognizing fractional parts
Science: Recognizing parts of skeletal structure, parts of an organism
Social Studies: Finding subdivisions of political divisions (wards, cities, counties, states, regions, countries, continents) on maps
Enrichment Areas: Identifying sections of orchestras or bands, recognizing verses and choruses of songs

TEACHING SUGGESTIONS

Help the students identify objects in the classroom which are wholes made up of geometric shapes (examples include ceiling tiles, floor tiles, brick or concrete block walls, window panes, etc.). Point out to students how the part appears within the whole.

MODEL LESSON

LESSON

Introduction

- Hold up a sleeve pattern piece over your own clothes.
  Q: Sometimes when parts are put together, they don't look quite as you would expect. This pattern piece becomes the part of the shirt (or dress) that you see here.
- Point to sleeve of shirt or dress.
  Q: It looks different as a piece than it does as part of a whole shirt.

Explaining the Objective to the Students

Q: In this exercise, you are going to practice recognizing what the whole looks like when the parts are put together.

Class Activity

- Use attribute blocks or handmade tangrams to show and identify each shape. First, show the shapes separately, then place them side by side, and finally, join them to make different figures. Describe the shape of the total figure, and name the separate shapes that make up the total figure.

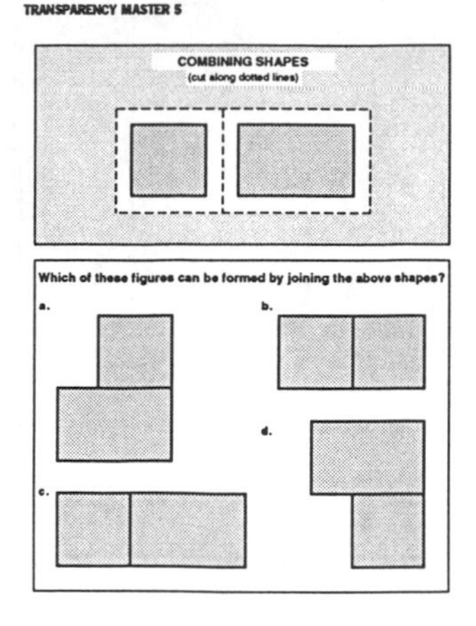

- Project the transparency TM 5 and have the students follow in the book.
  Q: This is a picture of the example on page 41. We need to see which of these figures can be made by putting these two shapes together.
- Show that the square and rectangular pieces which you have cut out can be easily moved around. Put the two pieces on figure *a* with edges matching.
  Q: All the edges match. This figure can be made by joining the shapes.
- Move the pieces over to figure *b*.
  Q: All the edges do not match. This figure can't be made by joining the shapes.
- Move the pieces down to figure *c*.
  Q: All the edges match. This figure can be made by joining the shapes.
- Move the pieces over to figure *d*.
  Q: All the edges match. This figure can be made by combining the pieces. In your workbook, three of the figures have been circled, showing that they can be made by putting the pieces together.

GUIDED PRACTICE

EXERCISE: **B-40**

- When students have had sufficient time to complete the exercise, check their answers by discussing correct choices and reasons for eliminating other choices.

INDEPENDENT PRACTICE

- Assign **B-41** and **B-42**.

THINKING ABOUT THINKING

Q: What did you pay attention to when you matched these figures?

1. I looked at the shapes to be joined.
2. I mentally moved the shapes to see if I could form the combination.
3. I checked that the combination I mentally formed matched the combination to be formed.
4. If I could not form a combination, I checked on the reason I could not.

PERSONAL APPLICATION

Q: Where else can you see how parts have been combined to make up a whole?

A: Examples include parts of a sandwich or pizza; sections of oranges or grapefruit; parts of a clock, gear, or motor; puzzles, tangrams, or other construction toys; pieces of furniture.

EXERCISES B-43 to B-47

FINDING AND TRACING PATTERNS

ANSWERS B-43 through B-47 — Student book pages 43–45
Guided Practice: B-43 b, d, The key pattern is a parallelogram slanted to the right. See below.
Independent Practice: B-44 a, c, "kite"-shaped quadrilateral; **B-45** a, c, two triangles; **B-46** b, d, triangle on top of a rectangle; **B-47** a, c, a small triangle inside a larger triangle See below.

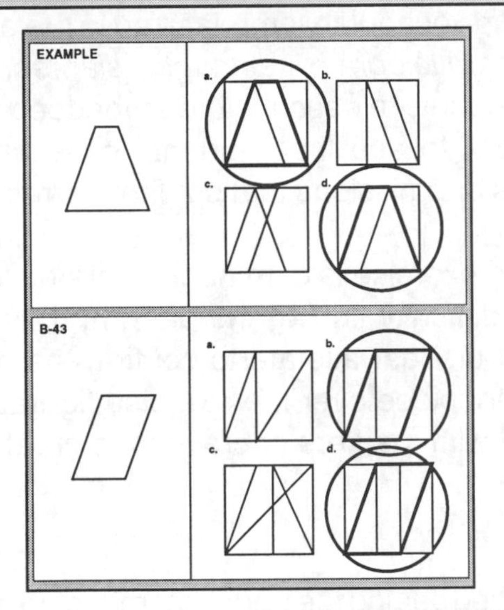

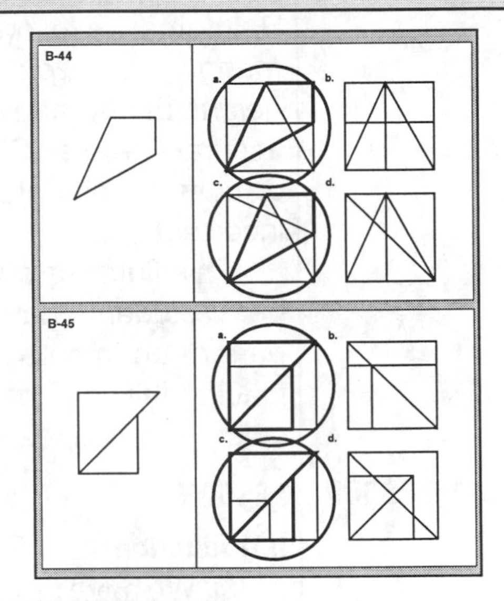

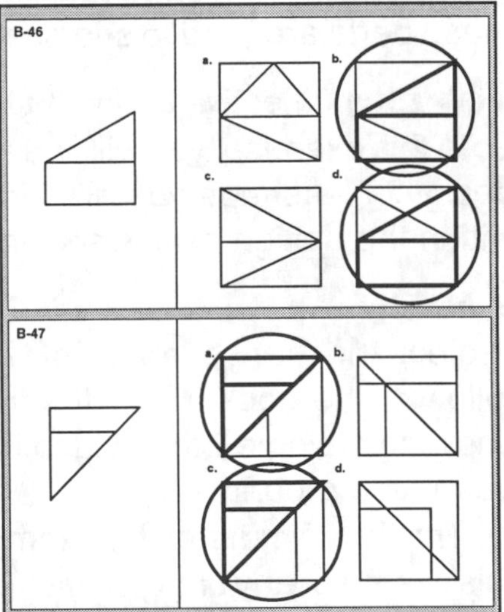

LESSON PREPARATION

OBJECTIVE AND MATERIALS

OBJECTIVE: Students will find and trace a pattern that is hidden in a larger design. The shape must face the same direction within the design as it is shown in the example.

MATERIALS: Transparency of TM 6 (page 254, top section, cut out trapezoid) • crayons or markers (optional)

CURRICULUM APPLICATIONS

Language Arts: Recognizing rhyme pattern in poetry, finding hidden words in larger words, finding repeated letter patterns in words, phonics instruction
Mathematics: Finding patterns or other shapes within geometric shapes
Science: Showing the path of blood through the body (or the path of any body system)
Social Studies: Indicating a historical trip on a map (the Oregon Trail, the Pony Express route, the travels of Lewis and Clark); locating the boundaries of national parks, counties, forest regions, etc.
Enrichment Areas: Finding repeated interval patterns in music, reproducing art projects, learning individual steps to form a dance pattern

TEACHING SUGGESTIONS

Reinforce the following vocabulary emphasized in the exercises: *side, angle, triangle, trapezoid, parallelogram, rectangle, slanted, diagonal, base,* and *height.* Encourage students to describe the embedded pattern in reasonably accurate terms. Check to see that students have answered correctly by discussing the details or omissions that confirm correct choices or eliminate incorrect ones.

This figure-ground exercise is commonly more difficult that it appears to be. If students have difficulty finding the pattern, they may trace the given pattern on thin paper or clear acetate to confirm the embedded pattern by placing the transparent piece over the exercise figures. Students may also color in the trapezoid with markers or crayons to create greater contrast.

MODEL LESSON

LESSON

Introduction

Q: We have practiced seeing the parts that make up the whole. Sometimes these parts are hard to see because other lines or shapes get in the way.

Explaining the Objective to the Students

Q: In this exercise, you will find and trace a pattern that is hidden in a larger design. The shape you are looking for must face in the same direction within the design as it is shown in the example.

Class Activity

- Project the transparency of TM 6 as students follow in the book. Move the transparency piece showing a single trapezoid to illustrate that it can be moved around.
  Q: This is a picture of the example on page 43. In this exercise, any time you move the trapezoid you must keep it facing the same direction. Don't turn the trapezoid as you move it, just slide it. We are going to use this shape to see if it is "hidden" in each figure.
- Move the trapezoid to cover figure *a*.
  Q: This trapezoid is the same size as the trapezoid in figure *a*, so this trapezoid is hidden in figure *a*. Trace the trapezoid so it can be seen in the figure.

- Demonstrate the tracing using a transparency marker. Move the trapezoid to cover figure *b*.
  Q: Notice that the left slanted part of the trapezoid is missing, so the trapezoid is not hidden in figure *b*.
- Move the trapezoid to cover figure *c*.
  Q: Notice that the right slanted side of the trapezoid is missing, so the trapezoid is not hidden in figure *c*.
- Move the trapezoid to figure *d* and cover the figure with the trapezoid so it fits.
  Q: All of the edges fit again, so the trapezoid is hidden in figure *d*.

GUIDED PRACTICE

EXERCISE: **B-43**

INDEPENDENT PRACTICE

- Assign exercises **B-44** through **B-47**.

THINKING ABOUT THINKING

Q: What did you pay attention to when you matched these figures?

1. I looked carefully at the details in the shape on the left (size of the figure, length of the sides, size of the angle, similarity to a common object, etc.).
2. I looked for matching equal parts (side, angle, etc.) in each of the four figures in the exercise.
3. I checked that all the sides and angles were the same.
4. I checked how those figures that didn't match were really different.

PERSONAL APPLICATION

Q: When might you have to find or follow a line or pattern in a larger design?

A: Examples include reading a map, simple electrical circuits, finding hidden figures in pictures.

EXERCISES B-48 to B-52

DIVIDING SHAPES INTO EQUAL PARTS—A

ANSWERS B-48 through B-52 — Student book pages 46–47
Guided Practice: B-48 Make sure each half has an equal number of cubes.
Independent Practice: B-49 to **B-52** Make sure each half has an equal number of cubes.

LESSON PREPARATION

OBJECTIVE AND MATERIALS

OBJECTIVE: Using interlocking cubes, students will divide patterns to show equal parts.

MATERIALS: Interlocking cubes (Multilink or Unifix cubes) • paper • crayons or markers (optional)

CURRICULUM APPLICATIONS

Language Arts: Dividing a long poem for several people to memorize

Mathematics: Showing equal fractional parts of a whole, basic symmetry exercises

Science: Identifying symmetry of body parts or natural phenomena (leaves, plants, shells, etc.)

Social Studies: Dividing historical time lines into intervals, making or interpreting charts and graphs

Enrichment Areas: Creating geometric art designs, dividing group art projects fairly

TEACHING SUGGESTIONS

Ask students to describe how they divided each of the shapes and how they confirmed that the resulting parts were equal.

MODEL LESSON

LESSON

Introduction

- Hold up a sheet of paper. (One time fold across, the other fold up and down.)
  Q: It is easy to fold a sheet of paper exactly in half.
- Hold up a second sheet of paper.
  Q: I can also fold a sheet of paper exactly in half a second way.

Explaining the Objective to the Students

Q: In this activity with interlocking cubes, you will divide patterns to show equal parts.

Class Activity

- Distribute a collection of interlocking cubes to each pair of students: ten cubes of one color and ten cubes of a second color.
  Q: Open your books to page 46. Build each of the patterns in the examples.
- Give students ample time to build each of the patterns.
  Q: How are each of the patterns alike?
  A: Both are ways of dividing a shape into equal parts.

  Q: How are the patterns different?
  A: One pattern is long (back and forth), the other pattern is short (up and down). OR One pattern is made up of two rectangles and the other is made up of two squares.

GUIDED PRACTICE

EXERCISE: **B-48**

INDEPENDENT PRACTICE

- Assign **B-49** through **B-52**.

THINKING ABOUT THINKING

Q: What did you pay attention to when you divided the shapes into equal parts?

1. I looked carefully to see if a similar paper shape could be folded into equal parts.
2. I built the two parts.
3. I checked to see that the parts were equal by pulling the halves apart and stacking them on top of one another.

PERSONAL APPLICATION

Q: When might you need to decide whether something is divided into equal parts?

A: Examples include sharing food or toys, cutting or folding paper.

EXERCISES B-53 to B-61

DIVIDING SHAPES INTO EQUAL PARTS—B

ANSWERS B-53 through B-61 — Student book pages 48–49
Guided Practice: B-53 yes; **B-54** no, rectangle E is taller than rectangle F
Independent Practice: B-55 yes; **B-56** yes; **B-57** no, rectangle K is narrower; **B-58** no, part M is larger; **B-59** yes; **B-60** no, part R is larger (Q is a convex half circle while R is a concave shape); **B-61** yes

LESSON PREPARATION

OBJECTIVE AND MATERIALS

OBJECTIVE: Students will recognize whether or not a figure has been divided exactly in half.

MATERIALS: Transparency of TM 6 (p. 254, bottom section) (One rectangle is to be left whole; the other is to be cut as indicated.) • crayons or markers (optional) • classroom items or shapes which can and cannot be divided into equal parts

CURRICULUM APPLICATIONS

Language Arts: Dividing a long poem for several people to read aloud or memorize

Mathematics: Showing equal fractional parts of a whole, basic symmetry exercises

Science: Identifying symmetry of body parts or natural phenomena (leaves, plants, shells, etc.)

Social Studies: Dividing historical time lines into intervals, making or interpreting charts and graphs

Enrichment Areas: Doing geometric art designs, dividing group art projects fairly

TEACHING SUGGESTIONS

If the figure has been divided unequally, encourage students to explain how the parts are unequal. Have students identify parts by position (upper, lower, left, right), compared sizes (larger, smaller), and compared lengths (longer, shorter, same). (If students have had prior instruction, use the following terms: *area*, *perimeter*, *radius*, *base*, *height*, etc.) It is helpful to cut apart the shapes to confirm whether one shape fits exactly on top of another.

MODEL LESSON

LESSON

Introduction

- Use a symmetrical object or an even number of blocks to show that some things can be divided exactly in half.

  Q: Sometimes it is easy to see what something looks like when it is divided exactly in half.

- Use an object which is divided slightly off center or an odd number of blocks to show that some things are not or cannot be divided equally. Point out the unequal parts.
  Q: Notice that in this object, one part is slightly larger than the other part.

Explaining the Objective to the Students

Q: In this activity, you will recognize whether or not a figure has been divided exactly in half.

Class Activity

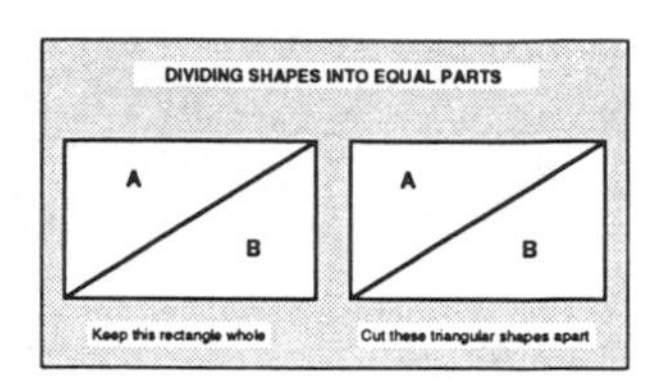

- Project the transparency of the whole rectangle (bottom of TM 6) on the top half of the projector. Lay separated triangles *A* and *B* on the lower half of the projector so that the students see that they can be moved around.
  Q: This is a picture of the example on page 48 in your book. We are going to test whether this rectangle has been divided into equal parts. If the parts are equal, then when I put them together all the edges will match exactly.
- Lay triangle *a* and triangle *b* on top of the whole rectangle.
  Q: These two triangles fit the rectangle exactly. If the rectangle has been divided exactly in half, then these triangles will match each other exactly when I compare them.
- Move the two triangles, one in each hand, to another part of the projector. Turn one triangle so that it can be stacked on the other to form a single triangle.
  Q: The triangles are equal because all the edges match exactly. Therefore, we know that the rectangle was divided into equal parts.

GUIDED PRACTICE

EXERCISES: **B-53**, **B-54**

INDEPENDENT PRACTICE

- Assign exercises **B-55** through **B-61**. Students may use colored markers to trace inside the outline of the two parts for better discrimination.

THINKING ABOUT THINKING

Q: What did you pay attention to in deciding if the shapes were divided into equal parts?

1. I looked to see if the pieces were the same shape.
2. I looked to see if the pieces were the same size.

PERSONAL APPLICATION

Q: When might you need to see whether something is divided into equal parts?
A: Examples include sharing food or toys, cutting or folding paper.

EXERCISES B-62 to B-72

DIVIDING SHAPES INTO EQUAL PARTS—C

ANSWERS B-62 through B-72 — Student book pages 50–52
Guided Practice: B-62 to **B-65** Count the spaces between the dots to confirm that the shapes are equal. See below.
Independent Practice: B-66 to **B-72** See below.

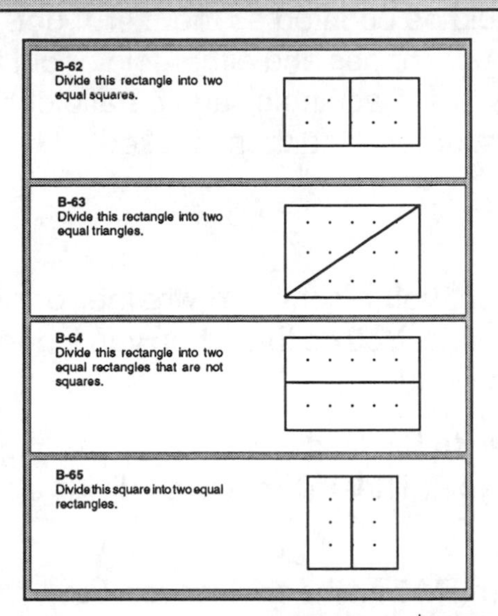

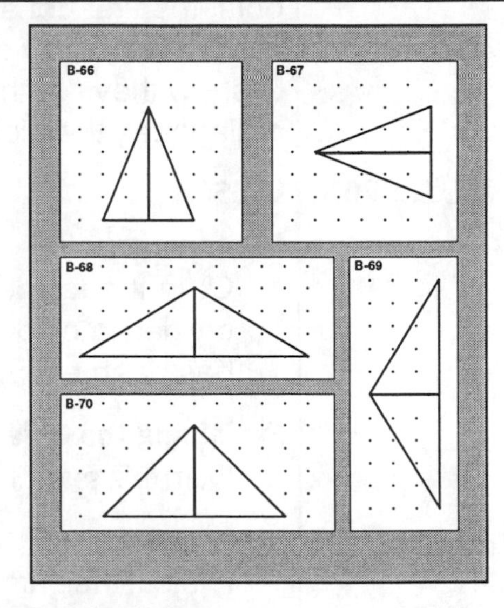

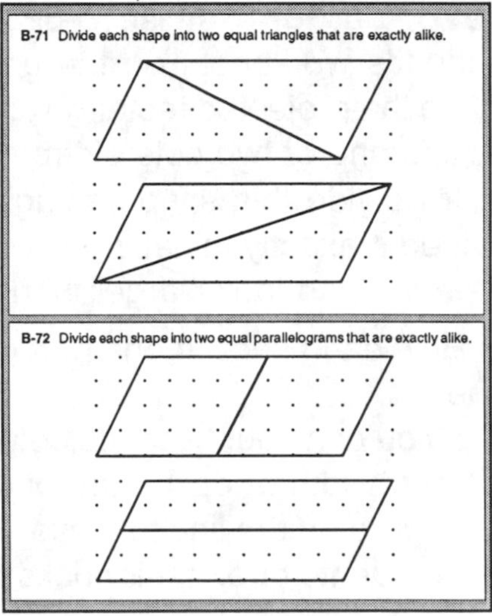

LESSON PREPARATION

OBJECTIVE AND MATERIALS

OBJECTIVE: Students will learn to divide a whole into equal parts.
MATERIALS: Transparency of TM 7 (p. 255)

CURRICULUM APPLICATIONS

Language Arts: Dividing a long poem for several people to memorize
Mathematics: Showing equal fractional parts of a whole, basic symmetry exercises
Science: Identifying symmetry of body parts or natural phenomena (leaves, plants, shells, etc.)
Social Studies: Dividing historical time lines into intervals, making or interpreting charts and graphs
Enrichment Areas: Doing geometric or symmetrical designs in art

TEACHING SUGGESTIONS

Learning the concept *equal* and confirming visually that two parts are equal are key to this lesson. Using words and phrases that mean *equal* (such as *same*, *congruent*, *symmetrical*, *as wide as*, and *as long as*) help reinforce this concept.

To confirm if shapes are equal, count spaces instead of dots. If one counts dots, the first dot should be counted as the "zero" dot.

Students may cut out shapes and either fold or cut them into equal parts to show they are the same. Rectangles and parallelograms can be cut along a diagonal, the pieces turned, and then stacked.

MODEL LESSON

LESSON

Introduction

Q: In the last lesson, you recognized whether or not a shape had been divided into equal parts. You do that when you judge whether a candy bar has been evenly divided.

Explaining the Objective to the Students

Q: In this exercise, you will learn to correctly divide a shape into equal parts.

Class Activity

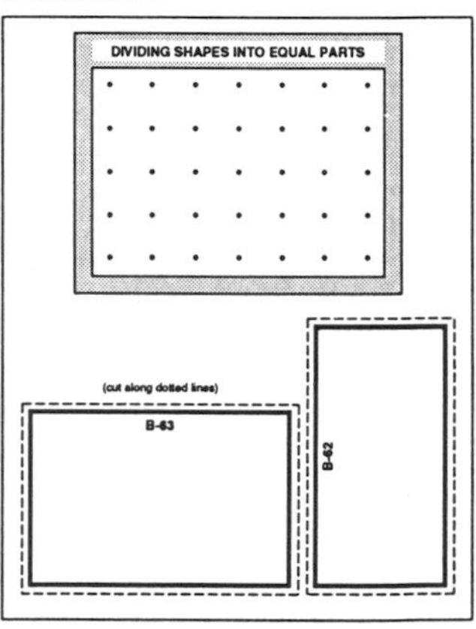

- Project the grid from TM 7 on the chalkboard. Lay the "cutout" of exercise **B-62** (page 50 in the student's workbook) on the grid. Use chalk to divide the projected rectangular grid with a vertical line one or two dots off from the center line.

  Q: If I divide this shape along this line, have I divided it exactly in half?

  A: One part looks larger than the other.

  Q: How can I know if one part is larger than the other?

  A: You can count spaces between dots to find the width of each part, or you can fold the figure on the line to make sure each part is the same size. Have the students open their books to page 50, exercise **B-62**.

- Erase the line you drew and redraw the line vertically down the center.

  Q: Take your pencil and divide the rectangle in exercise **B-62** into two equal squares the way I did here.

- Point to the division.

  Q: There are also other ways that I can divide the rectangle into equal parts. In the next two exercises (**B-63** and **B-64**), divide the rectangle into equal parts according to the instructions. Count the spaces to confirm your answer.

GUIDED PRACTICE

EXERCISES: **B-62** through **B-65**

- Check to see that students have divided the shapes correctly. Discuss how to confirm whether the shapes are equal by counting the spaces between the dots. Project the "cutout" of **B-63** from TM 7 and have students show how they divided the rectangle.

INDEPENDENT PRACTICE

- Assign exercises **B-66** through **B-72**.

THINKING ABOUT THINKING

Q: What did you pay attention to in deciding how to divide shapes into equal parts?

1. I counted to see if there were an even number of spaces in a line segment.
2. If there were an even number of spaces, then I knew I could divide the line in half.

PERSONAL APPLICATION

Q: Why is it sometimes important to divide things equally?

A: Examples include so that my friends don't get angry or feel cheated when we share something, so projects turn out as they are designed.

EXERCISES B-73 to B-80

WHICH SHAPE COMPLETES THE SQUARE?

ANSWERS B-73 through B-80 — Student book pages 53–55
Guided Practice: B-73 c; **B-74** b
Independent Practice: B-75 d; **B-76** b; **B-77** a; **B-78** a; **B-79** d; **B-80** d

LESSON PREPARATION

OBJECTIVE AND MATERIALS

OBJECTIVE: Students will visualize the missing part of a whole by choosing a shape to complete a square.

MATERIALS: Transparency of TM 8 (p. 256, cut apart as indicated) • markers or crayons for students (optional)

CURRICULUM APPLICATIONS

Language Arts: Deciding which letter correctly completes a word
Mathematics: Completing partial geometric shapes
Science: Deciding which bone completes a skeletal part
Enrichment: Choosing pattern pieces in sewing or woodworking, deciding note value to complete a measure of music, balancing artwork on a page, completing artwork that has been started

TEACHING SUGGESTIONS

The key concepts are *completes* and *equals.* Make a list of the words that students use to explain why each shape does or does not complete the square ("fits, fills, finishes, covers, etc.").

As an optional strategy, students may draw and color the missing area of the square and the areas of the four shapes using the same color. The similarity of colored areas is a helpful visual cue for young children. If some students still have difficulty seeing the correct completion shape, they may trace and cut out the pieces, putting them together like a puzzle. When the students have had sufficient time to complete the exercises, check answers by discussing correct choices and reasons for eliminating incorrect choices.

MODEL LESSON

LESSON

Introduction

Q: We have practiced dividing shapes. You have seen how important it is to be able to estimate visually whether something has been divided in half.

Explaining the Objective to the Students

Q: In this exercise, you will learn to visualize the missing part of a whole by choosing a shape to complete a square.

Class Activity

- Project the bottom half of TM 8 with the four pieces cut apart but arranged next to the large incomplete square in the original order.
  Q: This is a square, but it has a piece missing. We need to see which of the given shapes will make the square complete.

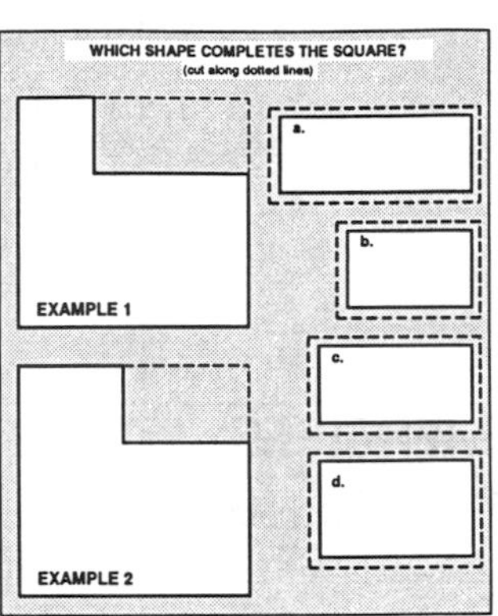

- Move shape *a* to the space on the square.
  Q: Does shape *a* complete the square?
  A: No.

  Q: Why not?
  A: It is too wide.

- Replace shape *a* with shape *b*.
  Q: Does shape *b* complete the square?
  A: No.

  Q: Why not?
  A: It is not wide enough.

- Replace shape *b* with shape *c*.
  Q: Does shape *c* complete the square?
  A: Yes.

  Q: How do you know?
  A: It just fits in the dotted outline.

- Replace shape *c* with shape *d*.
  Q: Does shape *d* complete the square?
  A: No, it is too tall.

  Q: We will put a circle around shape *c* because it completed the square.

- Optional: Repeat a similar procedure with example 2 on TM 8.

GUIDED PRACTICE

EXERCISE: **B-73**, **B-74**

INDEPENDENT PRACTICE

- Assign **B-75** through **B-80**. In exercises **B-75** through **B-80**, students may sketch in the completed square if they cannot do the exercises otherwise.

THINKING ABOUT THINKING

Q: What did you pay attention to deciding what shape completes the square?

1. I looked at the shape of the missing part.
2. I looked at the four shapes for similar shapes.
3. When I found a similar shape, I checked to see if it was the same size as the missing shape.
4. If I was uncertain, I checked the size of the piece with a ruler.

PERSONAL APPLICATION

Q: When might you have to decide what shape is needed to complete an object?

A: Examples include deciding how many pieces of cake are missing, which puzzle piece you need, or which part of a model or pattern is necessary to complete an object.

EXERCISES B-81 to B-83

WHICH SHAPES MAKE SQUARES?

ANSWERS B-81 through B-83 — Student book pages 56–58
Guided Practice: B-81 See below.
Independent Practice: B-82 to **B-83** See below.

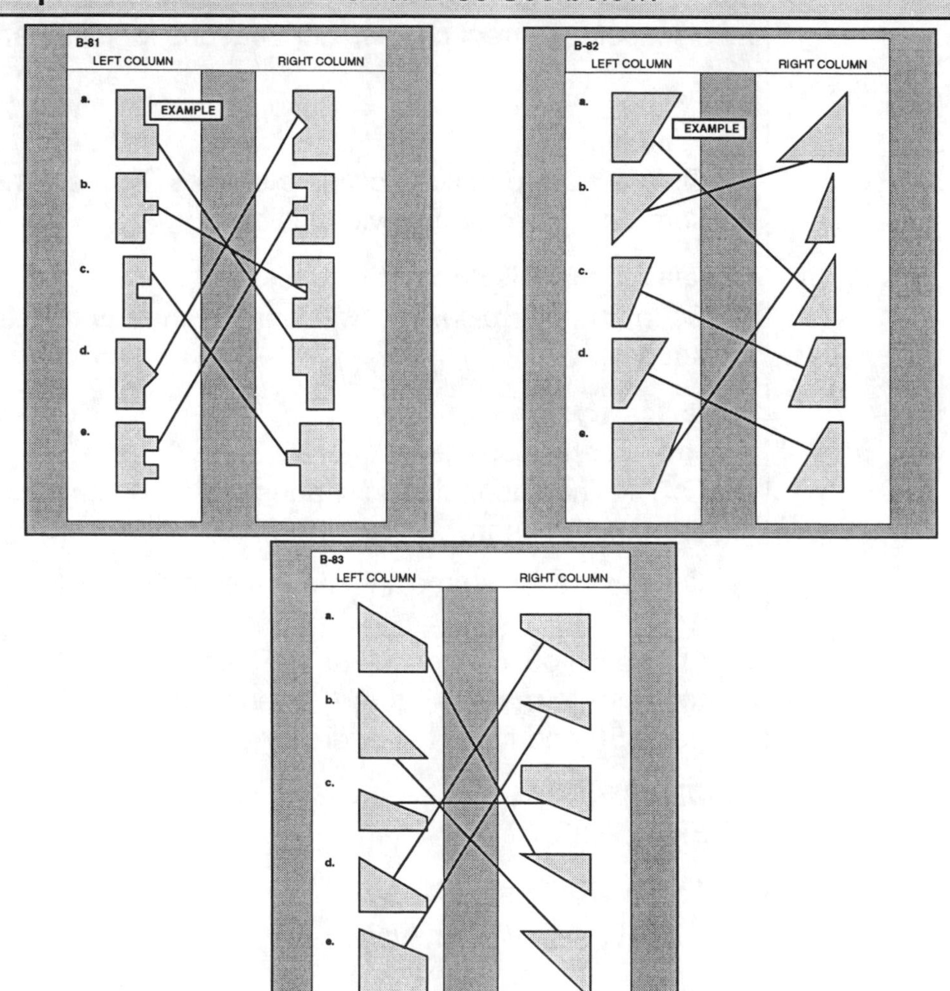

LESSON PREPARATION

OBJECTIVE AND MATERIALS

OBJECTIVE: Students will visualize the missing part of a whole by matching shapes that fit together to make a square.
MATERIALS: Transparency of student book page 56 • scissors, markers, or crayons for students (optional)

CURRICULUM APPLICATIONS

Language Arts: Deciding which letter correctly completes a word
Mathematics: Completing partial geometric shapes
Science: Deciding which bone completes a skeletal part, matching leaf or insect halves for symmetry
Enrichment: Choosing pattern pieces in sewing or woodworking, deciding note value to complete a measure of music, balancing artwork on a page, completing unfinished artwork

TEACHING SUGGESTIONS

Reinforce the following key terms emphasized in this lesson: *completes* and *equals*. Make a list of the words that students use to explain why each shape does or does not complete the square.

If some students still have difficulty seeing the correct completion shape, they may cut out the pieces, putting them together like a puzzle. When the students have had sufficient time to complete the exercises, check answers by discussing correct choices and reasons for eliminating incorrect choices.

MODEL LESSON

LESSON

Introduction

Q: We have practiced dividing shapes. You estimated visually whether something had been divided in half.

Explaining the Objective to the Students

Q: In this exercise, you will identify shapes that fit together to make a square.

Class Activity

- Project a transparency of page 56.
  Q: The line connects two shapes that fit together to make a square.
- Cut out the top shape and move it so it fits into its mate.
  Q: Now you can see that the two shapes fit together to make a square. Open your books to page 56 and finish the exercise.

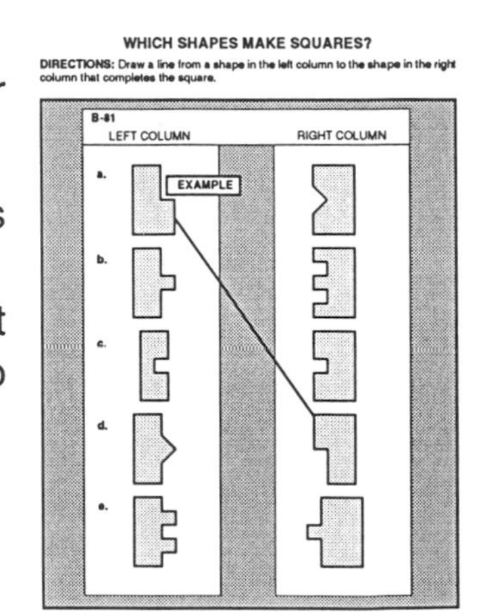

GUIDED PRACTICE

EXERCISE: **B-81**

INDEPENDENT PRACTICE

- Assign **B-82** and **B-83**.

THINKING ABOUT THINKING

Q: What did you pay attention to when you selected the shapes that make squares?

A: I imagined a square around each piece in the left column. I imagined how the remaining piece should look and then looked for that missing shape in the right column.

PERSONAL APPLICATION

Q: When might you have to decide what shape is needed to complete an object?

A: Examples include deciding how many pieces of cake are missing, which puzzle piece you need, or which part of a model or pattern is necessary to complete an object.

EXERCISES B-84 to B-100

COPYING A FIGURE/DRAWING IDENTICAL SHAPES

ANSWERS B-84 through B-100 — Student book pages 59–63
Guided Practice: B-84 to **B-92** Match the constructed figures to the figures on the worksheet. **B-93** to **B-94** Count the dots to confirm that the shapes are equal.
Independent Practice: B-95 to **B-100** Count the dots to confirm that the shapes are equal.

LESSON PREPARATION

OBJECTIVE AND MATERIALS

OBJECTIVE: Students will copy a shape, first by using interlocking cubes and then by drawing.

MATERIALS: Interlocking cubes (Unifix or Multilink cubes) • transparency of TM 9 (p. 257, cut apart as indicated) • markers or crayons for students (optional)

CURRICULUM APPLICATIONS

Language Arts: Preparing diagrams for reports or displays
Mathematics: Working problems involving shapes
Science: Drawing diagrams in science reports
Social Studies: Drawing maps
Enrichment: Copying works of art or music

TEACHING SUGGESTIONS

This lesson may take two sessions (**B-84** through **B-92** in the first session, **B-93** through **B-100** in the second).

MODEL LESSON

LESSON

Introduction

Q: We have divided shapes and estimated visually whether something has been divided in half.

Explaining the Objective to the Students

Q: It is important to be able to copy a shape. In this exercise, you will copy a shape first by building with interlocking cubes and then by drawing.

Class Activity

- Distribute at least twelve cubes to each pair of students.
  Q: Open your books to page 59. With your cubes, build the shapes shown in exercises **B-84** and **B-85**.
- Check students' progress and the accuracy of their matching.
  Q: Now you are ready to work on exercises **B-86** through **B-92**.
- Check students' progress.
- Project transparency of TM 9 on the chalkboard.
  Q: Now we will draw some shapes using dots as units instead of blocks. Notice that the rectangle is 2 units high and 3 units long.

TRANSPARENCY MASTER 9

DRAWING IDENTICAL SHAPES

(cut along dotted lines)

- Point to the dimensions.
  Q: In order to copy this figure, I need to pick a starting point. This is my "zero" point. When I count 2 units, I don't count the starting zero point. Count: zero, one, two. Draw a line from the "zero" point to point "two." I will do the same thing to make the other side of the rectangle.
- Repeat the process to finish copying the figure.
  Q: Connect the dots to make squares within the rectangle to resemble the interconnecting cubes.
- Optional: Repeat the process with the transparent 3 unit by 4 unit rectangle.
  Q: Try exercises **B-93** and **B-94** and I'll check your work.
- Check students' progress.
  Q: Now you are ready to work on exercises **B-95** through **B-100**.
- Check students' progress.

GUIDED PRACTICE

EXERCISES: **B-84** through **B-94**

INDEPENDENT PRACTICE

- Assign exercises **B-95** through **B-100**.

THINKING ABOUT THINKING

Q: What did you pay attention to in order to draw identical shapes?

1. I counted the spaces to find out the size of each side of the shape.
2. I counted the same number of spaces on the drawing grid and drew each side.
3. I checked to see that the shapes were the same.

PERSONAL APPLICATION

Q: When might you have to copy a shape, figure, or picture?

A: Examples include drawings needed for reports or projects or for fun.

EXERCISES B-101 to B-106

COMPARING SHAPES—SELECT

ANSWERS B-101 through B-106 — Student book pages 64–66
Guided Practice: B-101B a; **B-102A** a; **B-102B** b
Independent Practice: B-103A a; **B-103B** c; **B-104A** b; **B-104B** c; **B-105A** b; **B-105B** a; **B-106A** b; **B-106B** c

LESSON PREPARATION

OBJECTIVE AND MATERIALS

OBJECTIVE: Students will describe shapes by comparing and contrasting them.
MATERIALS: Transparency of student book page 64

CURRICULUM APPLICATIONS

Language Arts: Visual discrimination exercises for reading readiness
Mathematics: Figural similarity exercises, direction of numeral formations (5, 7, etc.)
Science: Recognition of similarly shaped leaves, insects, or shells
Social Studies: Matching puzzle sections to geographic features on map puzzles
Enrichment Areas: Recognizing shapes of road signs, discerning different patterns in art

TEACHING SUGGESTIONS

Reinforce the following vocabulary emphasized in the lesson: *base*, *height*, *width*, *sides*, and *corners*. *Angle* may be introduced if appropriate to your students. Students may practice the comparative and superlative forms of adjectives used to describe figures, i.e., tall, taller, tallest, etc.

MODEL LESSON

LESSON

Introduction

Q: When we divided and drew shapes, we also had to compare the parts.

Explaining the Objective to the Students

Q: In this exercise, you will describe shapes by comparing their characteristics.

Class Activity

- Project transparency of page 64.
  Q: Open your books to page 64. Notice figures *a*, *b*, and *c*. The height of each rectangle is shown. Which is the shortest?
  A: Rectangle *c*

- Point to **B-101.**
  Q: Which of the rectangles is the tallest?
  A: Rectangle *a*

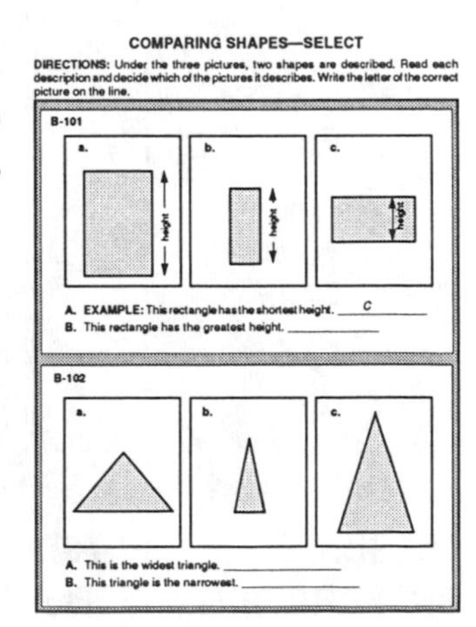
COMPARING SHAPES—SELECT

DIRECTIONS: Under the three pictures, two shapes are described. Read each description and decide which of the pictures it describes. Write the letter of the correct picture on the line.

B-101

a. b. c.

A. EXAMPLE: This rectangle has the shortest height. c

B. This rectangle has the greatest height. ______

B-102

a. b. c.

A. This is the widest triangle. ______

B. This triangle is the narrowest. ______

GUIDED PRACTICE

EXERCISES: **B-101**, **B-102**

INDEPENDENT PRACTICE

- Assign exercises **B-103** through **B-106**.

THINKING ABOUT THINKING

Q: What did you pay attention to when you compared shapes?

1. I paid attention to the important words (characteristics) in the description.
2. I looked for those specific characteristics in the three shapes.
3. I picked the shape that has those characteristics.
4. I checked to see why the other shapes don't fit the description.

PERSONAL APPLICATION

Q: When might you have to compare shapes at home?

A: Examples include putting away toys or tools, matching building blocks, matching parts or sections from construction toys, putting away dishes or silverware.

EXERCISES B-107 to B-113

COMPARING SHAPES—EXPLAIN

ANSWERS B-107 through B-113 — Student book pages 67–70
Guided Practice: B-107 Triangle 3 is taller and wider than triangle 4. **B-108** Trapezoid 5 is shorter and wider than trapezoid 6.
Independent Practice: B-109 Shape 7 is a pentagon which has a base and height the same as triangle 8. **B-110** Shape 9 is a pentagon which has a base and height the same as pentagon 10. The pentagon has been turned. **B-111** Shape 11 is a diamond (rhombus) which is smaller than diamond 12.
Guided Practice: B-112 to **B-113** See below.

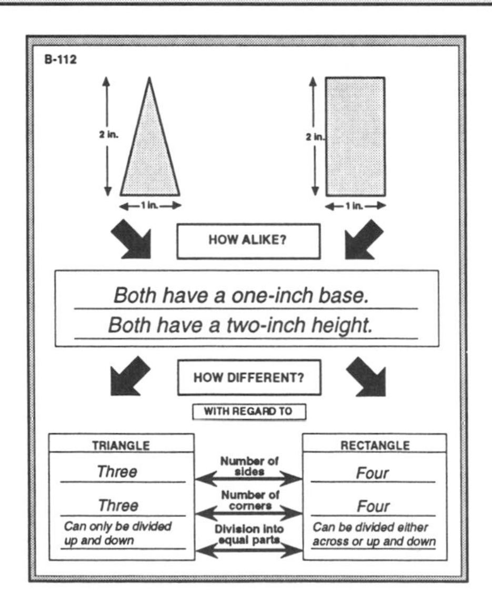

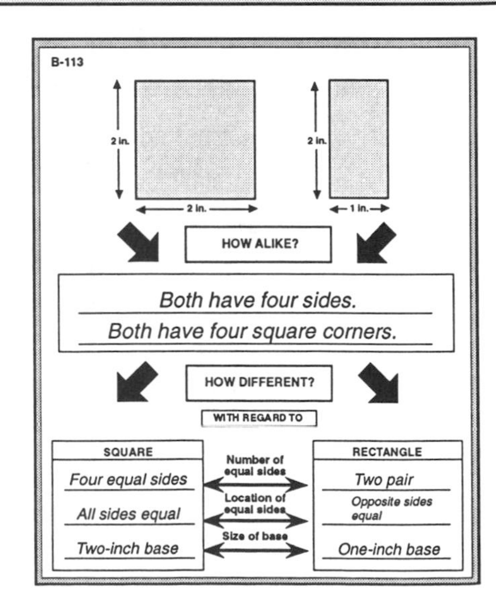

LESSON PREPARATION

OBJECTIVE AND MATERIALS

OBJECTIVE: Students will practice explaining how shapes are alike and different.

MATERIALS: Transparency of student book page 67, TM 10 (p. 258)

CURRICULUM APPLICATIONS

Language Arts: Descriptive writing to compare objects, noting similarities and differences in words

Mathematics: Describing geometric figures, writing numerals correctly

Science: Comparing scientific properties, weather conditions, states of matter

Social Studies: Comparing cultures and cultural artifacts

Enrichment Areas: Describing different types of music and instruments, discerning different patterns in art

TEACHING SUGGESTIONS

Reinforce the following vocabulary emphasized in the lesson: *triangle*, *rectangle*, *trapezoid*, *pentagon*, *diamond* (*rhombus*), *base*, *height*, *width*, *sides*, and *corners*. Encourage students to use these terms to describe and compare the specific characteristics of different shapes.

This lesson has two parts. In exercises **B-107** to **B-111**, students write a description comparing and contrasting two shapes. In exercises **B-112** and **B-113**, students use a graphic organizer to compare and contrast shapes. To extend this lesson, ask students to compare and contrast two objects in the classroom that contain the same geometric shapes.

MODEL LESSON

LESSON

Introduction

Q: We have practiced comparing shapes.

Explaining the Objective to the Students

Q: In this activity, you will describe shapes by comparing and contrasting them, first, by writing a description and second, by using a graphic organizer.

Class Activity

- Project transparency of page 67.

  Q: Open your books to page 67. Look at figures 1 and 2. Which is the tallest?

  A: Rectangle 1

  Q: Which is the thinnest?

  A: Rectangle 1

  Q: How can you combine these two statements into a comparison statement?

  A: Rectangle 1 is taller and thinner than rectangle 2.

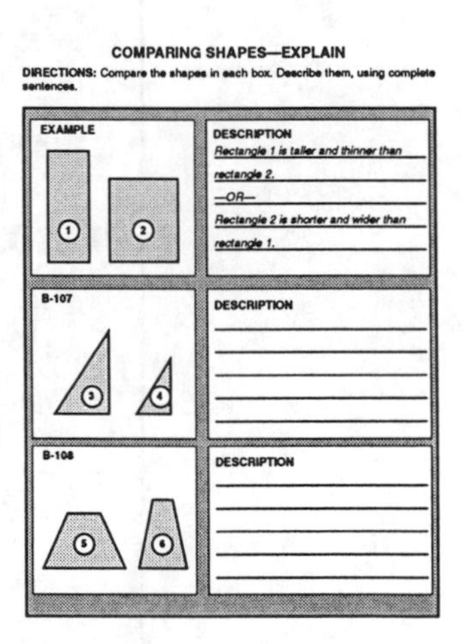
COMPARING SHAPES—EXPLAIN

DIRECTIONS: Compare the shapes in each box. Describe them, using complete sentences.

EXAMPLE	DESCRIPTION *Rectangle 1 is taller and thinner than rectangle 2.* *—OR—* *Rectangle 2 is shorter and wider than rectangle 1.*
B-107	DESCRIPTION
B-108	DESCRIPTION

- Point to **B-107**.

  Q: Which of the triangles is the tallest?

  A: Rectangle 3

  Q: Which of the triangles is widest?

  A: Rectangle 3

  Q: Write a comparison description in the space provided.

- Point to **B-108** and follow the same format you used with **B-107**.
- Have students do exercises **B-109** to **B-111** independently.
  Q: Do exercises **B-109** to **B-111** in your book.
- Project transparency master 10. Draw two shapes or place two cutouts on the page.
  Q: Now, we are going to use a graphic organizer to compare and contrast two shapes. Look at the two shapes at the top of the page. How are they alike?
  A: Write two similarities.
  Q: How are the two shapes different?
  A: Write three differences. Under the heading "with regard to," specify the type of characteristic that is different, i.e., number of sides, size, etc.

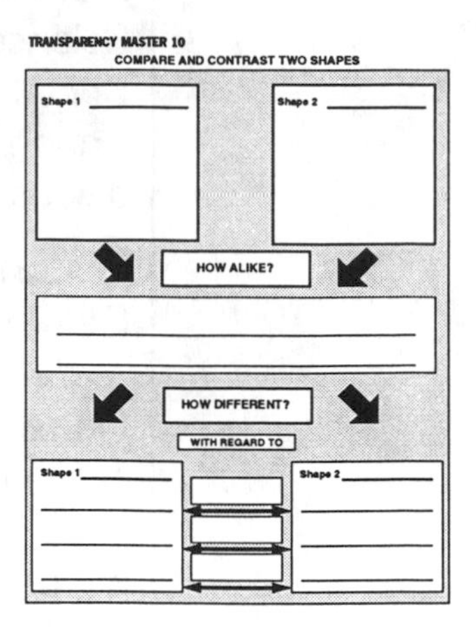

- Use the blank graphic organizer to compare other shapes: rectangle and trapezoid, trapezoid and rhombus, trapezoid and parallelogram, etc.

GUIDED PRACTICE

EXERCISES: **B-107** to **B-108** and **B-112** to **B-113**

INDEPENDENT PRACTICE

- Assign exercises **B-109** through **B-111**.

THINKING ABOUT THINKING

Q: What did you pay attention to when you compared shapes?
1. I paid attention to the important characteristics in the two shapes.
2. I gave names to those characteristics.
3. I combined the names and characteristics into a description.
4. I checked to see that my description was complete.

PERSONAL APPLICATION

Q: When might you have to describe shapes?
A: Examples include making requests; giving directions; describing buildings, vehicles, or household objects.

Chapter 3

FIGURAL SEQUENCES
(Student book pages 71–103)

EXERCISES C-1 to C-11

COPYING A PATTERN

ANSWERS C-1 through C-11 — Student book pages 72–73
Guided Practice: C-1 to **C-6** Check students' patterns.
Independent Practice: C-7 to **C-11** Check students' patterns.

LESSON PREPARATION

OBJECTIVE AND MATERIALS

OBJECTIVE: Students will build a pattern of blocks.
MATERIALS: Interlocking cubes (Unifix or Multilink cubes) to model creating patterns with cubes

CURRICULUM APPLICATIONS

Mathematics: Tesselations
Science: Seeing patterns in leaves, shells, and life cycles
Social Studies: Showing patterns of geographic changes or types of topography on maps
Enrichment Areas: Art exercises involving patterns, repeating patterns in music or movement

TEACHING SUGGESTIONS

Reinforce the following vocabulary emphasized in this lesson: *pattern, interlocking*, and *cube.* Encourage students to use these terms in their discussion.

MODEL LESSON

LESSON

Introduction

Q: Patterns are all around us. Where in this room can you see examples of regular or repeating patterns?

A: Examples might include fabric in clothing, brick or cement block walls, floor tiles, ceiling tiles, Venetian blinds, leaf arrangements on plants, etc.

Explaining the Objective to the Students

Q: In this lesson, you will build patterns with interlocking cubes.

Class Activity

- Using interlocking cubes, build a two-color, three-cube pattern different from the ones in the student book.
  Q: Use your set of blocks to build a pattern like this one.
- Check that the students can duplicate a simple pattern.
- Build a three-color, four-cube pattern.
  Q: Use your set of blocks to build a pattern like this one.
- Check that the students can duplicate the more complex pattern.

GUIDED PRACTICE
EXERCISES: **C-1** through **C-6**

INDEPENDENT PRACTICE
- Assign exercises **C-7** through **C-11**.

THINKING ABOUT THINKING

Q: What did you pay attention to when you matched a pattern?
1. I looked carefully at the pattern.
2. I used colored cubes to match the pattern in the book.
3. I checked that the two patterns matched.

PERSONAL APPLICATION

Q: Where do you remember seeing patterns?

A: Examples might include designs on fabric, decorative trims on buildings, brick or cement block walls, floor tiles, ceiling tiles, Venetian blinds, leaf arrangements on plants, etc.

EXERCISES C-12 to C-26

WHICH COLOR COMES NEXT?—SELECT

ANSWERS C-12 through C-26 — Student book pages 74–77
Guided Practice: C-12 d; **C-13** b
Independent Practice: C-14 b; **C-15** d; **C-16** c; **C-17** color 2; **C-18** color 2; **C-19** color 3; **C-20** color 4; **C-21** color 1; **C-22** color 2, 1; **C-23** color 1, 2; **C-24** color 3, 4; **C-25** color 1, 1; **C-26** color 2, 2

LESSON PREPARATION

OBJECTIVE AND MATERIALS
OBJECTIVE: Students will build a pattern of blocks and find the next block in the pattern.
MATERIALS: Interlocking cubes (Unifix or Multilink cubes)

CURRICULUM APPLICATIONS
Language Arts: Sequencing sounds in spelling, alphabetizing, dictionary practice
Mathematics: Recognizing number sequences, learning place value with manipulatives
Science: Seeing patterns in seasonal changes and weather conditions
Social Studies: Showing geographic changes or types of topography on maps
Enrichment Areas: Art exercises involving patterns, repeating patterns in music or movement

TEACHING SUGGESTIONS
Reinforce the following vocabulary emphasized in this lesson: *pattern*, *interlocking*, and *cube*.

When asked what will come next in a pattern, some students simply "see" the pattern and intuitively know what color comes next. Other students name the colors from left to right and use the word pattern as a clue.

MODEL LESSON

LESSON

Introduction

Q: Can you find more examples of patterns in this room?

A: Examples might include furniture arrangement, rows of books, borders on bulletin boards, floor tiles, ceiling tiles, calendars, leaf arrangements on plants, etc.

Explaining the Objective to the Students

Q: In this activity, you will build a pattern of blocks and find the next block in the pattern.

Class Activity

- Build the pattern shown in the example on page 74.

  Q: Use your set of blocks to build a pattern like this one.

- Check that the students can duplicate a long pattern.

  Q: What color comes next?

  A: Color *d*

  Q: How did you know what color came next?

  A: Answers will vary.

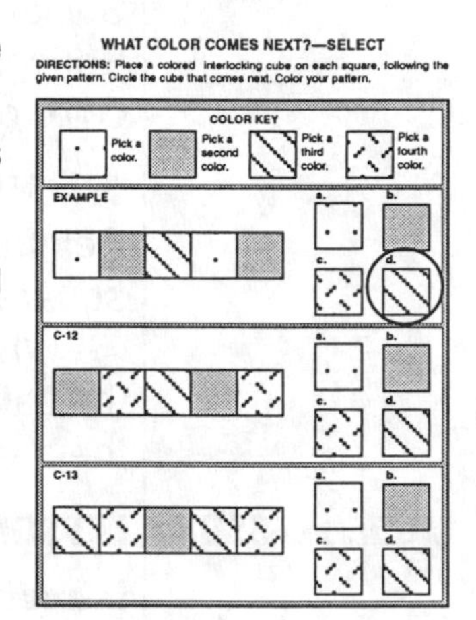

GUIDED PRACTICE

EXERCISES: **C-12, C-13**

INDEPENDENT PRACTICE

- Assign exercises **C-14** through **C-26**. The class may complete this lesson in two sessions.

THINKING ABOUT THINKING

Q: What did you pay attention to when you decided which figure came next?

1. I looked carefully at the colors of the cubes.
2. I looked for a pattern of changes.
3. I figured out, if the pattern continued, what the next cube would be like.
4. I checked that the other cubes didn't fit the pattern.

PERSONAL APPLICATION

Q: What are some other kinds of patterns we see around us?

A: Examples might include wallpaper patterns, roof tiles, strings of beads or holiday lights, fences, rows of trees or plants, etc.

EXERCISES C-27 to C-38

WHICH SHAPE COMES NEXT?—SELECT

ANSWERS C-27 through C-38 — Student book pages 78–81
Guided Practice: C-27 b; **C-28** c

Independent Practice: C-29 d; **C-30** c; **C-31** a; **C-32** c; **C-33** a; **C-34** d; **C-35** b; **C-36** c; **C-37** e; **C-38** a

LESSON PREPARATION

OBJECTIVE AND MATERIALS

OBJECTIVE: Students will find the next figure in a series by identifying the pattern of the series.
MATERIALS: TM 11 (p. 259) • attribute or design blocks (optional)

CURRICULUM APPLICATIONS

Mathematics: Counting patterns, pattern block activities
Science: Seeing patterns in leaves, shells, and life cycles
Social Studies: Recognizing patterns in ethnic decorations or artifacts
Enrichment Areas: Repeating rhythm patterns in music, repeating patterns of color or shape in art exercises such as printmaking or stenciling

TEACHING SUGGESTIONS

Reinforce the following terms emphasized in this lesson: *circle*, *rectangle*, *square*, *triangle.* Encourage students to use these terms in their discussion.

When asked what will come next in a pattern, some students simply "see" the pattern and intuitively know what shape comes next. Other students name the shapes from left to right and use the word pattern as a clue.

MODEL LESSON

LESSON

Introduction

Q: Sometimes changes in characteristics, such as shape or color, become regular patterns or sequences. A sequence is a repeating pattern, such as the days of the week on a calendar. Where in this room can you see examples of regular or repeating patterns?

A: Answers will vary.

Explaining the Objective to the Students

Q: In this lesson, you will find the next shape in a repeating pattern.

Class Activity

- Set up the sequence shown below on the chalk tray. (You might also draw the sequence on the board with colored chalk or use transparency figures.)

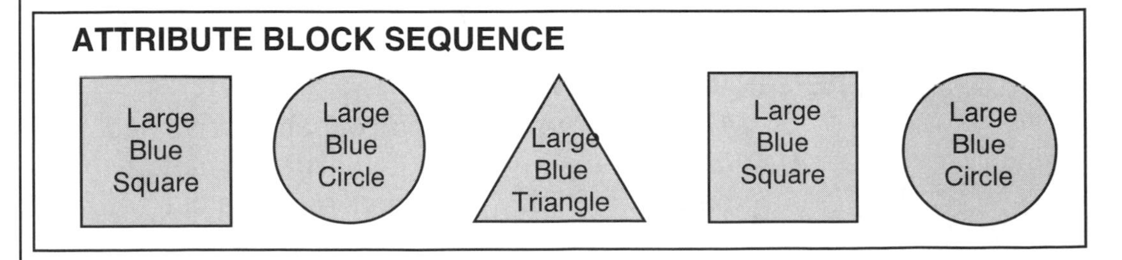

Use this set first. Separated to the right of this group, set up a series of red and blue blocks that includes a large blue triangle.

- Point to the set of blue blocks on the left.

Q: Here you see a row of shapes. Look for a pattern to find the next figure in the series. Compare the first and second figures. What was changed?
A: The shape changed, but the color stayed the same.

Q: Now compare the second and third figures. What was changed?
A: The shape changed, but the color stayed the same.

Q: In the first and second figures, the color did not change but the shape did. Look at the rest of the row. Can anyone see a pattern in the row?
A: The shape is changing from square to circle to triangle, but the color is staying the same.

Q: What figure do you think should come next?
A: Large blue triangle

- Take a large blue triangle from the red/blue set on the right and place it next to the blue circle.
  Q: How did you know that the large blue triangle would come next?
  A: The large blue square signals that the pattern is beginning again.

- Reinforce the pattern by pointing to the shapes as you say them.
  Q: Square, circle, triangle, square, circle, triangle
- Point to the set of red and blue shapes on the right.
  Q: What other sequences can you make using these shapes?
- Allow time for students to create several patterns.
- Project TM 11.
  Q: Several kinds of sequences use shapes. One kind of sequence is changing only shape.

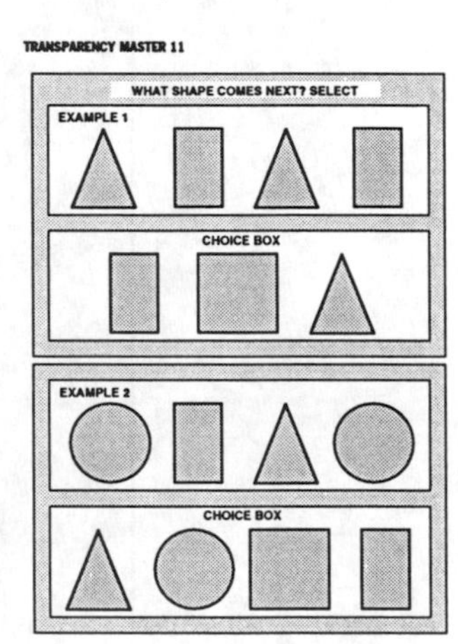

- Write "changing only shape" on the board.
  Q: That is what we just did with the blue shapes.
- Point to the blue shapes on the left.
  Q: We found a repeating pattern and continued it.
- Point to the shapes on TM 11 as you explain.
  Q: In this sequence, which block should come next? The pattern we are given is triangle, rectangle, triangle, rectangle. If that is so, the next figure should be a triangle.
- Draw a circle around the triangle. Note: An alternate solution might be (triangle, rectangle), (triangle, rectangle, rectangle), in which case the rectangle would be correct.
- A second changing-only-shape sequence example is given on TM 11. Use this sequence in a similar fashion for reinforcement.

GUIDED PRACTICE

EXERCISES: **C-27**, **C-28**

INDEPENDENT PRACTICE

- Assign exercises **C-29** through **C-38**.

THINKING ABOUT THINKING

Q: What did you pay attention to when you decided which figure should come next?

1. I looked carefully at the shapes.
2. I looked for a pattern of changes.
3. I figured out what the next shape would be if the pattern continued.
4. I checked that the other shapes didn't fit the pattern.

PERSONAL APPLICATION

Q: When would you need to select the next shape or design to continue a repeating pattern?

A: Examples might include locating brands of foods that are grouped together on grocery shelves or in freezers, matching edges on wallpaper or borders, knitting a pattern in a sweater.

EXERCISES C-39 to C-51

TUMBLING

ANSWERS C-39 through C-51 — Student book pages 82–86
Guided Practice: C-39 to C-42 See below.

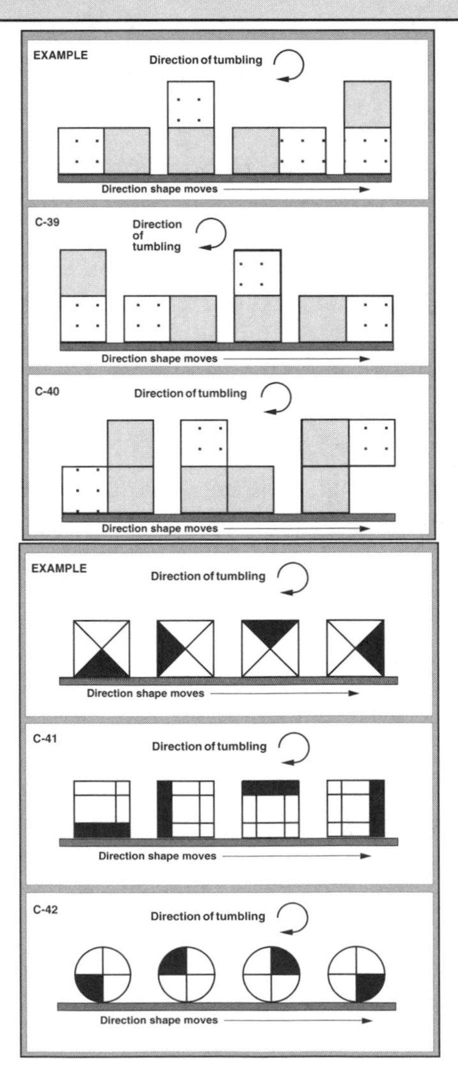

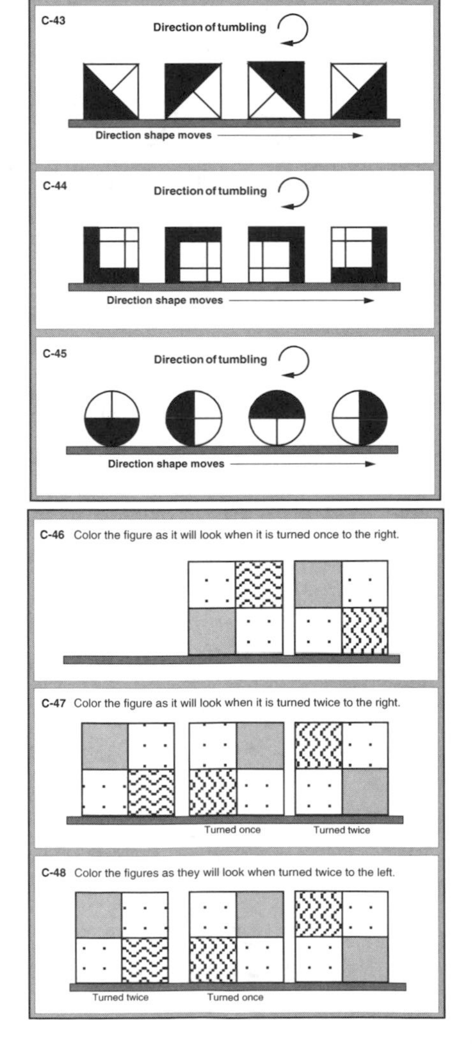

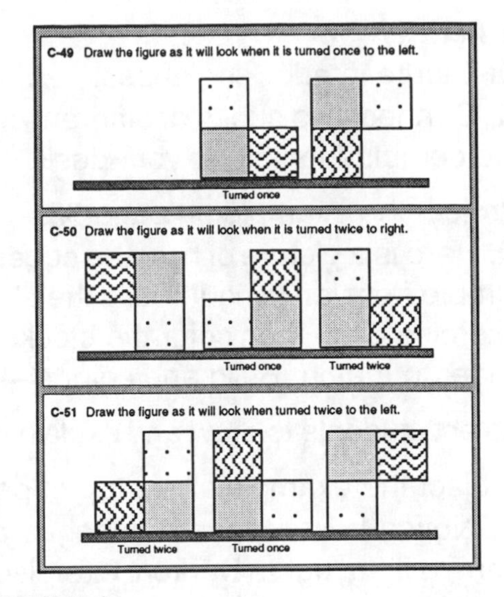

LESSON PREPARATION

OBJECTIVE AND MATERIALS

OBJECTIVE: Students will show how objects look when they have been tumbled along a surface.

MATERIALS: Interlocking cubes: 2 of one color and one of a second color • transparencies of student book pages 82 and 83 and TM 12 (p. 260) • washable transparency marker

CURRICULUM APPLICATIONS

Language Arts: Mirror-image words (mom, wow, dad, pop)
Mathematics: Rotating geometric shapes
Science: Explaining rotation of the earth, experiments with mirrors or gears
Social Studies: Map orientation, redrawing charts or graphs
Enrichment Areas: Folk or square dancing patterns; quilting, sewing, and needlework; making prints of rotating patterns

TEACHING SUGGESTIONS

Depending on the language level of your students, you may introduce additional terms. Tumbling is also known as *rotation*. Rotation to the right is also called *clockwise rotation*. Rotation to the left is called *counterclockwise rotation*.

Students may wish to color the pictures to match the interlocking cubes in order to emphasize the rotation. If students have difficulty recognizing rotation, use transparency master 12.

MODEL LESSON

LESSON

Introduction

- Write "changing only shape" on the board.
  Q: When you are putting a puzzle together, you often have to turn the pieces to see if they fit.

Explaining the Objective to the Students

Q: In this lesson, you will figure out what a shape will look like after it has been turned.

Class Activity

- Distribute interlocking cubes.
  Q: Connect two cubes of different colors. Tumble the combination along your desk.
- Project the example on page 82.
  Q: Here is a picture of how the cubes look as they tumble from left to right. Build the blocks shown in exercise **C-39** then color the blocks on the worksheet to match. Build and color **C-40**.
- Select students to draw and explain their answers.
- Project the example at the top of page 83.
  Q: Notice how a pattern changes as it tumbles from left to right. Darken exercises **C-41** and **C-42** to show how they look as they tumble. Another word for tumbling is rotating.

GUIDED PRACTICE

EXERCISES: **C-39** through **C-42**

- After students have time to complete their practice, project page 83 and select students to draw and explain their answers.

INDEPENDENT PRACTICE

- Assign exercises **C-43** through **C-51**.

THINKING ABOUT THINKING

Q: What did you pay attention to when you decided how a figure would look when it was turned?

1. I looked carefully at the details (color of cubes, pattern) to decide what to look for as the object tumbled.
2. I looked for a pattern of tumbling to the right (clockwise) or to the left (counterclockwise).
3. I determined what the next figure would look like if the pattern continued.

PERSONAL APPLICATION

Q: When might it be helpful to predict what a shape or pattern will look like when it has been turned?

A: Examples include puzzle pieces, Rubik's Cube, putting together models or clothing patterns.

EXERCISES C-52 to C-56

WHICH FIGURE COMES NEXT?—SELECT (FLIPS)

ANSWERS C-52 through C-56 — Student book pages 87–88
Guided Practice: C-52 a (flipping up and down about a horizontal axis)
C-53 c (flipping left and right about a vertical axis)
Independent Practice: C-54 c (flipping up and down about a horizontal

axis) **C-55** b (flipping left and right about a vertical axis) **C-56** c (flipping left and right about a vertical axis) See below.

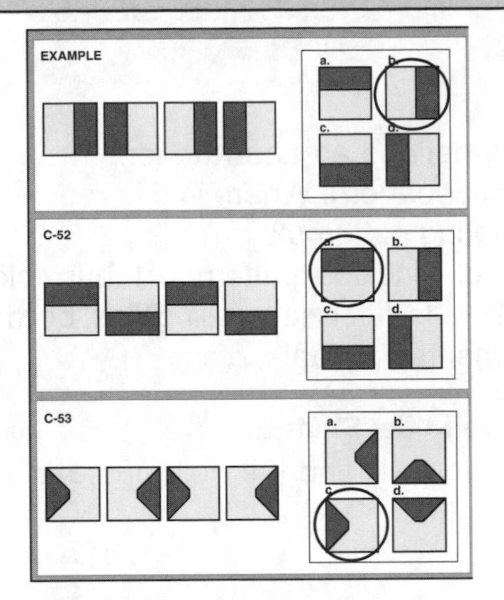

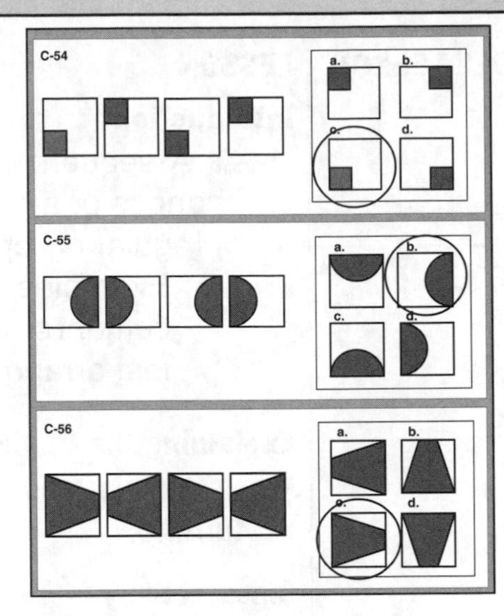

LESSON PREPARATION

OBJECTIVE AND MATERIALS

OBJECTIVE: Students will find the next figure in a series by identifying the pattern that the series follows.

MATERIALS: TM 13 (p. 261) • attribute or design blocks (optional)

CURRICULUM APPLICATIONS

Language Arts: Discriminating between letters (b/d, p/q), inverting letter order in words

Mathematics: Recognizing geometric changes in position (flips, turns, and slides), symmetry, arranging basic problems (e.g., horizontal or vertical addition)

Science: Recognizing alternating patterns in leaves or shells; showing symmetry in plants or animals; mirror-image exercises; understanding gears, wheels, motors, or kaleidoscopes; recognizing the rotation of planets

Social Studies: Showing geographic changes or types of topography on maps, using or making graphs that show negative and positive changes, reading maps

Enrichment Areas: Art exercises involving rotation, reflection, or positive and negative space; repeating interval patterns in music; playing games that call for taking turns or doing rotations (musical chairs, relay races, positions on a volleyball court, etc.)

TEACHING SUGGESTIONS

Reinforce the following terms emphasized in this lesson: *repeating*, *flip*, *rotation*, *sequence*. Encourage students to use these terms in their discussion. You may wish to include the terms *horizontal*, *vertical*, and *axis* if you wish to teach those concepts. The proper term for flipping is *reflection*.

When a child uses the word *turn*, he or she may be describing either turning in the same plane (rotation) or flipping on an axis (reflection). Help students express the difference between rotation in the last lesson and reflection in this one. Note: Students who disagree with the answers given in the manual should be allowed to explain their selection. If they "see" a different sequence and if their selection follows the sequence they see, then their answers should also be accepted as correct.

MODEL LESSON

LESSON

Introduction

Q: A sequence is defined as an arrangement of things having regular changes or a regular pattern. Where in this room can you see examples of regular or repeating patterns?

A: Examples might include patterned fabric, clothing trims, brick or cement block walls, floor tiles, ceiling tiles, borders on bulletin boards, leaf arrangements on plants, etc.

Explaining the Objective to the Students

Q: In this lesson, you will find the next figure in a series that involves turning.

Class Activity

- Project the transparency of TM 13. Cover example 2 so that only example 1 and the choice box are visible.

  Q: What pattern do you see in this sequence?

  A: Possible answers include reverses, flips, or alternating left and right. The arrows seem to be flipping back and forth. (Students may accurately recognize that the arrow could be rotating 180°.)

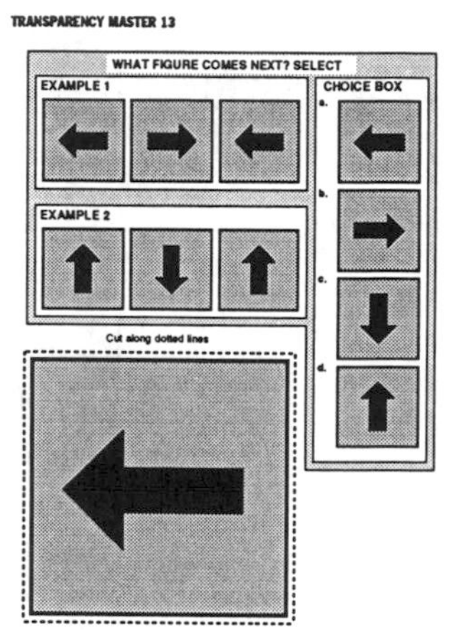

- Replace the transparency with the large square cut from the bottom of TM 13. Put a straw or pencil along the vertical axis and flip the arrow two or three times.

  Q: How would you describe how this arrow is turning?

  A: The black arrow seems to be flipping about the straw (or pencil).

- Flip the large pattern two or three times about the vertical axis, ending with the arrow pointing to the left.

  Q: The first figure points to the left.

- Flip the arrow so that it points to the right.

  Q: The second figure points right.

- Flip the arrow so that it points back to the left.

  Q: The third figure is in the same position as the first.

- Flip the arrow so that it points to the right.

  Q: The fourth figure, then, should be in this position. I need to find the figure that looks like this one.

- Point to the figures in the choice box.

Q: Which figure in the choice box looks like the one that should come next in the sequence?
A: Figure *b*

- Draw a line from the end of row 1 to figure *b*.
  Q: We call this kind of sequence flipping.
- Write "flipping" on the board.
  Q: We can also flip the arrow up and down.
- Project example 2 of TM 13. Cover example 1 so that only example 2 and the choice box are visible.
  Q: Look at the arrow's position this time. The black arrow seems to be flipping again, but this time it seems to be flipping up and down.
- Project the large square from TM 13; place it in the same orientation as the first square in row two (arrow pointing upward). Lay a pencil or straw along the horizontal axis.
  Q: In the first position, the arrow points up.
- Flip so that the arrow points down.
  Q: In the second position, the arrow points down.
- Flip so the arrow points up again.
  Q: In the third position, the arrow points up again. What should the fourth figure look like?
- Pause for student answers.
  A: Figure *c* (the arrow points down)
- Place TM 13 on the projector.
  Q: What is the difference between flips and rotations? When you flip the square, one side stays still and the other sides move. You have to pick up part of the square and turn it over.
- Flip back and forth.
  Q: I can flip it left and right...
- Flip twice about a vertical axis.
  Q: Or I can flip it up and down,...
- Flip twice about a horizontal axis.
  Q: ...but each time I have to move it through space. I have to pick it up and put it down. If I am doing a rotation sequence, I don't have to pick up the figure. I can turn (rotate) it to the right...
- Turn the large square to the right through four positions.
  Q: ...or I can turn it to the left.
- Turn the large square to the left through four positions.

GUIDED PRACTICE
EXERCISES: **C-52, C-53**

INDEPENDENT PRACTICE

- Assign exercises **C-54** through **C-56**.

THINKING ABOUT THINKING

Q: What did you pay attention to when you decided which figure came next?

1. I looked carefully at the details (position of the figure, pattern or color) to decide what characteristic was being changed.
2. I looked for a pattern of changes like tumbling (rotating) or flipping (reflecting).
3. I figured out what the next figure would look like if the pattern continued.
4. I selected the next figure in the pattern.

PERSONAL APPLICATION

Q: Where do you remember seeing repeating patterns of shapes?

A: Examples include clothing fabric, brick or concrete block walls, floor tiles, ceiling tiles, Venetian blinds, leaf arrangements on plants, beads, etc.

Q: When might you have to decide if something has been flipped?

A: Examples include when a letter has been reversed, seeing words in a mirror, handling dress patterns which have to be reflected to cut the second piece, print making, negatives in photography.

Q: When might recognizing rotation be important?

A: Examples include visualizing how puzzle pieces fit, turning maps to read route markers.

EXERCISES C-57 to C-62

WHICH FIGURE COMES NEXT?—SELECT (CHANGES IN DETAIL)

ANSWERS C-57 through C-62 — Student book pages 89–90
Guided Practice: C-57 d (subtracting detail) **C-58** b (adding detail)
Independent Practice: C-59 b (subtracting detail) **C-60** d (adding detail)
C-61 a (subtracting detail) **C-62** b (subtracting detail) See below.

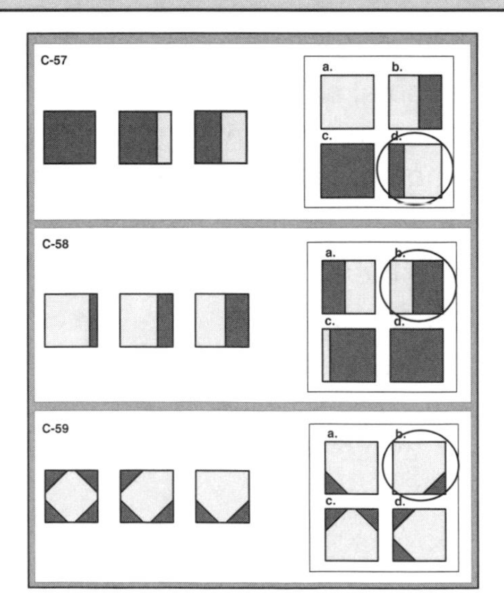

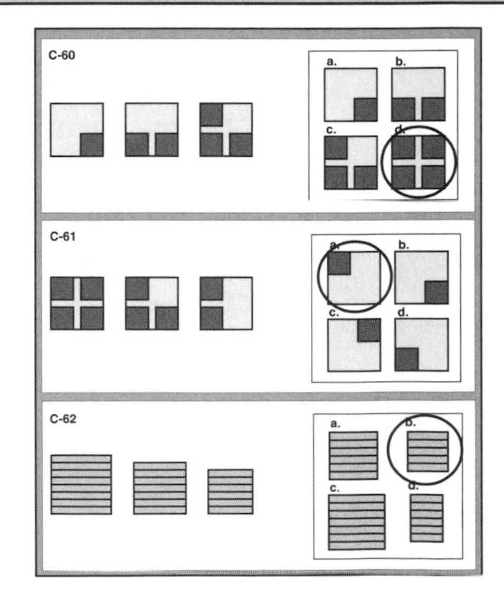

LESSON PREPARATION

OBJECTIVE AND MATERIALS

OBJECTIVE: Students will find the next figure in a series by identifying the pattern of added or subtracted detail.

MATERIALS: TM 14 (p. 262)

CURRICULUM APPLICATIONS

Language Arts: Forming new words by adding or subtracting prefixes and suffixes

Mathematics: Making charts and graphs, recognizing enlargement or reduction of the proportions of geometric figures

Science: Recognizing the operation of gears, wheels, motors; recognizing growth or injury to organisms

Social Studies: Identifying geographic changes or types of topography on maps, comparing different types of maps that show the same geographic area, comparing photographs to maps

Enrichment Areas: Repeating interval patterns in music; art exercises involving adding detail; playing games that call for adding additional details, tokens, playing pieces, etc.

TEACHING SUGGESTIONS

Reinforce the following vocabulary emphasized in this lesson: *adding detail, subtracting detail.* Encourage students to use these terms in their discussion.

MODEL LESSON

LESSON

Introduction

Q: We have been looking carefully at sequences that show changes in shape or changes by rotation or flipping.

Explaining the Objective to Students

Q: In this lesson, you will decide if a shape shows a pattern of increasing or decreasing detail.

Class Activity

- Project figure 1, cut from TM 14.
  Q: In this figure, there is a black bar on the left side of the square.
- Add figure 2, putting it directly over the first figure.
  Q: Now the black bar has grown.
- Add figure 3, putting it directly over the first two figures.
  Q: In this figure, the black part covers about half of the square.
- Add figure 4, putting it directly over the first three figures.
  Q: Now, most of the square is black. This type of sequence is known as adding detail.

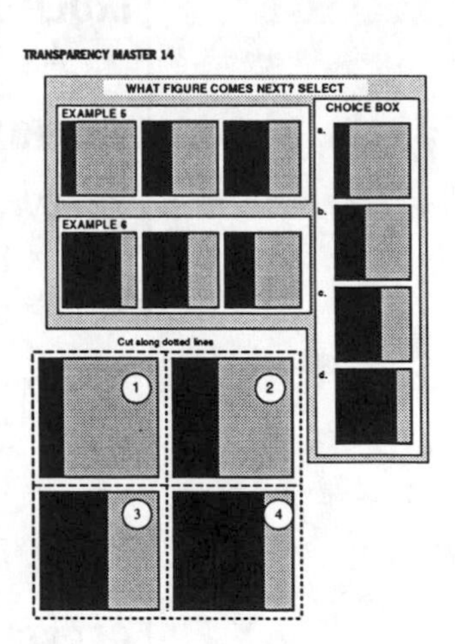

- Write “adding detail” on the board. Project example 5 and the choice box from TM 14.

Q: In example 5, which figure in the choice box will continue the sequence?
A: Figure *d*

- Draw a line from the end of row 5 to figure *d.*
  Q: If we reverse the process and take some of the shaded parts away, it is called subtracting detail, much in the same way as we subtract numbers. This kind of sequence starts with a lot of color or lines and each figure in the sequence has less. Let's use this figure as an example. In the first position, most of the square is black.
- Remove figure 4.
  Q: Now I have subtracted some of the black part.
- Remove figure 3.
  Q: I have removed more of the color, leaving only part of the square black.
- Remove figure 2.
  Q: Now only the bar at the side remains black. I have subtracted an equal amount of the black part each time. I have removed detail. The detail in this example is the amount of black space. (Note: You may have an intelligent, creative student who recognizes the difference in positive and negative space and perceives that in the first demonstration you were subtracting gray and in the second you were adding gray. That perception is also correct and should be encouraged.)
- Write "subtracting detail" on the board. Project example 6 from TM 14.
  Q: Which figure in the choice box seems to continue this sequence?
  A: Figure *a*
- Draw a line from the end of row 6 to figure *a.*

GUIDED PRACTICE

EXERCISES: **C-57**, **C-58**

INDEPENDENT PRACTICE

- Assign exercises **C-59** through **C-62**.

THINKING ABOUT THINKING

Q: What did you pay attention to when you decided which figure came next?

1. I looked carefully at the details to decide what characteristic was being changed.
2. I looked for a pattern of changes like tumbling (rotating), flipping (reflection), or adding or subtracting details.
3. I figured out what the next figure would look like if the pattern continued.
4. I selected the next figure in the pattern.

PERSONAL APPLICATION

Q: Where else might you notice added or subtracted detail?
A: Examples include doing puzzles where you need to find a figure that is different from the others or find the two figures that are the same, finding hidden figures in pictures.

EXERCISES C-63 to C-74

WHICH FIGURE COMES NEXT?—DRAW IT!

ANSWERS C-63 through C-74 — Student book pages 91–93
Guided Practice: C-63 repeating patterns of shapes; **C-64** repeating patterns of shapes
Independent Practice: C-65 flipping from lower left to upper right corner, or reflection about the diagonal; **C-66** turning to the right, or clockwise rotation; **C-67** turning to the left, or counterclockwise rotation; **C-68** subtracting detail; **C-69** turning to the right, or clockwise rotation; **C-70** flipping left and right, or reflection about a vertical axis; **C-71** flipping from lower left to upper right corner, or reflection about the diagonal; **C-72** flipping from top to bottom, or reflection about a horizontal axis; **C-73** adding detail—1/4 shaded, 1/2 shaded, 3/4 shaded, all shaded; **C-74** adding detail—no shading, 1/4 shaded, 1/2 shaded, 3/4 shaded See below.

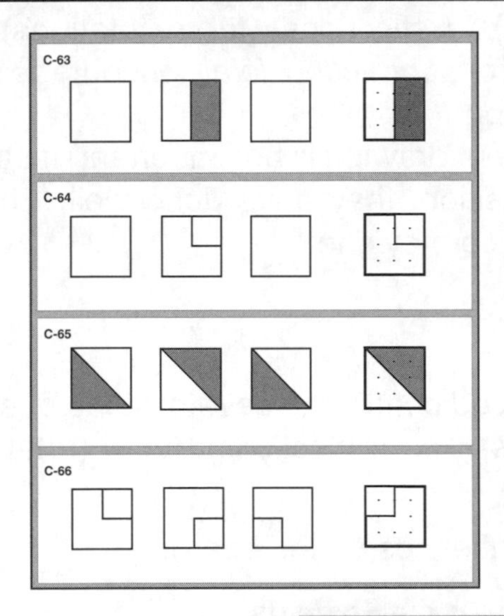

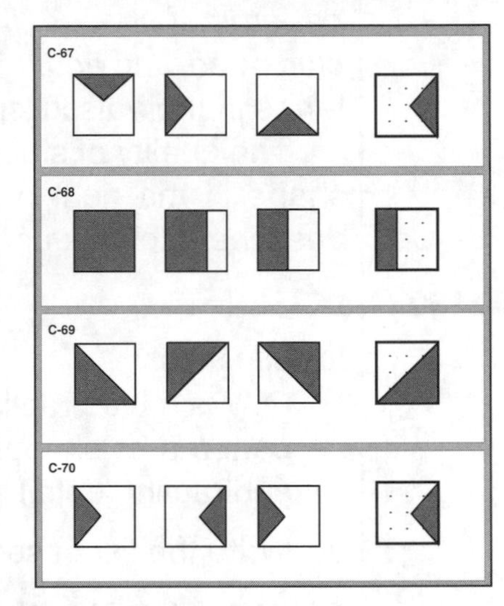

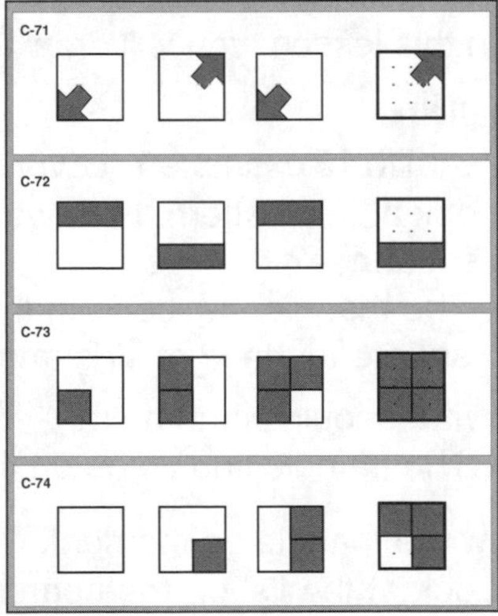

LESSON PREPARATION

OBJECTIVE AND MATERIALS

OBJECTIVE: Students will continue a sequence by drawing the last shape on a grid of dots.

MATERIALS: TM 15 (p. 263) • attribute blocks (optional)

CURRICULUM APPLICATIONS

Language Arts: Choosing the right picture(s) to fill in or complete a story sequence, filling in the missing sections on a partially completed outline format
Mathematics: Filling in the missing item(s) in a sequence of shapes, completing mathematical tables
Science: Biological stages in the development of animals or plants, duplicating steps in laboratory experiments
Social Studies: Recognizing repeating causal relationships; developing a time line pattern; predicting changes in population, production output, etc. by finding past statistical patterns; reading and interpreting bar and line graphs
Enrichment Areas: Art projects requiring construction of repeated patterns, memorizing music or dance patterns

TEACHING SUGGESTIONS

Reinforce the following vocabulary emphasized in this lesson: *repeating patterns of shapes*, *flips* (reflections), *turns* (rotations), *clockwise*, *counter-clockwise*, and *adding* or *subtracting detail*. Encourage students to use these terms in their discussion.

The quality of student drawing is not important. If students have the right shape in the right position, they have done well. Encourage students to describe each sequence they see.

MODEL LESSON

LESSON

Introduction

Q: We have practiced continuing several kinds of sequences: repeating patterns of shapes, flips (reflections), turns (rotations), and adding or subtracting detail.

- Write the list of sequences on the board.

Explaining the Objective to the Students

Q: In this lesson, you will draw the next shape in a sequence.

Class Activity

- Project TM 15 example 7 (cover examples 8–10).
  Q: Look at this pattern. How would you describe this sequence?
  A: The triangle moves from the left side of the square to the right side and then to the left.
  Q: What should come next?
  A: The triangle should be on the right.

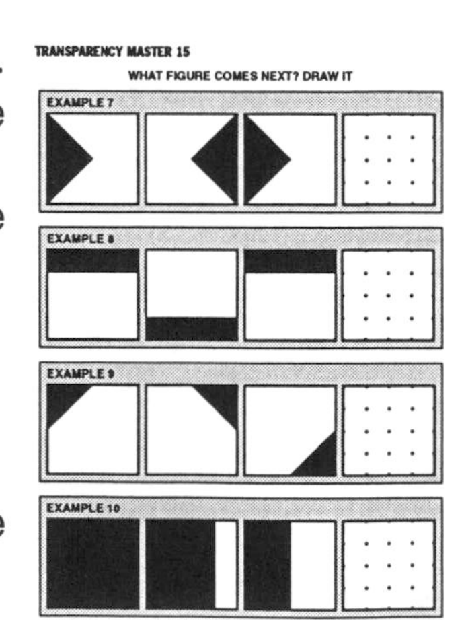

- Draw the triangle on the right.
  Q: Look at the list on the board. Which of these terms describes this sequence?
  A: Flips
  Q: How did the figure flip?
  A: From left to right

- Project TM 15 example 8 (cover examples 9–10).
  Q: Look at this pattern. How would you describe this sequence?
  A: The black part moves from the top to the bottom and then back to the top.
  Q: What should come next?
  A: The black part should be on the bottom.
- Draw the black rectangle on the bottom.
  Q: Look at the list on the board. Which of these terms describes this sequence?
  A: Flips
  Q: How did the figure flip?
  A: From top to bottom
- Project TM 15 example 9 (cover example 10).
  Q: Look at this pattern. How would you describe this sequence?
  A: The triangle moves around the corners of the square from top left, to top right, to bottom right.
  Q: What should come next?
  A: The triangle should be on the bottom left.
- Draw the triangle on the bottom left.
  Q: Look at the list on the board. Which of these terms describes this sequence?
  A: Turns
  Q: How did the figure turn?
  A: From left to right, or clockwise
- Project TM 15 example 10.
  Q: Look at this pattern. How would you describe this sequence?
  A: The black part is getting smaller.
  Q: What should come next?
  A: The black rectangle should be narrow and on the left.
- Draw the rectangle on the left.
  Q: Look at the list on the board. Which of these terms describes this sequence?
  A: Subtracting detail

GUIDED PRACTICE

EXERCISES: **C-63**, **C-64**

INDEPENDENT PRACTICE

- Assign exercises **C-65** through **C-74**.

THINKING ABOUT THINKING

Q: What did you pay attention to when you decided which figure came next?

1. I looked carefully at the details (position of the figure, added or subtracted detail, pattern or color) to decide what characteristic was changed.
2. I looked for a pattern of changes like rotating, reflecting, more detail, less detail, etc.
3. I figured out what the next figure would look like if the pattern continued.
4. I drew the next figure in the pattern.

PERSONAL APPLICATION

Q: When might you have to make or figure out the next item in a sequence?

A: Examples include knitting, carving, weaving, art activities. Examples like **C-63** and **C-64** have the appearance of a column of bricks or blocks being put in place. The repeating sequences may remind students of decorative designs.

EXERCISES C-75 to C-83

DESCRIBE A SEQUENCE/SEQUENCE OF POLYGONS AND ANGLES

ANSWERS C-75 through C-83 — Student book pages 94–99
Guided Practice: C-75 This is a sequence of circles that gets larger. **Independent Practice: C-76** This is a sequence of adding rectangles to a stack (adding detail). **C-77** In this sequence, the arrow turns in a counterclockwise rotation. **C-78** In this sequence, the half circle flips right to left (reflection about a vertical axis). **C-79** In this sequence, the triangle flips up and down (reflection about a horizontal axis). **C-80** In this sequence, the stack of rectangles gets smaller (subtracting detail). **C-81** In this sequence, the pattern flips from corner to opposite corner (reflection about a diagonal). **C-82** to **C-83** See below.

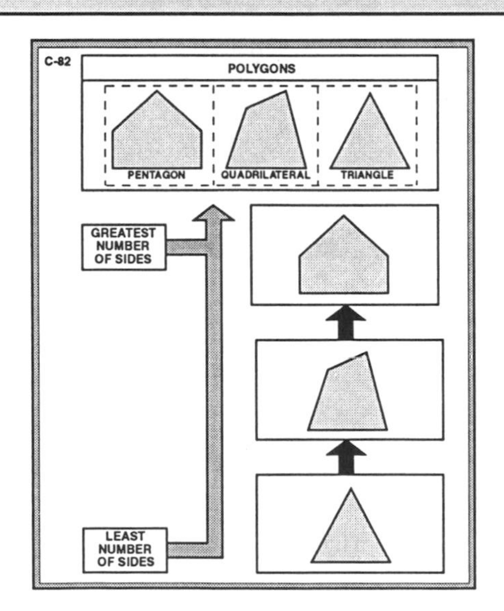

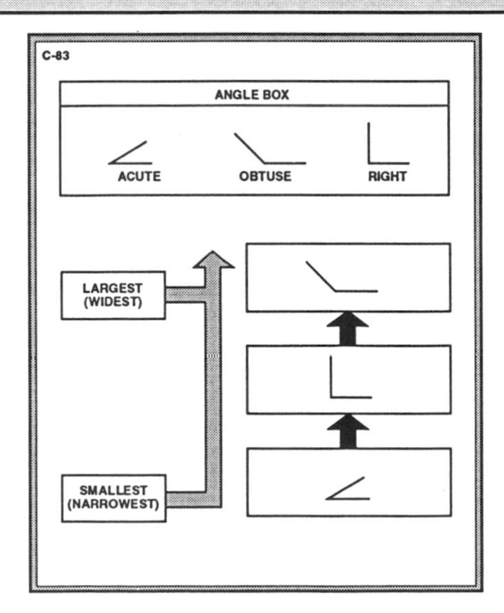

LESSON PREPARATION

OBJECTIVE AND MATERIALS

OBJECTIVE: Students will write a description of a sequence.
MATERIALS: Transparency of page 94 of the student book

CURRICULUM APPLICATIONS

Language Arts: Choosing the right picture(s) to fill in or complete a story sequence
Mathematics: Filling in the missing item(s) in a sequence of shapes, completing mathematical tables
Science: Tracing biological stages in the development of animals or plants, following steps in science activities
Social Studies: Developing a time line pattern, reading and interpreting bar and line graphs
Enrichment Areas: Art projects requiring construction of repeated patterns, memorizing music or dance patterns

TEACHING SUGGESTIONS

Reinforce the following key concepts emphasized in this lesson: *changing only shape*, *flips*, *turns*, *adding or subtracting detail, rotation, reflection.* Encourage students to use these terms in their discussion and to write complete sentences. The example on page 94 illustrates color reversal, while **C-75** illustrates size change.

MODEL LESSON

LESSON

Introduction

Q: Name the kinds of sequences we have seen.
A: Changing only shape, flips, turns, adding or subtracting detail

Explaining the Objective to the Students

Q: In this lesson, you will write a description of a sequence.

Class Activity

- Project transparency of page 94.

Q: How is the second figure in the sequence different from the first?
A: The gray and black areas are exchanged.

Q: Now, look at the next two figures; are they changed in the same way?
A: Yes, again the gray and black areas are exchanged.

Q: After checking the last two figures, you are ready to write a description.

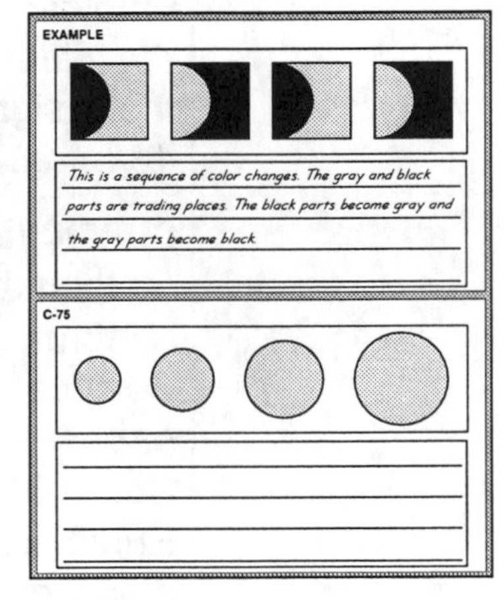

GUIDED PRACTICE

EXERCISE: **C-75**

INDEPENDENT PRACTICE

- Assign exercises **C-76** through **C-83**.

THINKING ABOUT THINKING

Q: What did you pay attention to when you decided which figure came next?

1. I looked carefully at the details (size of the figure, number of sides, position of the figure, size of the angle, pattern or color, added or subtracted detail) to decide what characteristic was changed.
2. I looked for a pattern of changes (getting larger, getting smaller, rotating, reflecting, more detail, less detail, etc.).
3. I figured out what the next figure would look like if the pattern continued.
4. I wrote a description of what I observed.

PERSONAL APPLICATION

Q: When might you need to describe a pattern sequence?

A: Examples include describing a series of events in a picture story or a cartoon strip, following the cooking instructions pictured on the side of packaged foods, sequencing events on a time line or a graph.

EXERCISES C-84 to C-89

PAPER FOLDING—SELECT

ANSWERS C-84 through C-89 — Student book pages 100–101
Guided Practice: C-84 c; **C-85** d; **C-86** b
Independent Practice: C-87 a; **C-88** b; **C-89** c

LESSON PREPARATION

OBJECTIVE AND MATERIALS

OBJECTIVE: Students will learn to recognize which pattern makes a particular design.

MATERIALS: 5" × 8" models of the items in the example and exercise **C-84** (p. 100) • Use a hole punch and brightly colored paper to make a model of the unfolded design and models of the four folded patterns *a*, *b*, *c*, and *d*. (Make the folded patterns in such a way that they can be unfolded and compared to the design.) • paper and scissors for teacher and students • photocopies of TM 16 (p. 264) and TM 17 (p. 265)

CURRICULUM APPLICATIONS

Mathematics: Visual perception skills, drawing geometric shapes to form patterns

Science: Reproducing crystal or snowflake patterns, predicting appearance or position

Social Studies: Seeing patterns in charts, graphs, or schedules

Enrichment Areas: Breaking down a dance routine or an art project into basic steps, creating symmetrical patterns in art, creating stencils, making masks

TEACHING SUGGESTIONS

Reinforce the following vocabulary emphasized in this lesson: *right*, *left*, *center*, *top*, and *bottom* (or *upper* and *lower*). Encourage students to use these terms in their discussion. *Inner*, *outer*, and *near* are relative terms that require some point for comparison. The concepts of curve and symmetry may be introduced at this time. (Note: In this lesson, *pattern* refers to the folded item; *design* refers to the unfolded item.)

Students must be able to perceive the folded edge and distinguish it from the outer edges in these exercises. If students have trouble visualizing the folded pattern of each design, they may use half sheets of paper and a hole punch to duplicate patterns and designs. Students may also wish to make up similar problems for their classmates to solve.

Note: In exercises **C-84** through **C-86**, students will select which pattern makes a particular design after it has been folded. In exercises **C-87** through **C-89**, students will reverse the process, selecting the design which results when a pattern is unfolded.

MODEL LESSON

LESSON

Introduction

Q: Have you ever cut a design from a folded paper? The design does not look the same when it is unfolded as it did when you were cutting the folded paper.

- As the class watches, cut a fancy, sharply curved pattern from an edge of a piece of scrap paper folded in half. Place the folded pattern on the overhead projector.
  Q: In earlier lessons, we used the word *pattern* to describe a repeating sequence. In this lesson, pattern refers to an outline for cutting. The folded paper you see on the projector is my pattern. Let's see what the design looks like when I unfold it.
- Unfold the paper.
  Q: When I unfold the pattern, I have a lovely design. Now two sides have the same curves as the one edge I cut.
- Fold the paper again. Select a noticeable curve and point as you explain.
  Q: This curve...
- Point to identify the curve as you slowly open the pattern.
  Q: ...becomes two curves.
- Point to both curves.
  Q: Let's practice. Cut the fanciest pattern you can make.
- Allow time for students to cut patterns and discuss their designs.
  Q: Because our patterns have such large curves, the unfolded design is lovely. You can see easily the two edges that result...
- Point to both edges of a sample student-cut pattern.
  Q: ...from the one that you cut.

Explaining the Objective to the Students

Q: In this lesson, you will select a pattern that makes a particular design.

Class Activity—Option 1

- Place the folded patterns marked *a*, *b*, *c*, and *d* on the chalkboard tray, tabletop, or overhead projector so that all students can see them. Unfold the unmarked pattern of *a*. Hold it up for students to see then place it to the left of the folded patterns.
  Q: Which pattern—*a*, *b*, *c*, or *d*—do you think will make this design when it is unfolded?
  A: Pattern *a*

Q: How can we be sure that *a* is the right choice?
A: Unfold pattern *a* and match it to the design.

- Point as you explain that the hole is near the upper edge of both pattern *a* and the design.
  Q: Why can't pattern *b* be the answer?
  A: The holes are too large and don't match up when we unfold the pattern and put it in front of the design.
  Q: Why can't pattern *c* be the answer?
  A: The hole in pattern *c* is in the center, not near the upper edge.
  Q: Why can't pattern *d* be the answer?
  A: The hole in pattern *d* is near the lower edge, not the upper edge.
  Q: Look at the example on page 100. The design at the top of the page looks like this one.

PAPER FOLDING—SELECT

DIRECTIONS: The figure on the left represents a sheet of paper with holes punched in it. How will the sheet look when folded along the dotted line? Circle the correct answer in the choice box.

EXAMPLE — Unfolded — CHOICE BOX — Folded — a. b. c. d.

C-84 — Unfolded — CHOICE BOX — Folded — a. b. c. d.

C-85 — Unfolded — CHOICE BOX — Folded — a. b. c. d.

C-86 — Unfolded — CHOICE BOX — Folded — a. b. c. d.

- Pick up design *a*.
  Q: This design, when unfolded, looks like the unfolded pattern. Notice that it has been circled. Now look at exercise **C-84** on the same page.
- Hold up the unfolded pattern corresponding to **C-84**. Hold up models. Allow time for student discussion.
  Q: Which pattern looks like this design?
  A: Pattern *c*
  Q: Circle *c* in the choice box.

Class Activity—Option 2

- Distribute hole punches and photocopies of TM 16 and TM 17. Ask students to create paper models of the patterns. Follow the dialogue in the model lesson above while students use the paper models.

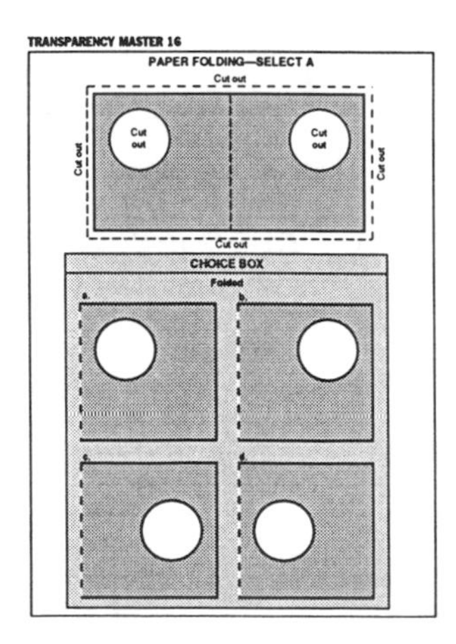

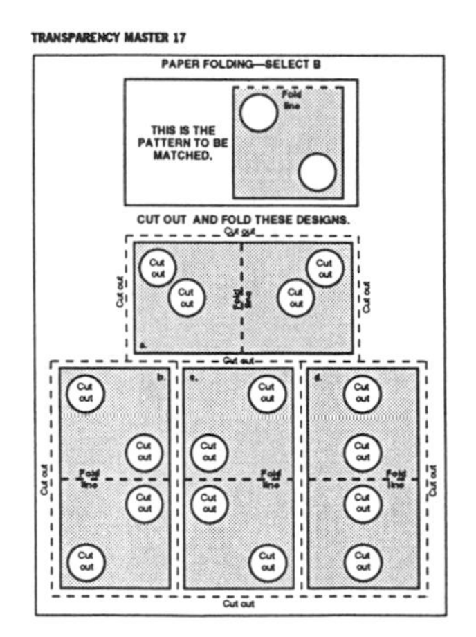

GUIDED PRACTICE

EXERCISES: **C-84** through **C-86**

INDEPENDENT PRACTICE

- Assign exercises **C-87** through **C-89.**

THINKING ABOUT THINKING

Q: What did you pay attention to when you decided which figure came next?

1. I looked carefully at the location of the holes.
2. In my mind, I lined up the fold lines.
3. I matched the pattern and design.

PERSONAL APPLICATION

Q: When might it be helpful if you were able to figure out which pattern created a design or which design is made from a pattern?

A: Examples include refolding a map, cutting a pattern to create a particular snowflake design.

EXERCISES C-90 to C-97

PAPER FOLDING—DRAW

ANSWERS C-90 through C-97 — Student book pages 102–3
Guided Practice: C-90 to **C-93** See below.
Independent Practice: C-94 to **C-97** See below.

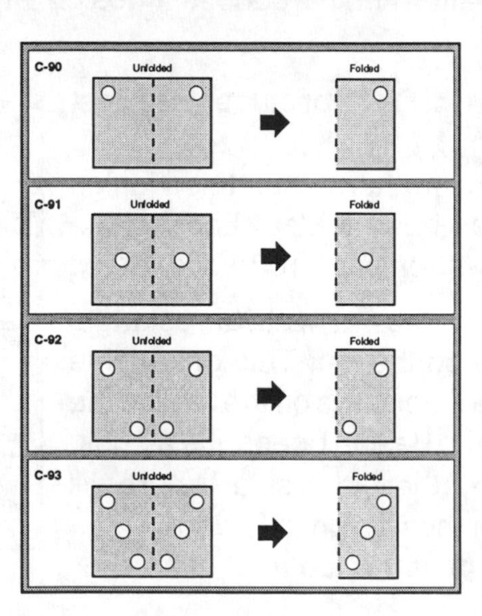

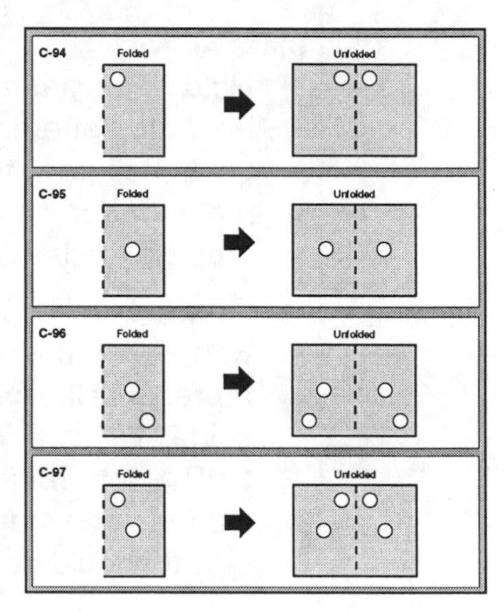

LESSON PREPARATION

OBJECTIVE AND MATERIALS

OBJECTIVE: Students will draw the pattern that results when a design is folded and the design that results when a pattern is unfolded.

MATERIALS: Paper model of **C-90**, page 102 of the student book • paper and scissors for teacher and students

CURRICULUM APPLICATIONS

Mathematics: Recognizing symmetrical figures
Science: Recognizing symmetry in natural forms
Social Studies: Map drawing skills, projecting a graph pattern
Enrichment Areas: Making a string of paper dolls, a paper airplane, or a kite; origami; drawing the unfinished side of a figure or picture

TEACHING SUGGESTIONS

Reinforce the following vocabulary emphasized in this lesson: *right/left/center, inner/outer, edge/fold, pattern* (folded)/*design* (unfolded), and *symmetry* (optional). (Note: The word *pattern* refers to the folded item; the word *design* refers to the unfolded item. It may be necessary to tell students that the circles represent holes that go all the way through both sides of the folded sheet.) Encourage students to use these terms in their discussion.

Students must perceive the folded edge and distinguish it from the outer edges in these exercises. Identify these edges and encourage students to use the directional and relative terms discussed in the previous lessons.

MODEL LESSON

LESSON

Introduction

Q: In previous lessons, you practiced recognizing which pattern creates which design when it is unfolded and what pattern results when a design is folded.

Explaining the Objective to the Students

Q: In this lesson, you will draw the design that results when a pattern is unfolded or the pattern that produces a design when it is folded.

Class Activity

- Hold up the pattern of **C-90** for students to see. Fold the pattern.
  Q: Let's draw the pattern that this design makes after it has been folded. Look at example exercise **C-90** on page 102 while I draw.

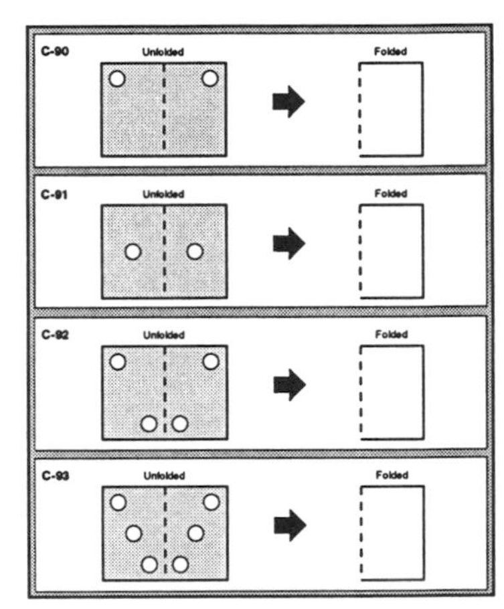

- On the blackboard, draw a vertical rectangle with a dotted line on the left. This dotted line represents the fold line and is on the left like the spine of a book. The design opens like a book.
  Q: The dotted line is the folded side. Where will the hole go when the design is folded?
  A: It should go near the top right corner.
- Draw in the hole as shown.

GUIDED PRACTICE

EXERCISES: **C-90** through **C-93**

INDEPENDENT PRACTICE

- Assign exercises **C-94** through **C-97**. Note: In the preceding exercises, students drew the pattern of a design when it is folded. In exercises **C-94** through **C-97,** students will draw the design which results when a pattern is unfolded.

THINKING ABOUT THINKING

Q: What did you pay attention to when you drew the folded pattern?

1. I identified the fold line of the design.
2. I decided where the hole would go when the design was folded.

3. I drew the pattern the design would make when it was folded.
4. I followed the same process to draw the design of the unfolded pattern.

PERSONAL APPLICATION

Q: When must you be able to recognize or draw what something will look like when it is folded or unfolded?

A: Examples include getting correct linens from a shelf or clothes from a drawer; folding napkins or clothes; paper cutting or paper folding; finding correct pattern pieces for constructing clothing, shop, or carpentry projects; visualizing completed models.

Chapter 4

FIGURAL CLASSIFICATIONS

(Student book pages 105–44)

EXERCISES D-1 to D-9

MATCH A SHAPE TO A GROUP

ANSWERS D-1 through D-9 — Student book pages 106–9
Guided Practice: D-1 a; **D-2** e; **D-3** b; **D-4** d
Independent Practice: D-5 b; **D-6** d; **D-7** a; **D-8** e; **D-9** c

LESSON PREPARATION

OBJECTIVE AND MATERIALS

OBJECTIVE: Students will practice classifying objects by shape.
MATERIALS: Transparency Master 18 (p. 266) • washable transparency markers or colored chalk in three colors • attribute blocks (optional)

CURRICULUM APPLICATIONS

Language Arts: Decoding skills in reading readiness, recognizing sentence types from end marks, learning to form letters in handwriting
Mathematics: Distinguishing between types of arithmetic problems (addition, subtraction, etc.) by sign and/or problem configuration
Science: Classifying natural objects (leaves, fish, shells, etc.) according to shape
Social Studies: Identifying the functions of road signs from their shapes, associating international signs and symbols with their meanings
Enrichment Areas: Distinguishing note values in printed music, classifying architectural structures by common shapes

TEACHING SUGGESTIONS

Make sure that students know the definition of and can identify a pentagon. Use the examples *tricycle* (three-wheeler) and *triangle* (three-angled) to illustrate how the prefix for a figure name is often related to the number of angles or sides it has. Explain that *penta* is Greek for five.

MODEL LESSON

LESSON

Introduction

Q: We call ourselves a "class" of students. What do all the students in this class have in common? In this class, everyone is about the same age, meets in the same place, studies the same things, and has me for a teacher. *Class* means more than just a school room; it also means a group which has a common characteristic. When we describe the group by naming that common characteristic, we are classifying.

Explaining the Objective to the Students

Q: In this lesson, you will classify objects by shape.

Class Activity—Option 1

- Demonstration using attribute blocks or cutout shapes: On a flat surface or chalk rail, arrange a group of three rectangles having different sizes and colors.

Q: What do all of these shapes have in common?

A: They are all rectangles.

- From the remaining shapes, invite a student to pick another shape that belongs to the group.

  Q: We said that a class is a group with a common characteristic and that the common characteristic of these shapes is that they are all rectangles. Therefore, we can say that these shapes all belong to the class "rectangle."

Class Activity—Option 2

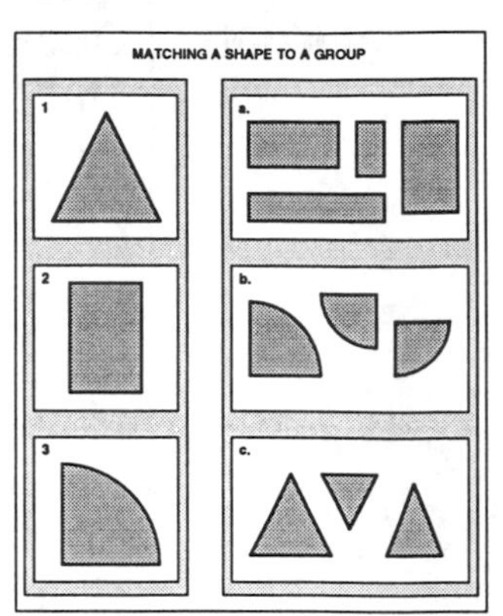

- Project transparency of TM 18. Use a colored marker to connect 1 to *c*. If you project on a chalkboard, use colored chalk.

  Q: Why can I connect this shape...
- Point to 1.

  Q: ...to this group?
- Point to group *c*.

  A: Group *c* is a group of triangles and box 1 has a triangle in it.

  Q: What is the name of the shape in box 2?

  A: Rectangle

  Q: To which group does it belong?

  A: Group *a*
- Use a second color to connect box 2 to group *a*.

  Q: What is the shape in box 3?

  A: It looks like a big piece of pie.

  Q: To which group does it belong?

  A: Group *b*
- Use a third color to connect box 3 to group *b*.

GUIDED PRACTICE

EXERCISES: **D-1** through **D-4**

- When students have had sufficient time to complete these exercises, check answers by discussing their choices and the reasons for eliminating other choices. Students should describe each shape and class as precisely as is appropriate for their vocabulary and developmental level.

INDEPENDENT PRACTICE

- Assign exercises **D-5** through **D-9**.

THINKING ABOUT THINKING

Q: What did you pay attention to when you classified the figures?

1. I looked carefully at the shape of the figure on the left.

2. I looked for a group of figures that had the same characteristic(s) as the first shape.
3. I drew a line connecting the shape to the group.

PERSONAL APPLICATION

Q: When might you want or need to classify objects by shape?

A: Examples include sorting eating or cooking utensils, sorting construction toys or tools, sorting edge pieces from interior pieces in a picture puzzle, organizing workshop materials at home or in school.

EXERCISES D-10 to D-18

MATCH A PATTERN TO A GROUP

ANSWERS D-10 through D-18 — Student book pages 108–9
Guided Practice: D-10 e; **D-11** b; **D-12** a; **D-13** c
Independent Practice: D-14 c; **D-15** e; **D-16** a; **D-17** b; **D-18** d

LESSON PREPARATION

OBJECTIVE AND MATERIALS

OBJECTIVE: Students will practice classifying objects by pattern or design.
MATERIALS: Transparency of TM 19 (p. 267) • washable transparency markers or colored chalk in three colors • attribute blocks (optional)

CURRICULUM APPLICATIONS

Language Arts: Classifying words by letter pattern, recognizing prefixes and suffixes
Mathematics: Classifying numbers according to place value, sorting geometric shapes
Science: Classifying leaves, insects, flowers, shells, birds, etc. by pattern
Social Studies: Using a legend to distinguish topographical areas on a map (mountains, deserts, etc.), interpreting graphs
Enrichment Areas: Classifying color families in art, classifying music by sound or rhythm pattern (classical, jazz, etc.)

TEACHING SUGGESTIONS

Students should describe each pattern and class as precisely as is appropriate for their vocabulary and developmental level.

MODEL LESSON

LESSON

Introduction

Q: Recall that *class* means more than just a school room; it also means a group which has a common characteristic. When we describe the group by naming that common characteristic, we are classifying.

Explaining the Objective to Students

Q: In this lesson, you will classify objects by pattern.

Class Activity

- Project transparency of TM 19. Use a colored marker to connect box 1 to *b*. If you project on a chalkboard, use colored chalk.

  Q: Why can I connect this pattern...

TRANSPARENCY MASTER 19

MATCHING A PATTERN TO A GROUP

- Point to box 1.
  Q: ...to this group?
- Point to group *b*.
  A: Box 1 is a square with a dark pie-shaped piece in one corner.
  Q: To which group does it belong?
  A: Group *b*
  Q: Describe the pattern in box 2.
  A: A square with a dark circle in one corner
  Q: To which group does it belong?
  A: Group *c*
- Use a second color to connect box 2 to group *c*.
  Q: Describe the pattern in box 3.
  A: A box with a dark triangle on one side
  Q: To which group does it belong?
  A: Group *a*
- Use a third color to connect box 3 to group *a*.

GUIDED PRACTICE

EXERCISES: **D-10** through **D-13**

- When students have had sufficient time to complete these exercises, check answers by discussing their choices and the reasons for eliminating other choices.

INDEPENDENT PRACTICE

- Assign exercises **D-14** through **D-18**.

THINKING ABOUT THINKING

Q: What did you pay attention to when you classified the patterns?

1. I looked carefully at the details of the pattern on the left (design, color, etc.).
2. I looked for a group of patterns that had the same characteristic(s) as the first figure.
3. I drew a line connecting the pattern to the group.

PERSONAL APPLICATION

Q: Where might you want or need to classify objects by pattern or color?
A: Examples include sorting art materials, fabrics, or puzzle pieces; sorting laundry by color.

EXERCISES D-19 to D-27

SELECT A SHAPE THAT BELONGS TO A GROUP

ANSWERS D-19 through D-27 — Student book pages 110–11
Guided Practice: D-19 c; **D-20** c
Independent Practice: D-21 b; **D-22** c; **D-23** d; **D-24** b; **D-25** c; **D-26** b; **D-27** d

LESSON PREPARATION

OBJECTIVE AND MATERIALS

OBJECTIVE: Students will find a figure that can belong to a group or class because of its shape or pattern.
MATERIALS: Transparency of TM 20 (p. 268)

CURRICULUM APPLICATIONS

Language Arts: Classifying words into pattern groups (silent-e words, double vowel words, "-ing" words)
Mathematics: Grouping fractions by common denominator, sorting geometric shapes
Science: Classifying animals into families (primates, marsupials, mammals, etc.), distinguishing between different food groups
Social Studies: Using legends to interpret maps and graphs; distinguishing between artifacts according to use, type, material, etc.
Enrichment Areas: Classifying types of music or forms of art and dance when given the classifying criteria

TEACHING SUGGESTIONS

Reinforce the names of the shapes used in this lesson and encourage students to use these names in their discussion. Use of the terms *parallelogram*, *pentagon*, and *obtuse* should depend on the grade and ability levels of students. Encourage students to be as precise as possible in describing figures and groups.

MODEL LESSON

LESSON

Introduction

Q: When we classified objects by shape or pattern, you matched a shape or pattern to a group by asking, "In which group does this shape belong?"

Explaining the Objective to the Students

Q: In this lesson, you will find another shape that belongs in a group. Now you are going to ask yourself, "Which one of these shapes belongs to the group?"

Class Activity

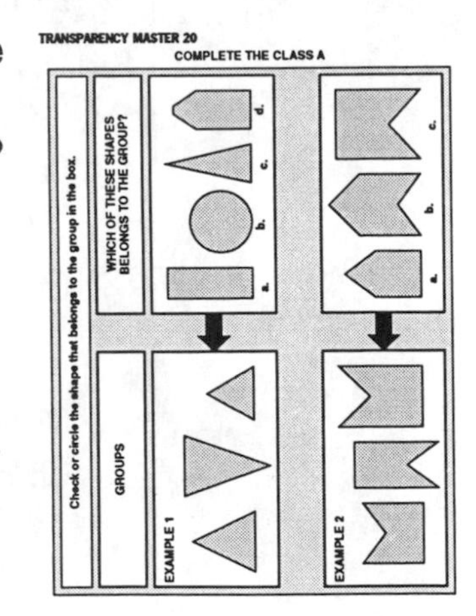

- Project TM 20 (example 1), keeping the right side covered.
  Q: How can we describe the class in example 1?
  A: It is a group of triangles.
- Move the covering to project figure *a*.
  Q: How can we describe figure *a?*
  A: It is a rectangle.
  Q: Does figure *a* belong to a class of triangles?
  A: No
  Q: Why not?
  A: It is not a triangle.
- Move the covering to project figure *b*.

Q: How can we describe figure *b?*
A: It is a circle.

Q: Can figure *b* belong to a class of triangles?
A: No

Q: Why not?
A: It is not a triangle.

- Move the covering to project figure *c.*
  Q: How can we describe figure *c?*
  A: It is a triangle.

  Q: This figure should be checked as the correct answer.
- Write a check mark on the projected image.
  Q: How can we describe figure *d?*
  A: It is a shape that looks like a house (a pentagon).

  Q: Can figure *d* belong to a class of triangles?
  A: It is not a triangle.

- Project example 2, keeping the right side covered.
  Q: We need to classify these figures by shape. How can we describe this class?
  A: The figures look like an M or a W. All the figures have five sides and two sides make a V shape.

- Move the covering to project figure *a.*
  Q: How can we describe figure *a?*
  A: It looks like a house.

  Q: Can figure *a* belong to this class of things?
  A: No

  Q: Why not?
  A: It does not look like an M.

- Move the covering to project figure *b.*
  Q: How can we describe figure *b?*
  A: It looks like an arrow.

  Q: Can figure *b* belong to this class of things?
  A: No

  Q: Why not?
  A: It does not look like an M.

- Move the covering to project figure *c.*
  Q: How can we describe figure *c?*
  A: It looks like an M.

  Q: This figure should be checked as the correct answer.
- Write a check mark on the projected image.

GUIDED PRACTICE

EXERCISES: **D-19** and **D-20**

INDEPENDENT PRACTICE

- Assign exercises **D-21** through **D-27**.

THINKING ABOUT THINKING

Q: What did you pay attention to when you classified the figures?

1. I looked carefully at the details of the shapes in the group box (number of sides, length of the sides, size of the angle).
2. I looked at each shape in the choices to see which had the same characteristic(s) as the shapes in the group box.
3. I checked the shape that belonged to the group.

PERSONAL APPLICATION

Q: When might you want or need to decide which item belongs to a particular class?

A: Examples include sorting laundry, deciding whether some article is washable or if it needs to be dry cleaned, sorting or retrieving tools and building materials.

EXERCISES D-28 to D-35

DESCRIBING CLASSES

ANSWERS D-28 through D-35 — Student book pages 112–13
Guided Practice: D-28 a, b; **D-29** a, b, c, d See below.
Independent Practice: D-30 a, c; **D-31** a, c; **D-32** d; **D-33** a, b, d; **D-34** c; **D-35** b, d See below.

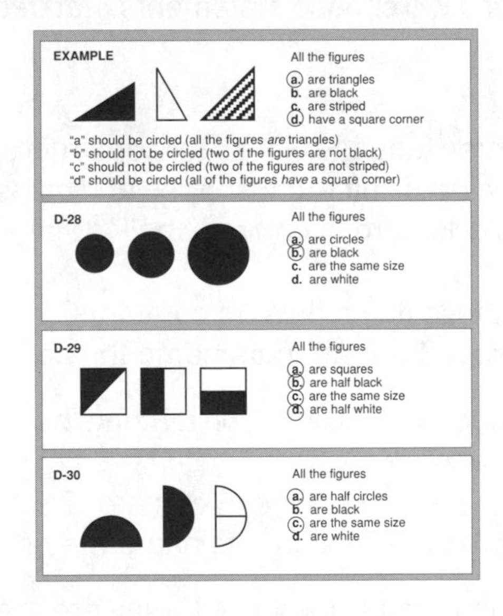

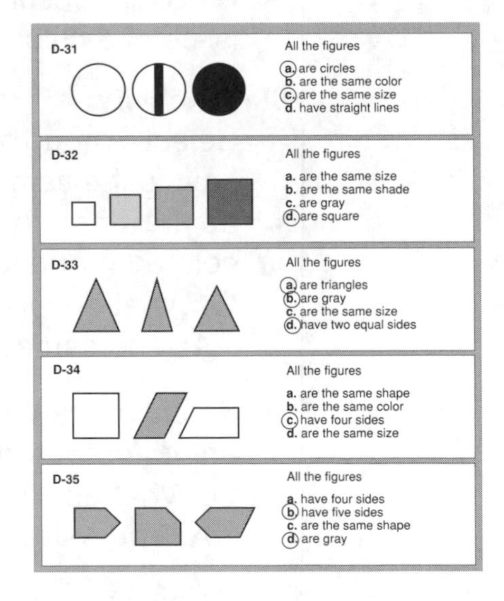

LESSON PREPARATION

OBJECTIVE AND MATERIALS

OBJECTIVE: Students will choose statements that describe the characteristics of a class.

MATERIALS: Attribute blocks or handmade equivalent

CURRICULUM APPLICATIONS

Language Arts: Describing the way things or people look, writing or giving directions for constructing something
Mathematics: Describing and reproducing geometric shapes
Science: Classifying organisms, land formations, or materials based on appearance; describing the results of an experiment or demonstration
Social Studies: Interpreting charts and graphs, drawing inferences from pictures
Enrichment Areas: Identifying common elements in a work of art or a dance

TEACHING SUGGESTIONS

Reinforce the following vocabulary used in this lesson: *characteristics, attributes, class.* Encourage students to use these terms in their discussion. Stress the word *all.* All members of a class must have a common characteristic or attribute. Since most of the classes possess more than one common attribute, the students must read each descriptive phrase. Next, the students decide if the statement is true for each member of the class. If the statement is false about any member of the class, then that characteristic or attribute cannot be used to describe the class. During discussion, students should support their choices and explain why the other answers are incorrect. This exercise promotes thoroughness in description and examination.

MODEL LESSON

LESSON

Introduction

Q: You have practiced putting shapes and patterns into classes. To do that, you had to identify the common characteristic of that class.

Explaining the Objective to the Students

Q: In this lesson, you will choose statements that describe the characteristics of a class.

Class Activity

- Select a group of three attribute blocks (or construct a set of three shapes) having the same color but different shapes and sizes. Note: Groups may be drawn on the blackboard in colored chalk or on a transparency using colored markers.
  Q: What do all of these items have in common?
  A: The same color (Have students name the color.)

- Select or draw a second group in which all items have the same shape but are different sizes and colors.
  Q: What do all of these items have in common?
  A: The same shape (Have students name the shape.)

- Select or draw a third group in which all items are the same size but have different colors and shapes.
  Q: What do all of these items have in common?
  A: The same size (Have students specify the size.)

Q: The important word in the questions that you have just answered is the word *all.* Before a sentence can describe a class, it must be true of all items in that class.

- Select the following group of attribute blocks: 2 large blue triangles, 1 large blue square, and 1 large red triangle.

  Q: Can we use the class "blue" to describe this group?
  A: No

  Q: Why not?
  A: It includes one red figure.

  Q: Can we use the class "triangles" to describe this group?
  A: No

  Q: Why not?
  A: It includes one square figure.

  Q: Can we use the class "large" to describe this group?
  A: Yes

  Q: Why?
  A: All of the figures are large.

  Q: As you do the exercises, be sure that the sentence is true of every item in the group.

GUIDED PRACTICE

EXERCISES: **D-28, D-29**

- When students have had sufficient time to complete these exercises, check their answers by discussing correct choices and reasons for eliminating incorrect choices. Encourage students to be as specific as possible when stating their rationale.

INDEPENDENT PRACTICE

- Assign exercises **D-30** through **D-35**.

THINKING ABOUT THINKING

Q: What did you pay attention to when you classified the figures?

1. I read the statements one at a time.
2. For each statement, I looked carefully at all the figures in the group to see if the statement was true for all members of the group.
3. I circled the true statements.

PERSONAL APPLICATION

Q: When might you be asked if a statement accurately describes the characteristics of what you see?

A: Examples include reporting an accident, injury, or fire to officials; describing an article of clothing to someone over the telephone; describing something to a blind person.

EXERCISES D-36 to D-49

MATCHING CLASSES BY SHAPE/PATTERN

ANSWERS D-36 through D-49 — Student book pages 114–16
Guided Practice: D-36 a; **D-37** d; **D-38** b; **D-39** c; **D-40** d; **D-41** c
Independent Practice: D-42 a; **D-43** e; **D-44** b; **D-45** c; **D-46** e; **D-47** d; **D-48** b; **D-49** a See below.

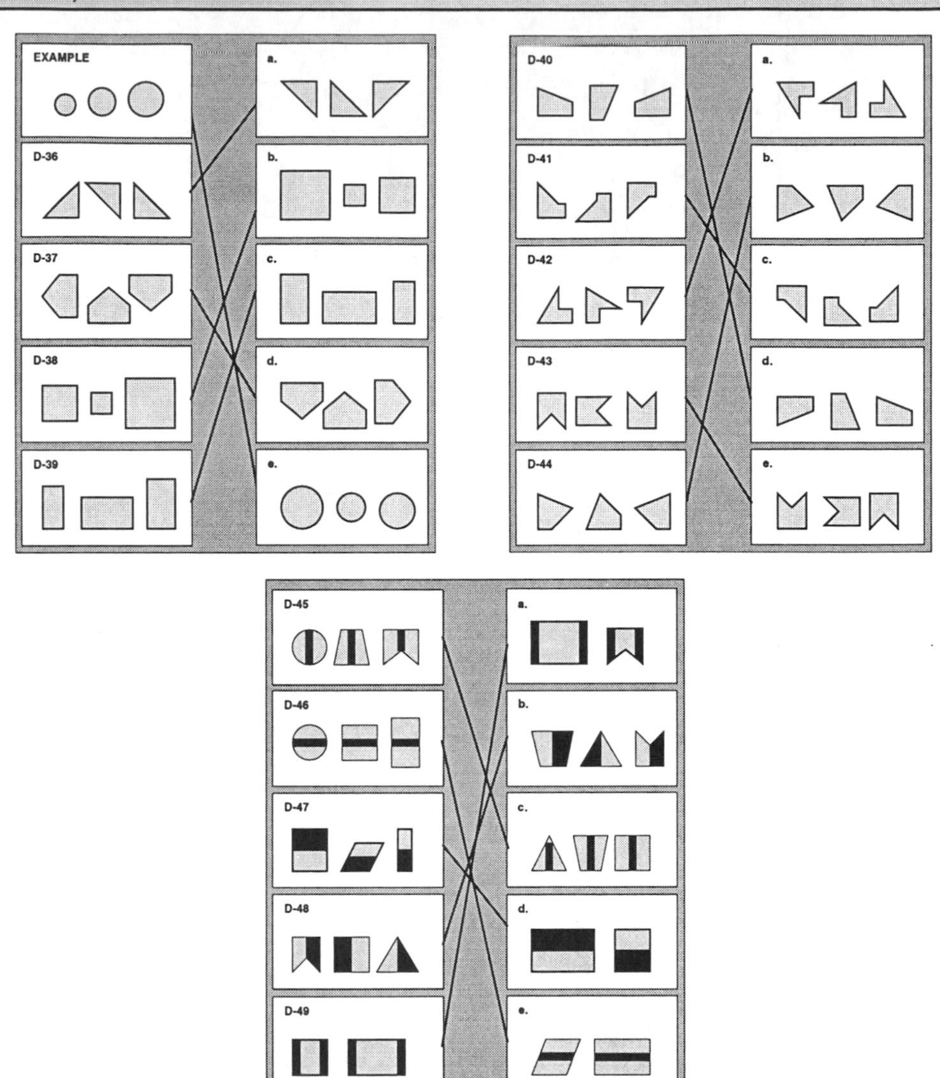

LESSON PREPARATION

OBJECTIVE AND MATERIALS

OBJECTIVE: Students will match groups which share the same characteristics.
MATERIALS: Transparency of student book page 114 • transparency marking pens or pencils in five colors • pattern blocks, attribute blocks, or handmade equivalent • 10 small boxes • 5 larger boxes

CURRICULUM APPLICATIONS

Language Arts: Matching pictures with letter sounds in reading readiness exercises
Mathematics: Comparing and modeling place values, using mathematical symbols to group arithmetic problems

Science: Classifying (leaves, insects, flowers, shells, birds) by the characteristics of shape or pattern, identifying symmetry and geometric forms found in the environment
Social Studies: Finding the same location on different types or sizes of maps; using a legend to read maps, graphs, or charts
Enrichment Areas: Locating types of books or items in a library; identifying families of musical instruments (winds, reeds, percussion) by sound, design, or shape; identifying similar shapes in different pictures

TEACHING SUGGESTIONS

Encourage students to use precise vocabulary in describing shape and pattern. Such geometric terms as *pentagon*, *right triangle*, *trapezoid*, and *parallelogram* should be used if the words are familiar to the students. When nonconventional shapes are used, encourage such imaginative similes as "arrowhead" or "home plate."

MODEL LESSON

LESSON

Introduction

Q: In the last few exercises, you have been matching objects with their proper classes and describing those classes.

Explaining the Objective to the Students

Q: In this exercise, you will match groups which share the same characteristics.

Class Activity

- For this demonstration using transparency of page 114, use a different colored marker for each line. Draw a line connecting the example to group *e*.
  Q: Why can I connect this group of shapes...

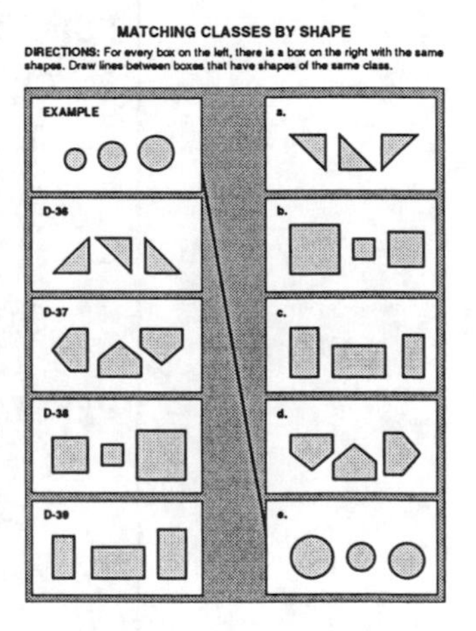

- Point to example.
  Q: ...to this group?
- Point to group *e*.
  A: Both the example and group *e* are groups of circles.

  Q: What are the shapes in group **D-36**?
  A: Triangles

  Q: To which group in the second column do these shapes belong?
  A: Group *a*
- Use a second color to connect **D-36** to *a*.
  Q: Name the shapes in group **D-37**.
  A: Pentagons (each looks like home plate or a small house)

  Q: Pentagon is the name of any five-sided closed figure. The Pentagon building in Washington, D.C., has five equal sides. To which group in the second column do the pentagons belong?
  A: Group *d*

- Use a third color to connect **D-37** to *d*.
  Q: Finish this exercise on your own then we will discuss the answers.
- Allow time for completion then continue the lesson.
  Q: What are the shapes in group **D-38**, and which group on the right goes with these shapes?
  A: Squares, group *b*
- Use a fourth color to connect **D-38** to *b*.
  Q: What are the shapes in group **D-39**, and which group on the right goes with these shapes?
  A: Rectangles, group *c*
- Use a fifth color to connect **D-39** to *c*.

GUIDED PRACTICE

EXERCISES: **D-36** through **D-41**

- When students have had sufficient time to complete these exercises, check answers by discussing correct choices and reasons for eliminating other choices. Encourage students to state their class descriptions and reasoning as specifically as possible.

INDEPENDENT PRACTICE

- Assign exercises **D-42** through **D-49**. Note: Point out to the students that in exercises **D-45** through **D-49** they are to match groups using patterns rather than shapes as the corresponding attributes.

THINKING ABOUT THINKING

Q: What did you pay attention to when you classified the figures?

1. I looked at a group of figures to see what all the figures had in common.
2. For each group, I looked for a matching group that had the same set of characteristics.
3. I drew a line connecting the pair of groups having similar characteristics.

PERSONAL APPLICATION

Q: When might you be asked to find matching groups of objects?
A: Examples include sorting (construction toys, eating utensils, clothing); locating products in grocery, department, or hardware stores.

EXERCISES D-50 to D-58

CLASSIFYING BY SHAPE—FIND THE EXCEPTION

ANSWERS D-50 through D-58 — Student book pages 117–18
Guided Practice: D-50 e (class: same quadrilateral rotated; exception: smaller quadrilateral) **D-51** f (class: parallelogram; exception: trapezoid) **D-52** b (class: pentagon; exception: triangle) **D-53** c (class: three-fourths circle; exception: half-circle)
Independent Practice: D-54 b (class: triangle with two sides equal [isosceles]; exception: triangle with a square corner [right triangle]) **D-55** c (class: rectangle; exception: square) **D-56** d (class: M-shaped

figures; exception: D-shaped figure) **D-57** b (class: half circle; exception: three-fourths circle) **D-58** c (class: trapezoid with two sides equal [isosceles]; exception: irregular trapezoid) See below.

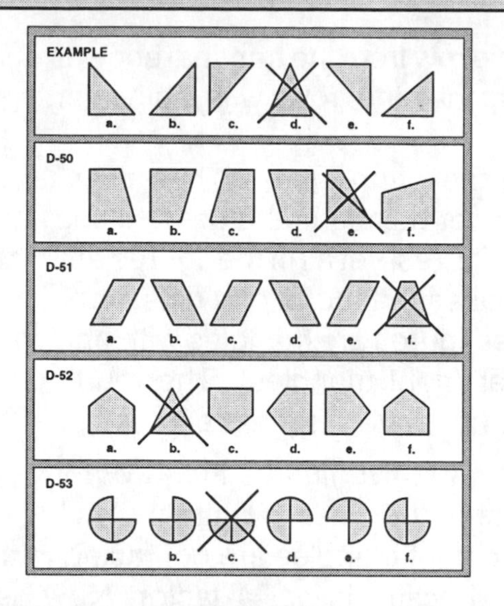

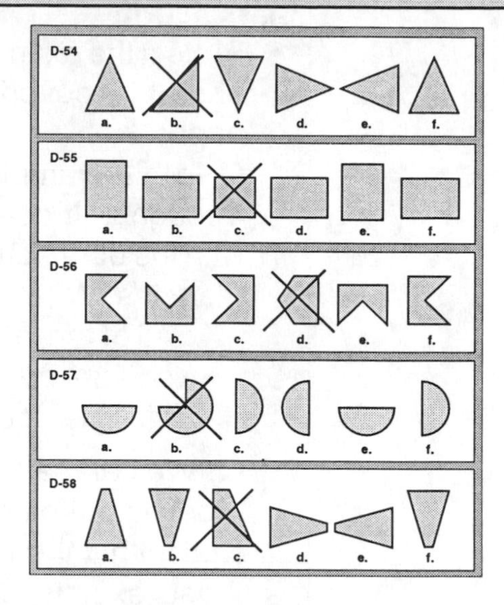

LESSON PREPARATION

OBJECTIVE AND MATERIALS

OBJECTIVE: Students will group figures by shape then find the figure that does not belong to this class.

MATERIALS: Transparency of student book page 117 • washable transparency marker

CURRICULUM APPLICATIONS

Language Arts: Distinguishing rhyming word patterns, recognizing exceptions to spelling or pronunciation rules

Mathematics: Distinguishing similar and dissimilar geometric patterns or shapes

Science: Classifying natural objects (leaves, fish, shells, etc.) according to shape

Social Studies: Reading a map

Enrichment Areas: Distinguishing between types of dance, art, or music

TEACHING SUGGESTIONS

Encourage students to use precise vocabulary when describing shapes. Geometric terms (*right triangle*, *rhombus*, *parallelogram*, *trapezoid*, and *pentagon*), names of fractional parts, and relationships of position (opposite, adjacent) should be used if the words and concepts are familiar to the students.

Imaginative similes should be encouraged when nonconventional shapes are used. Students should always name and/or describe both the class and the exception for each exercise.

MODEL LESSON

LESSON

Introduction

Q: In previous lessons, you matched groups which had the same characteristics.

Explaining the Objective to the Students

Q: In this lesson, you will decide which shapes go together to form a class and which shape does not belong to this class.

Class Activity

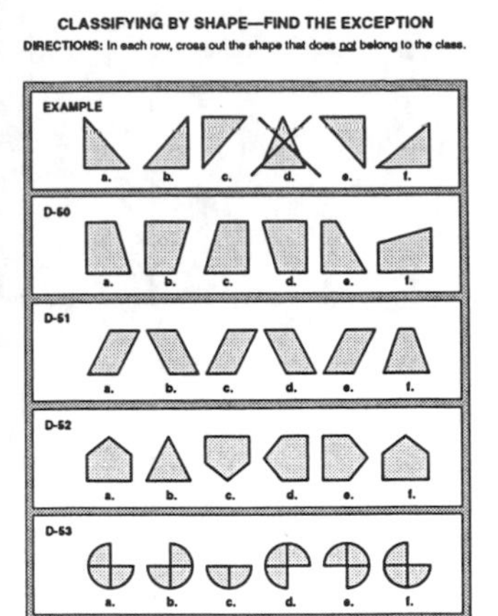

- Project the example row from the transparency of page 117, covering the other rows with a piece of paper.

  Q: In this example, the fourth figure (*d*) has an "X" through it. It does not belong in the same class with the other four figures, yet all of the figures are triangles. What makes the fourth one different?

  A: All of the other figures are triangles with one square corner (right triangles). The fourth triangle does not have a square corner.

  Q: We can say that the five figures that are not crossed out belong to the class "right triangles."

  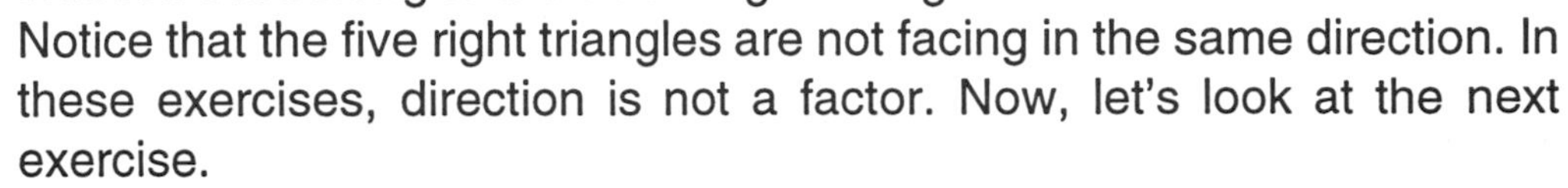

  Notice that the five right triangles are not facing in the same direction. In these exercises, direction is not a factor. Now, let's look at the next exercise.

- Move the cover down to project exercise **D-50**.

  Q: How are all of these shapes alike?

  A: They all have four sides; they are all gray; they all have two square corners; etc.

  Q: Which four shapes are most alike?

  A: Shapes *a*, *b*, *c*, *d*, and *f*

  Q: How is shape *e* different from the others?

  A: It is smaller at the top than the other five.

  Q: We will show that this is the figure that does not belong to this class by putting an "X" through figure *e*.

GUIDED PRACTICE

EXERCISES: **D-50** through **D-53**

- When students have had sufficient time to complete the exercises, check their answers by discussion. Ask students to name and describe the class each time and tell why the exception does not belong.

INDEPENDENT PRACTICE

- Assign exercises **D-54** through **D-58**.

THINKING ABOUT THINKING

Q: How did you decide which shapes went together and which shape did not belong?

1. I decided how all the shapes were alike.
2. I looked for one shape that was different from the others.
3. I put an "X" through the figure that was an exception.

PERSONAL APPLICATION

Q: When might you be asked to find the thing that doesn't belong to a group or class?

A: Examples include sorting anything (books, toys, silverware, dishes), reshelving misplaced items in a store or library.

EXERCISES D-59 to D-67

CLASSIFYING BY PATTERN—FIND THE EXCEPTION

ANSWERS D-59 through D-67 — Student book pages 119–20

Guided Practice: D-59 b (class: one-quarter shaded; exception: adjacent quarters shaded) **D-60** d (class: one corner shaded; exception: two corners shaded) **D-61** e (class: one-half shaded along the diagonal; exception: shading not along diagonal) **D-62** c (class: opposing portions [two parts] shaded; exception: one portion shaded)

Independent Practice: D-63 c (class: trapezoid-shaped section shaded; exception: triangle-shaped section shaded) **D-64** b (class: two sections [opposing quarters] shaded; exception: one section [quarter] shaded) **D-65** b (class: inner portion shaded; exception: outer portion shaded) **D-66** c (class: inner portion shaded; exception: outer portion shaded) **D-67** b (class: shaded triangles tip to tip; exception: shaded triangles base to base) See below.

Note: Occasionally a student will notice the rotation series within some of the pattern-classification exercises and use that as a classification technique. Such answers should be accepted and encouraged.

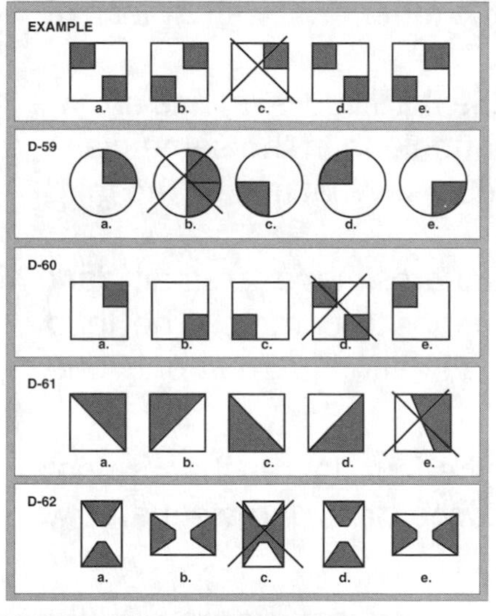

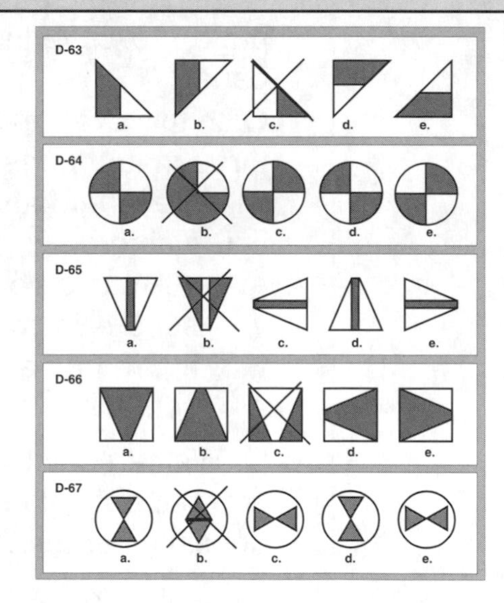

LESSON PREPARATION

OBJECTIVE AND MATERIALS

OBJECTIVE: Students will group figures by pattern then find the figure that does not belong to the group.

MATERIALS: Transparency of student book page 119 • washable transparency marker

CURRICULUM APPLICATIONS

Language Arts: Classifying words by letter pattern, recognizing prefixes and suffixes

Mathematics: Sorting geometric patterns
Science: Classifying leaves, insects, flowers, shells, birds, etc. by pattern
Social Studies: Using a legend to distinguish topographical areas (mountains, plains, etc.) on a map, interpreting graphs
Enrichment Areas: Classifying color families in art, classifying music by sound or rhythm pattern (classical, rock-and-roll, opera, etc.)

TEACHING SUGGESTIONS

Reinforce geometric terms (*right triangle, rhombus, parallelogram, trapezoid,* and *pentagon*), names of fractional parts (one-half, one-quarter, three-fourths), and relationships of position (inner, outer) if the words and concepts are familiar to students. Encourage students to use precise vocabulary when describing shapes.

Imaginative similes should be encouraged when nonconventional shapes are used. Students should always name and/or describe both the class and the exception for each exercise.

MODEL LESSON

LESSON

Introduction

Q: In the last lesson, you found the shape that did not belong in a class.

Explaining the Objective to Students

Q: In this lesson, you will decide which patterned figures go together to form a class and which patterned figure does not belong.

Class Activity

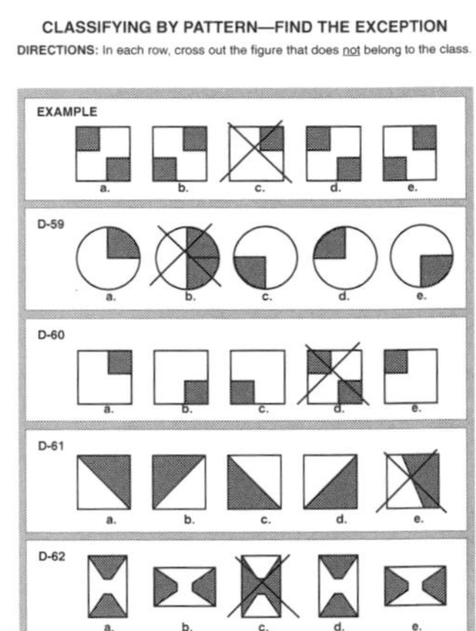

- Project the example row from the transparency of page 119, covering the other rows with a piece of paper.
  Q: In this example, the third figure *(c)* has an "X" through it. It does not belong in the same class with the other four figures. What makes the third figure different?
  A: All the other squares have a small dark square in two opposite corners. The third figure has a small dark square in only one corner.
  Q: We can say that the four figures that are not crossed out belong to the class "large squares with small dark squares in opposite corners."
- Move the cover down to project exercise **D-59**.
  Q: How are all these patterned figures alike?
  A: They are circles that are partly shaded.
  Q: Which four shapes are most alike?
  A: Shapes *a, c, d,* and *e*
  Q: How is shape *b* different from the others?
  A: It is half shaded while the other four are one-quarter (one-fourth) shaded.

Q: We can say that the four patterned figures that are not crossed out belong to the class "circles that are one-quarter shaded." We will show that figure *b* does not belong to the class by putting an "X" through it.

GUIDED PRACTICE

EXERCISES: **D-59** through **D-62**

- When students have had sufficient time to complete these exercises, check answers by discussion. Ask students to name and describe each class and tell why the exception does not belong.

INDEPENDENT PRACTICE

- Assign exercises **D-63** through **D-67**.

THINKING ABOUT THINKING

Q: How did you decide which patterned figures went together and which patterned figure did not belong?

1. I decided how all the patterned figures were alike.
2. I looked for one patterned figure that was different from the others.
3. I put an "X" through the figure that was the exception.

PERSONAL APPLICATION

Q: When might you be asked to find some pattern or color that doesn't belong to a group or class?

A: Examples include sorting art materials, fabrics, or puzzle pieces; sorting laundry by color.

EXERCISES D-68 to D-76

CLASSIFYING MORE THAN ONE WAY

ANSWERS D-68 through D-76 — Student book pages 121–22

Guided Practice: D-68 a (half-black vertically) and d (circles) **D-69** b (squares) and c (half-black along the diagonal) **D-70** c (rectangles) and d (gray shapes)

Independent Practice: D-71 c (circles) and d (gray figures) **D-72** a (squares) and c and f (black figures) **D-73** a (squares) and b (horizontal half-black figures) **D-74** a (checkered pattern) and c (circles) **D-75** f (triangles) and e (vertical half-black figures) **D-76** f (triangles) and b (horizontal half-black figures)

LESSON PREPARATION

OBJECTIVE AND MATERIALS

OBJECTIVE: Students will classify a given figure in more than one way.

MATERIALS: Transparency of student book page 121 • washable transparency markers • attribute blocks or facsimiles (optional)

CURRICULUM APPLICATIONS

Language Arts: Arranging groups of words using various categories (e.g., initial letters, alphabetical order, words that end in "-ing")

Mathematics: Sorting geometric shapes and patterns

Science: Classifying and reclassifying objects into different categories using shape, pattern, or color

Social Studies: Making different types of charts or graphs; grouping artifacts according to material, use, design, etc.
Enrichment Areas: Selecting proper materials for an art project from available supplies, classifying music using different categories (type, instrument, purpose, etc.)

TEACHING SUGGESTIONS

Reinforce the following vocabulary emphasized in this lesson: *horizontal, vertical, diagonal, characteristics, attributes*. Encourage students to use these terms as they explain why figures belong in some classes and not in others. Accurate descriptions of figures and classes should be the goal.

MODEL LESSON

LESSON

Introduction

Q: We have been classifying objects by shape or pattern. By doing such activities, we have found out that the same object can sometimes be classified more than one way.

Explaining the Objective to the Students

Q: In this lesson, you will place a figure in more than one class.

Class Activity—Option 1

- Demonstration using attribute blocks: Give each student an array of attribute blocks (buttons, shells, or other small objects can also be used), and ask them to sort the array using any characteristic they choose. After students have sorted the blocks, have them identify the characteristic they used. Then ask them to sort the same group another way using a different characteristic. When they have completed this task, discuss how they were able to fit the same items into two different arrangements.

Class Activity—Option 2

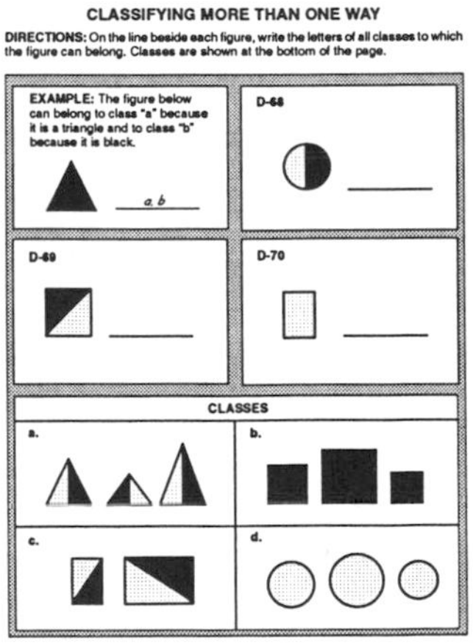

- Demonstration using transparency of page 121

CLASSIFYING MORE THAN ONE WAY

DIRECTIONS: On the line beside each figure, write the letters of all classes to which the figure can belong. Classes are shown at the bottom of the page.

EXAMPLE: The figure below can belong to class "a" because it is a triangle and to class "b" because it is black. a b

D-68

D-69

D-70

CLASSES

a. b. c. d.

Q: Look at the example. The answer tells us that the black triangle can belong to class *a* and to class *b*. The triangle belongs in class *a* because it is a triangle and all of the figures in class *a* are triangles. The triangle can also belong in class *b* because it is black and all of the figures in class *b* are black. The answer also tells us that the triangle doesn't fit into classes *c* or *d*. Let's find out why not. What could we call the class in group *c?*

A: Half-shaded (half-black) rectangles

Q: Why won't the black triangle fit into that class?

A: It is not a rectangle and it is not half-black.

Q: What could we call the class shown in group *d?*

A: Gray circles

Q: Why won't the black triangle fit into that class?

A: It is not a circle and it is not gray.

GUIDED PRACTICE

EXERCISES: **D-68** through **D-70**

- When students have had sufficient time to complete these exercises, check their answers by discussion. Have students name/identify the characteristics of each class and explain why the given figure fits or does not fit that class.

INDEPENDENT PRACTICE

- Assign exercises **D-71** through **D-76**.

THINKING ABOUT THINKING

Q: What did you pay attention to when you classified the figures?

1. I looked carefully at the details of the given figure (shape of the figure, number of sides, pattern or color, portion shaded).
2. I checked all the groups' figures to see if any had the same characteristic(s) as the given figure.
3. I listed the groups having the characteristic(s) of the given figure.

PERSONAL APPLICATION

Q: When might it be necessary for you to sort a group of objects into more than one category?

A: Examples include sorting tools or materials for different projects, arranging books or toys on shelves, arranging collections of objects (e.g., using country of origin, denomination, or date of issuance to sort stamps or money).

EXERCISES D-77 to D-78

CLASSIFYING BY COLOR/SHAPE—SORTING

ANSWERS D-77 through D-78 — Student book pages 123–24
Guided Practice: D-77 See below.
Independent Practice: D-78 See below.

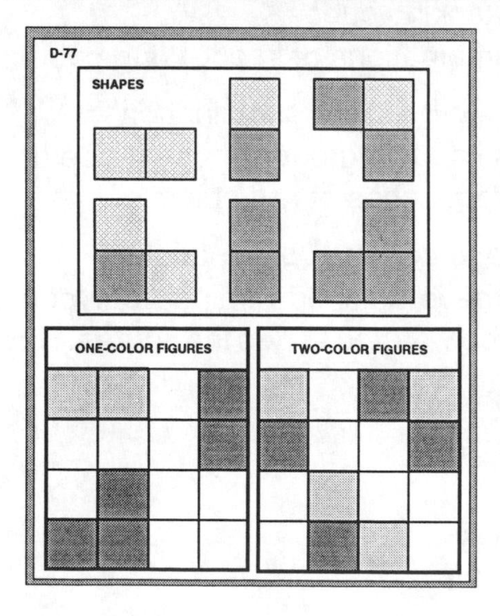

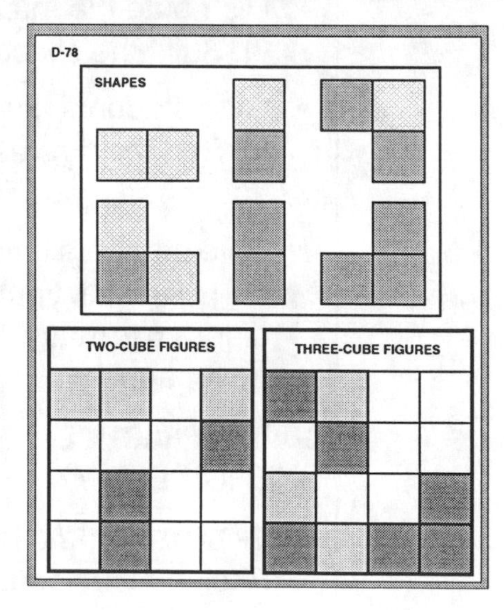

LESSON PREPARATION

OBJECTIVE AND MATERIALS

OBJECTIVE: Students will build groups of interlocking cubes and then sort the groups of cubes first by color and then by shape.

MATERIALS: Interlocking cubes (Multilinks or Unifix), 20 cubes per 2 students (10 each of two colors)

CURRICULUM APPLICATIONS

Language Arts: Decoding words, classifying words by letter pattern, recognizing prefixes and suffixes, recognizing sentence types from end marks

Mathematics: Distinguishing types of arithmetic problems by sign and/or problem configuration, classifying numbers according to place values, sorting geometric shapes

Science: Classifying natural objects (leaves, fish, shells, etc.) according to shape or pattern

Social Studies: Identifying road signs by shape, using a map legend, interpreting graphs

Enrichment Areas: Distinguishing between note values in printed music, classifying color families in art, classifying music by sound or rhythm pattern

TEACHING SUGGESTIONS

Encourage students to explain why the groups of cubes belong in one group and not the other.

MODEL LESSON

LESSON

Introduction

Q: We have been using pattern or shape as the identifying characteristic of a class when deciding which classes different figures fit into.

Explaining the Objective to the Students

Q: In this lesson, you will build groups of interlocking cubes and then sort the groups first by color then by shape.

Class Activity

- Distribute the interlocking cubes.
  Q: Build the six combinations of interlocking cubes.
- Give students time to build the combinations.
  Q: Sort the figures into two groups: those made from cubes of one color and those made from cubes of two colors.
- Give students time to sort the combinations.
  Q: Now that you have finished sorting your combinations, color your diagram to match the way you sorted the figures. Repeat the sorting on the next page.

GUIDED PRACTICE

EXERCISE: **D-77**

INDEPENDENT PRACTICE

- Assign exercise **D-78**.

THINKING ABOUT THINKING

Q: What did you pay attention to when you classified the figures?

1. I looked carefully at the details to decide how the figures were alike (number of colors or number of cubes).
2. I checked to be sure that all the figures had the same characteristic(s).
3. I moved members of the same class into their answer box.

PERSONAL APPLICATION

Q: Where else might you classify something by pattern or color?

A: Examples include sorting art materials, fabrics, or puzzle pieces; sorting laundry by color.

EXERCISES D-79 to D-94

COMPLETE THE CLASS

ANSWERS D-79 through D-94 — Student book pages 125–26
Guided Practice: D-79 8 (squares) **D-80** 9 (black figures) **D-81** 9 (triangles) **D-82** 2, 7, 9 (small figures)
Independent Practice: D-83 7 (squares) **D-84** 5 (gray) **D-85** 6 (black) **D-86** 8 (rectangles) **D-87** 9 (half-black) **D-88** 9 (triangles) **D-89** 3 (gray) **D-90** 9 (squares) **D-91** 9 (black) **D-92** 7 (circles) **D-93** 8 (six-sided figures—hexagons) **D-94** 6 (half-black)

LESSON PREPARATION

OBJECTIVE AND MATERIALS

OBJECTIVE: Students will determine how given shapes are alike then supply additional shapes that will fit into that class.

MATERIALS: Transparency of TM 21 (p. 269, cut apart as indicated)

CURRICULUM APPLICATIONS

Language Arts: Using end marks to identify questions and exclamations, choosing rhyming words by finding letter patterns

Mathematics: Sorting shapes, learning set theory and determining sets

Science: Classifying organisms, land formations, or equipment by known attributes

Social Studies: Making or using a map legend to gain information (types of roads, distances between cities, sizes of cities, state capitals and county seats, etc.)

Enrichment Areas: Classifying musical instruments according to type, shape, or typical use (marching band, rock band, orchestra); classifying art or dance according to given characteristics

TEACHING SUGGESTIONS

Students should name each class and explain why certain figures fit the class and other figures do not. Their explanations will reinforce the concept of class and categories. Younger children might call the half-black figures "black and gray figures" if they are not familiar with fractional parts.

MODEL LESSON

LESSON

Introduction

Q: We have been identifying characteristics of a class by color or shape.

Explaining the Objective to the Students

Q: In this lesson, you will determine how a group of shapes are alike and then find other shapes that fit that class.

Class Activity

- Project the shapes and example A from TM 21. Move shapes 1 and 4 to the bottom of the screen under the example.
  Q: In this exercise, we need to fill in the blank with another shape that goes with shapes 1 and 4. What do these two shapes have in common?
  A: Shapes 1 and 4 are both circles.

  Q: Which of the other figures can fit into this class?
  A: Shape 7

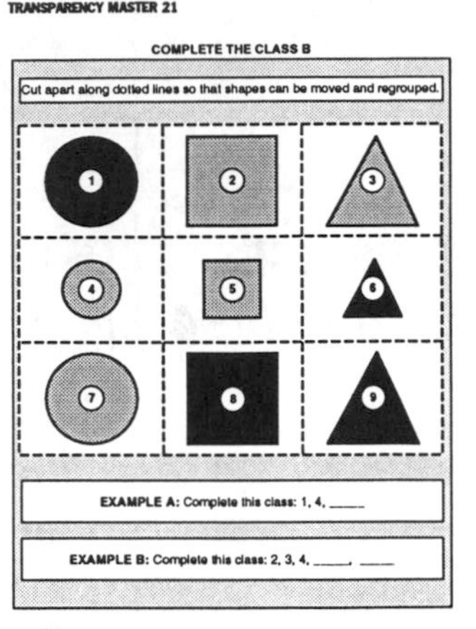

- Move shape 7 to the set at the bottom.
  Q: Since none of the other shapes fit, the class "circles" is complete. We will put the numeral 7 on the blank line to show that the class is complete.
- Write 7 on the blank then move shapes 1, 4, and 7 back to their original positions. Move the covering down to project example B. Move shapes 2, 3, and 4 to the bottom of the screen under the example.
  Q: What do these three shapes have in common?
  A: They are all gray.

  Q: Which of the other shapes can belong in this class?
  A: Other shapes in this class are 5 and 7.
- Move shapes 5 and 7 to the set at the bottom of the screen.
  Q: If there are no other gray objects left above, we can say that the class "gray shapes" is complete. What numbers must we put on the blanks in the second exercise to show that the class is complete?
  A: 5 and 7

GUIDED PRACTICE

EXERCISES: **D-79** through **D-82**

- Note: In **D-79**, you may need to point out that if we were looking for gray figures, the figures numbered 3 and 4 would also have been included between the given numbers 2 and 5.
- When students have had sufficient time to complete these exercises, check their answers by discussion. Have students name the class each time.

INDEPENDENT PRACTICE

- Assign exercises **D-83** through **D-94**.

THINKING ABOUT THINKING

Q: What did you pay attention to when you classified the figures?

1. I looked carefully at the details of the given figures to decide how they were alike (color, shape, or size).
2. I checked in the shape box to see if other figures had the same characteristic(s).
3. I listed the shape number of each member of the class on the answer blanks.

PERSONAL APPLICATION

Q: When could you be asked to decide which additional objects can fit into a given class?

A: Examples include sorting tools, books, or toys; putting away clothes, dishes, or silverware; sorting laundry; deciding arrangements of collections.

EXERCISES D-95 to D-103

FORM A CLASS

ANSWERS D-95 through D-103 — Student book page 127
Guided Practice: D-95 2, 4; **D-96** 1, 2, 5, 7; **D-97** 3, 5, 7, 9; **D-98** 3, 4, 6, 8, 9
Independent Practice: D-99 1, 5, 6; **D-100** 3, 7; **D-101** 2, 5, 7, 8; **D-102** 2, 9; **D-103** 4, 8

LESSON PREPARATION

OBJECTIVE AND MATERIALS

OBJECTIVE: Students will give the name of a class and will choose the shapes that can fit into that class.
MATERIALS: Transparency of TM 22 (p. 270, cut apart as indicated)

CURRICULUM APPLICATIONS

Language Arts: Sorting words into various categories (e.g., words with prefixes or suffixes, words with doubled consonants or vowels, alphabetical arrangements)
Mathematics: Using set theory; regrouping items, numerals, or geometric shapes
Science: Organizing plants, animals, or minerals; organizing a collection for a science project
Social Studies: Organizing artifacts or pictures of historic periods using various categories (time, type, culture, etc.)
Enrichment Areas: Rearranging pictures to form a collage, rearranging known dance steps to create a new dance

TEACHING SUGGESTIONS

Encourage students to form and name additional sets using the same shapes.

MODEL LESSON

LESSON

Introduction

Q: We have identified characteristics of classes and decided which figures go into a certain class.

Explaining the Objective to the Students

Q: In this lesson, you are given the name of a class and will choose the shapes that can fit into that class.

Class Activity

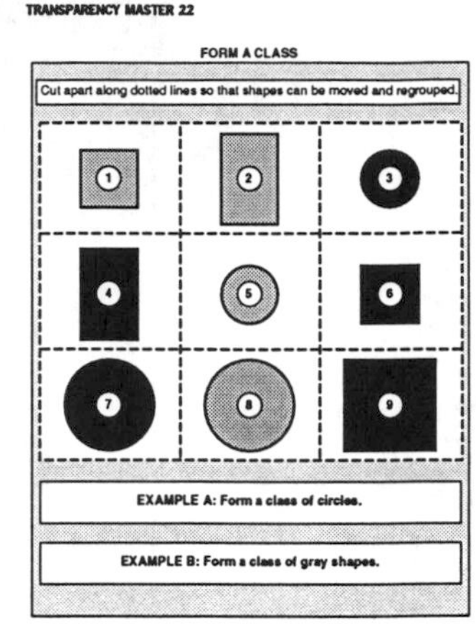

- Project figures 1, 2, and 3 (top row) of TM 22 on the overhead projector, covering the bottom rows with two separate strips of paper.
  Q: Let's see if we can form different classes from the same group of figures. Let's say we want to form a class of squares. Could we use any shapes in this row?
  A: Shape 1 is a square.
- Move shape 1 to the bottom of the screen. Move the strip of paper covering shapes 4, 5, and 6 (row two) up so that the second row is exposed and the top row is covered.
  Q: Do any shapes in this row fit into the class of squares?
  A: Shape 6 is a square.
- Move shape 6 to the bottom of the screen beside shape 1. Move the strip of paper covering shapes 7, 8, and 9 (row three) so that the top two rows are now covered and only the third row is exposed.
  Q: Do any shapes in this row fit into the class of squares?
  A: Shape 9 is a square.
- Move shape 9 to the bottom of the screen beside shapes 1 and 6. Uncover the entire group of shapes.
  Q: Here are the shapes we considered. From this group, we have made a set, or class, of squares. These shapes at the bottom, numbers 1, 6, and 9, make up that class.
- Move the shapes back to their original places.

GUIDED PRACTICE

EXERCISES: **D-95** through **D-98**

- Note: Explain to students that squares are rectangles with 4 equal sides. When students have had sufficient time to complete these exercises, check their answers by discussion. Students should explain why each figure does or does not fit into each particular class.

INDEPENDENT PRACTICE

- Assign exercises **D-99** through **D-103**.

THINKING ABOUT THINKING

Q: What did you pay attention to when you classified the figures?

1. I read the description of the class.
2. I checked in the shape box to see which figures had the same characteristic(s) as those described.
3. I listed the shape number of each member of the class on the answer blanks.

PERSONAL APPLICATION

Q: When might you need to sort a group of objects in more than one way?

A: Examples include sorting tools, utensils, or supplies; organizing collections of shells, stamps, rocks, dolls, etc.

EXERCISES D-104 to D-117

DRAW ANOTHER

ANSWERS D-104 through D-117 — Student book pages 128–30
Guided Practice: D-104 See below.
Independent Practice: D-105 to **D-117** (**D-113–116** Accept any of the 4 possible shapes in each class. **D-117** Accept any half-black shape.)

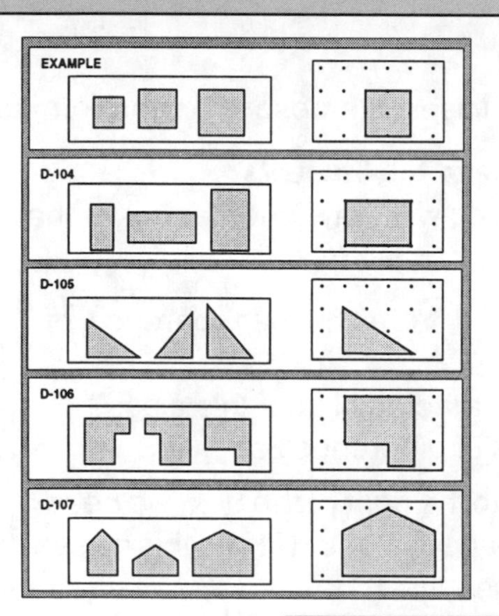

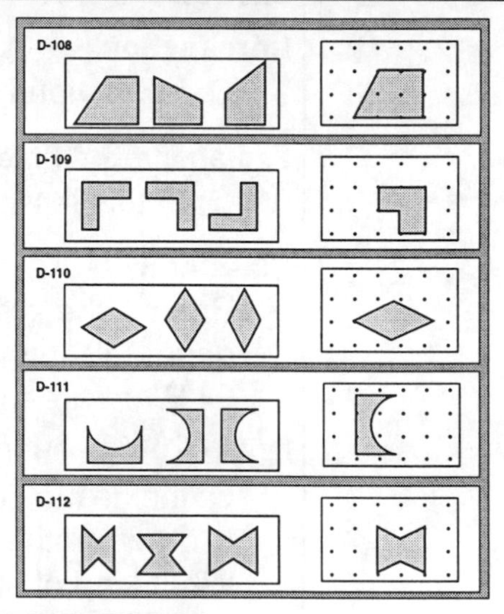

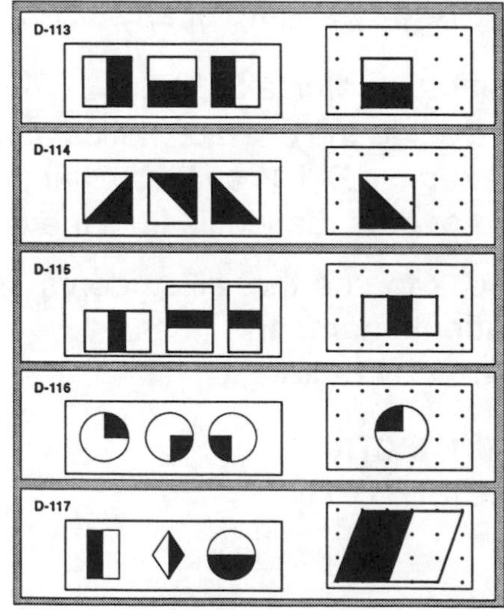

LESSON PREPARATION

OBJECTIVE AND MATERIALS

OBJECTIVE: Students will draw another figure that could belong in a given class.

MATERIALS: Transparency of student workbook page 128 • washable transparency markers

CURRICULUM APPLICATIONS

Language Arts: Identifying punctuation and proofreading marks

Mathematics: Adding numbers or shapes to given sets, constructing geometric shapes

Science: Naming another item that could belong to a classification, identifying cloud formations

Social Studies: Naming states by geographic region or other given category, using an atlas, adding additional information to an existing chart or graph

Enrichment Areas: Creating different art projects using the same type of material, adding to lists of instrument types (percussion, woodwinds, brass, strings, etc.)

TEACHING SUGGESTIONS

Students should accurately describe the shape and pattern of each class.

MODEL LESSON

LESSON

Introduction

Q: In previous exercises, we described and formed classes of figures.

Explaining the Objective to the Students

Q: In this lesson, you will draw another figure that belongs to a class.

Class Activity

- Project the example from the transparency of page 128.

  Q: What are the characteristics of this class?

  A: Gray squares of different sizes

  Q: Inside the dots to the right of the box, we need to draw another shape that could fit into this class. What kind of shape must it be?

  A: A gray square of any size

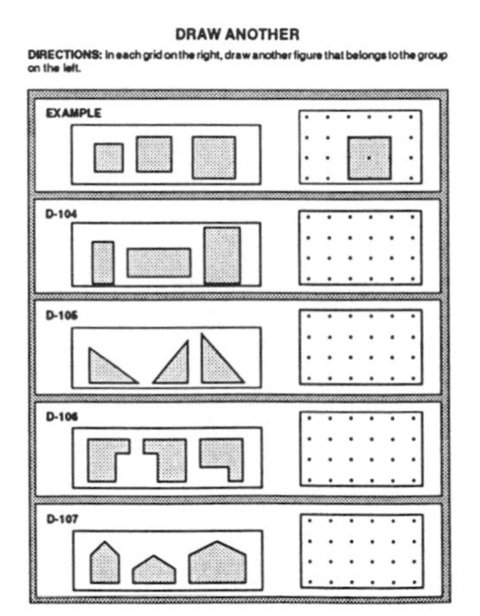

- Point to exercise **D-104**.

  Q: Decide the characteristics of the class shown in exercise **D-104** and draw another shape on the grid of dots that could fit in the class.

- Check to see that all students drew a gray rectangle of any size. Have students name the class.

  A: Gray rectangle

GUIDED PRACTICE

EXERCISE: **D-104**

INDEPENDENT PRACTICE

- Assign exercises **D-105** through **D-117**.

THINKING ABOUT THINKING

Q: How did you decide what figure to draw in the box?

1. I determined the characteristics of the class.
2. I drew a figure that fit those specific characteristics.

PERSONAL APPLICATION

Q: When might you be asked to provide another object that belongs to a given class?

A: Examples include sorting items for a library, finding or buying parts for models or art projects, finding common items that show different geometric shapes.

EXERCISES D-118 to D-120

CLASSIFYING BY SHAPE/COLOR/SIZE—SORTING

ANSWERS D-118 through D-120 — Student book pages 131–33
Guided Practice: D-118 See below.
Independent Practice: D-119 to **D-120** See below.

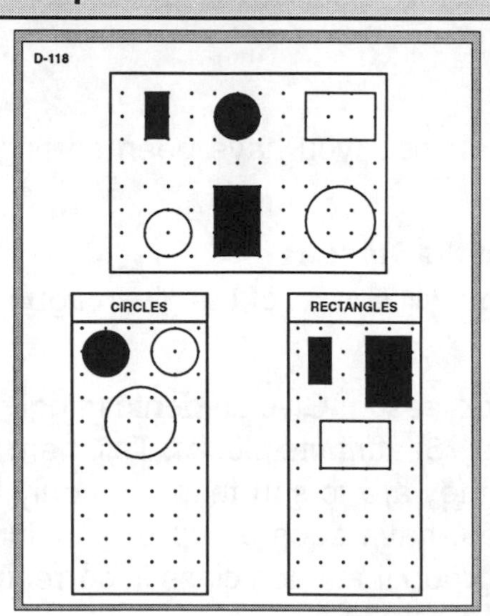

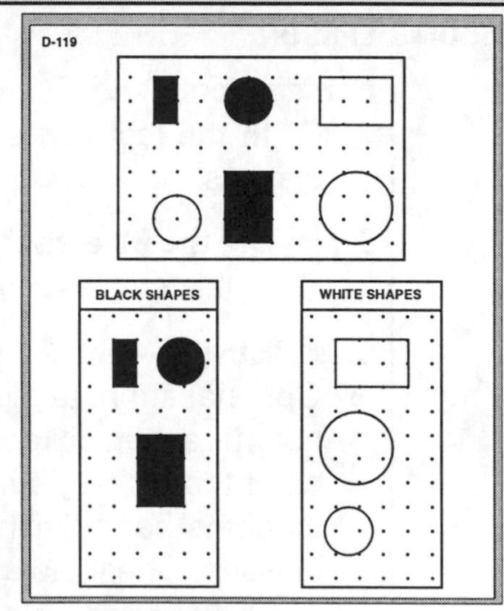

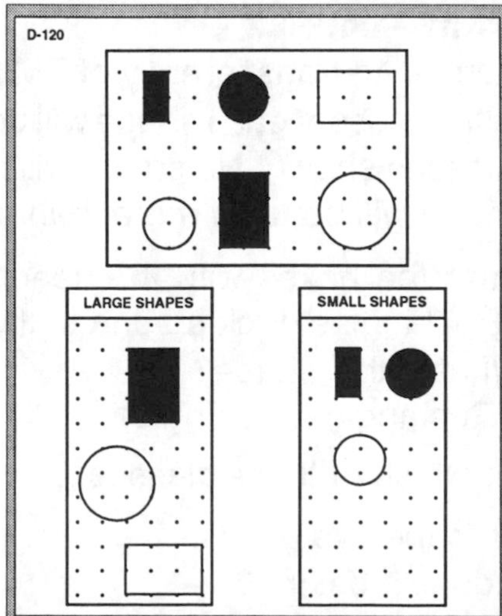

LESSON PREPARATION

OBJECTIVE AND MATERIALS

OBJECTIVE: Students will sort all of the given figures into classes.
MATERIALS: Transparency of TM 23 (p. 271, cut apart as indicated) • attribute blocks or facsimiles (optional)

CURRICULUM APPLICATIONS

Mathematics: Placing numerical information onto charts or graphs
Science: Classifying objects or organisms into given categories
Social Studies: Deciding which information to include and how to arrange it on a chart, map, or graph
Enrichment Areas: Classifying crayons, paints, or markers according to color family or other given category; classifying music or art mediums according to given categories

TEACHING SUGGESTIONS

Encourage students to use precise vocabulary in describing shape and pattern. If students have difficulty, it may help them to "color code" the shapes (e.g., coloring all of the rectangles red, all of the triangles blue, etc.). Color sometimes helps younger students see shape similarities. If students have not drawn the shapes to comparative scale (larger and smaller shapes), it may be necessary to explain the differences in size.

MODEL LESSON

LESSON

Introduction

Q: In the last few exercises, you have been sorting some figures into classes.

Explaining the Objective to the Students

Q: In this exercise, you will sort all of the given figures into classes.

Class Activity—Option 1

- Optional attribute block lesson: Give students (singly or in small groups) not more than sixteen (16) attribute blocks. Tell them that they are to use all of the blocks, but they are to sort them into only two classes. When students have finished, have each group or individual identify the two classes. Check as a group that each class is correctly constructed.

Class Activity—Option 2

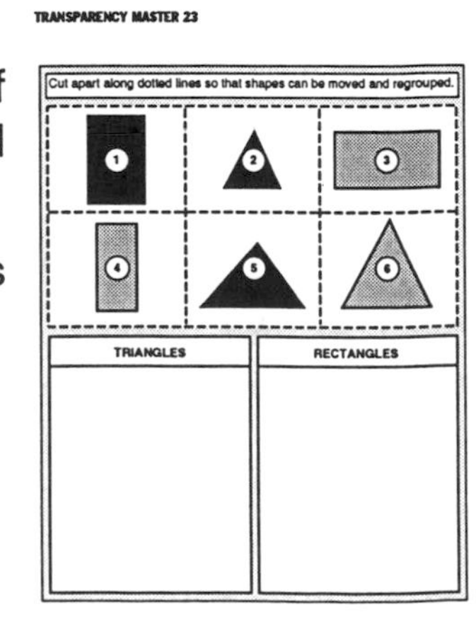

- Lesson using transparency of TM 23
  Q: In this lesson, each shape will belong in one of the classes shown. No shape will be left out and no shape will be used more than once.
- Point to the boxes with the headings Triangles and Rectangles. Pick up one of the shapes.
  Q: What is this shape?
  A: Rectangle (or triangle)
  Q: To which of these classes...
- Point to the two boxes.
  Q: ...does it belong?

- After the students determine the class, place the shape inside the correct class box. Repeat the above questions until all of the shapes are sorted into a class.

GUIDED PRACTICE

EXERCISE: **D-118**

- Check to be certain that students have placed each shape and that all shapes are in the correct classification.

INDEPENDENT PRACTICE

- Assign exercises **D-119** and **D-120**.

THINKING ABOUT THINKING

Q: What did you pay attention to when you classified the figures?

1. I read the heading on the sorting box.
2. I found all the figures in the shape box that had the given characteristic(s).
3. I drew the shapes in the correct sorting box.

PERSONAL APPLICATION

Q: When might you need to sort several objects into different groups, boxes, or containers?

A: Examples include sorting toys, books, eating utensils, tools, or supplies; sorting items into boxes for moving or packing.

EXERCISES D-121 to D-136

OVERLAPPING CLASSES—INTERSECTION

ANSWERS D-121 through D-136 — Student book pages 135–38
Guided Practice: D-121 to **D-123** See below.
Independent Practice: D-124 to **D-136**
See below.

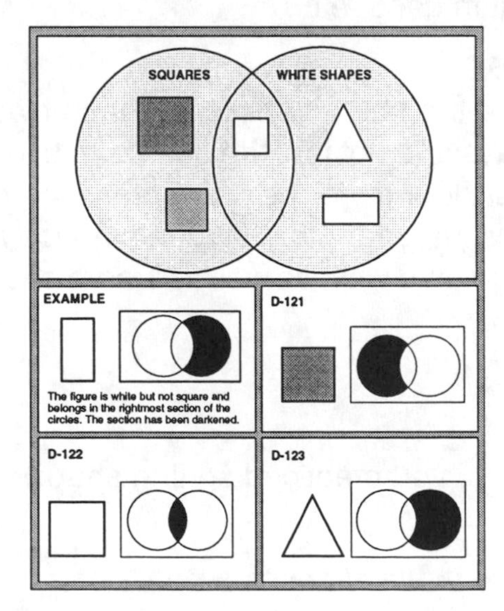

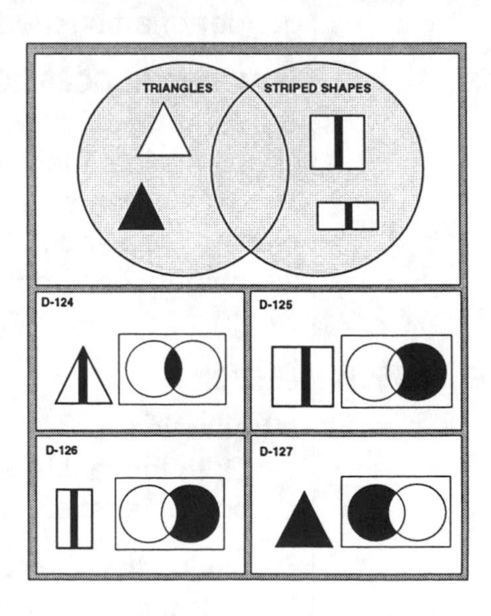

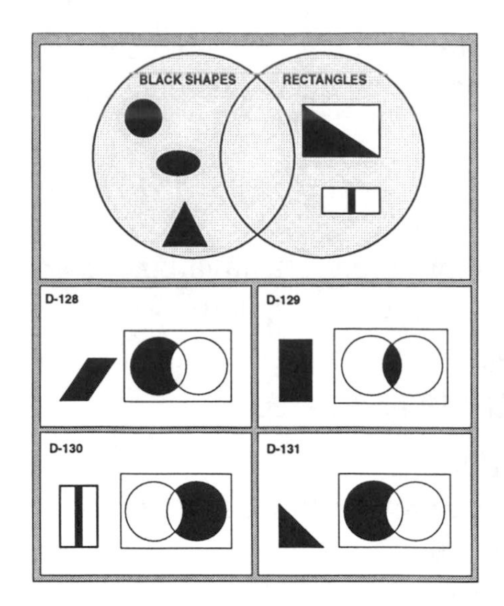

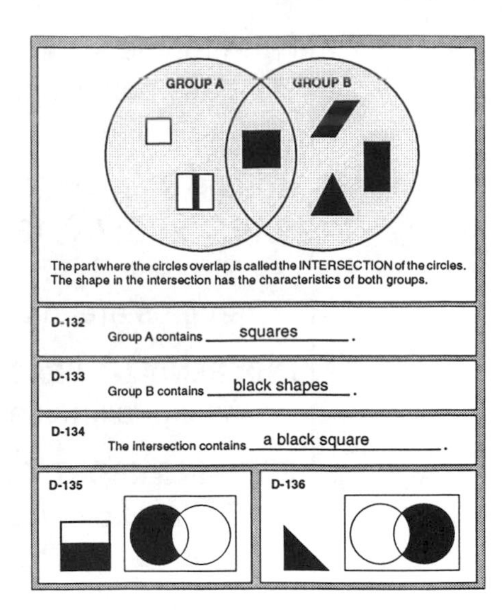

LESSON PREPARATION

OBJECTIVE AND MATERIALS

OBJECTIVE: Students will sort shapes using one or two characteristics.

MATERIALS: Transparencies of student workbook pages 134 and 135 • attribute blocks or facsimiles (optional) • two movable circles, hoops, or large rings of some type (optional)

CURRICULUM APPLICATIONS

Language Arts: Describing objects using *all*, *some*, and *not*

Mathematics: Set theory exercises, attribute block exercises, sorting geometric shapes

Science: Sorting natural objects into overlapping classes, classifying days according to different weather patterns (morning rain, afternoon rain, all-day rain, night rain, morning sun, etc.)

Social Studies: Interpreting graphic information, using a legend to read a map

Enrichment Areas: Classifying musical instruments into different types (marching band/orchestra), classifying works of art into different categories (era, gender of artist, medium used, etc.)

TEACHING SUGGESTIONS

Students should name the characteristics of each class and tell why each shape fits into that class and not into the others. In exercises **D-132** through **D-134**, students must determine the characteristics of each class before placing the shape. It is important that students notice and talk about where each figure does not go, and why, as well as where each figure does go, and why.

MODEL LESSON

LESSON

Introduction

Q: In the last lesson, you practiced sorting shapes using more than one characteristic.

Explaining the Objective to the Students

Q: In this lesson, you will sort some shapes by one characteristic and some shapes by two.

Class Activity—Option 1

- Optional attribute block lesson: Place one ring or circle representing red shapes and one representing triangles on a table. Select the following blocks from the attribute block set: small blocks—blue triangle, red square, red rectangle; large blocks—blue triangle, red hexagon, yellow triangle, red triangle
- Point to the ring representing red shapes.
  Q: This class is going to be red shapes.
- Point to the ring representing triangles.
  Q: This class is going to be triangles. We are going to pick out some blocks and put each one into the class that best describes it.
- Pick up the small blue triangle.
  Q: Which class should this shape fit in, the red class or the triangle class?
    A: Triangle class
- Put the shape into the triangle ring. Saving the large red triangle until last, continue until all other figures have been placed. Hold up the red triangle.
  Q: Where does this one belong?
    A: In the red circle (or in the triangle circle)

  Q: Yes, it is red (or a triangle), but that doesn't describe its shape (color), does it?
    A: No, put it in the triangle (red) circle.

  Q: Yes, it is a triangle (red), but that doesn't describe its color (shape), does it? We have a real problem here. We have one shape and two places to put it. Does anybody see a way to show that the shape is both red and a triangle?
- Students will probably try to move it back and forth between the red class and the triangle class or place it in the space between the two rings. Encourage students to move the rings. If they do not move the rings, you should move the rings so that they overlap as shown on page 134.
  Q: Does this show the different classes? Are all of the shapes in the triangle circle triangles? Are all of the shapes in the red circle red? Is the red triangle part of both circles? We call this a diagram of overlapping classes. *Overlapping* means that some of the shapes or objects fit into more than one class. This diagram is also sometimes called a Venn diagram.

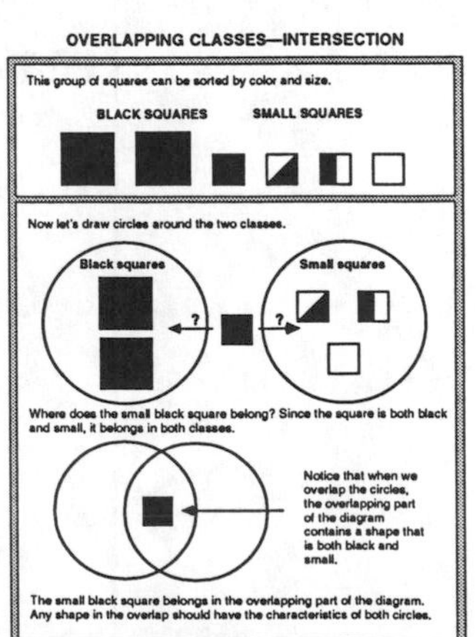

Class Activity—Option 2

- Demonstration using transparencies of student workbook pages 134 and 135: Project the transparency of page 134.
  Q: This is a picture of page 134 in your book. You may wish to follow along as we do this example.
- Following the instructions on the page, work through the example so that students will understand the method for showing overlapping classes.

Project the transparency of student workbook page 135.

Q: Look at the overlapping classes diagram on page 135. What are the characteristics of the classes inside each circle?

A: The class on the left is squares and the class on the right is white shapes.

Q: What can we call the class where these two circles overlap?

A: White squares

Q: Look at the shapes inside each of those classes. Why is each placed where it is on the diagram?

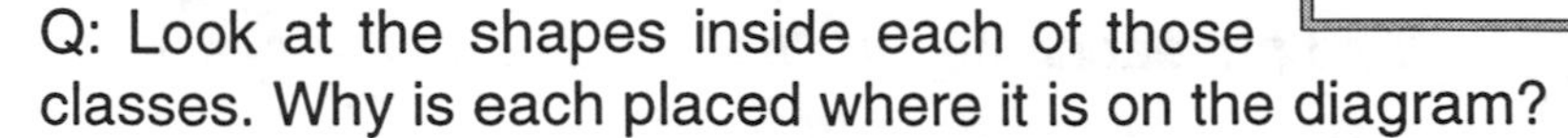

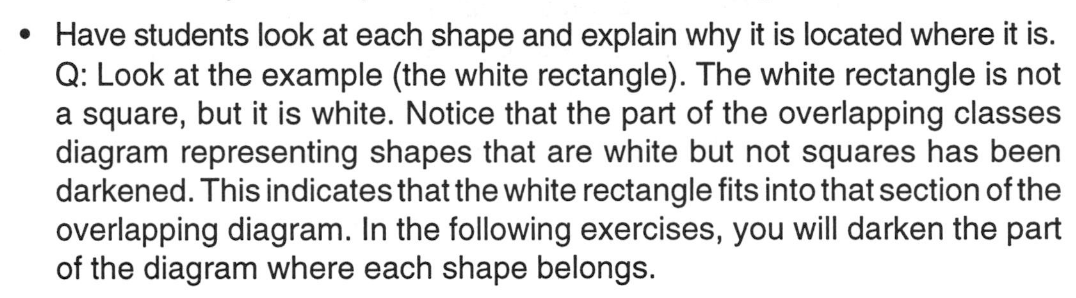

- Have students look at each shape and explain why it is located where it is.

  Q: Look at the example (the white rectangle). The white rectangle is not a square, but it is white. Notice that the part of the overlapping classes diagram representing shapes that are white but not squares has been darkened. This indicates that the white rectangle fits into that section of the overlapping diagram. In the following exercises, you will darken the part of the diagram where each shape belongs.

GUIDED PRACTICE

EXERCISES: **D-121** through **D-123**

- When students have had sufficient time to complete these exercises, check their answers by discussion.

INDEPENDENT PRACTICE

- Assign exercises **D-124** through **D-136**.

THINKING ABOUT THINKING

Q: What did you pay attention to when you classified the figures?

1. I looked carefully at the details to decide how the figures in a sorting circle were alike (size of the figure, shape, pattern, or color).
2. I checked to be sure that any figure added to the circle had the same characteristic(s).
3. If a figure has the characteristics of both sorting circles, it belongs in the overlapping part of the diagram (intersection).

PERSONAL APPLICATION

Q: When might you need to know if something fits into more than one class?

A: Examples include finding books in a library, locating products in a grocery store, finding different uses for tools (e.g., using one side of a claw hammer to drive nails and the other side to pull them).

EXERCISES D-137 to D-140

OVERLAPPING CLASSES—MATRIX

ANSWERS D-137 through D-140 — Student book pages 140–41
Guided Practice: D-137 See below.
Independent Practice: D-138 to **D-140** See below.

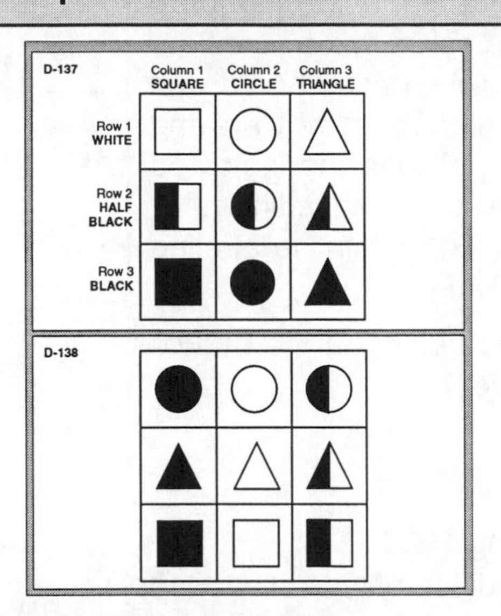

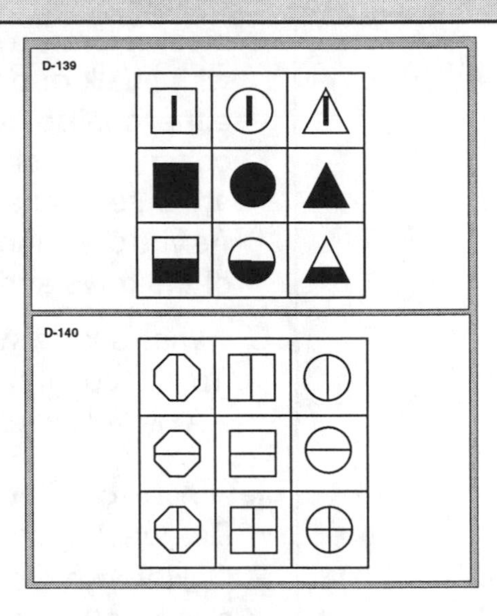

LESSON PREPARATION

OBJECTIVE AND MATERIALS

OBJECTIVE: Students will classify figures by more than one characteristic using a diagram called a matrix.

MATERIALS: TM 24 (p. 272) • washable transparency marker • attribute blocks or facsimiles (optional)

CURRICULUM APPLICATIONS

Language Arts: Reading graphs, tables, or schedules
Mathematics: Making and using arithmetic charts
Science: Classifying organisms or equipment using more than one characteristic (eye color/hair color, for example)
Social Studies: Making a graph or chart of survey results
Enrichment Areas: Comparing or contrasting works of art or pieces of music using multiple categories, comparing or contrasting works of two artists or composers

TEACHING SUGGESTIONS

Reinforce the following vocabulary emphasized in this lesson: *characteristic, matrix.* Encourage students to use precise vocabulary when describing shape and pattern.

MODEL LESSON

LESSON

Introduction

Q: In the last lesson, we learned that some figures could be classified by more than one characteristic. We used an overlapping-circles diagram, or a Venn diagram, to show that relationship.

Explaining the Objective to the Students

Q: In this lesson, you will use a diagram called a matrix to classify figures by two characteristics.

Class Activity—Option 1

- Optional demonstration using attribute blocks: Draw a nine-cell matrix on the floor with chalk or on a piece of cardboard with either chalk or a marker. Label the rows and columns as shown, placing the indicated attribute blocks in their corresponding cells. As you draw the matrix, explain the labeling of the rows and columns.

	Column 1 squares	Column 2 circles	Column 3 triangles
Row 1 red	red square	red circle	red triangle
Row 2 yellow	yelllow square		
Row 3 blue	blue square		

Q: What shape would belong in the middle (row 2, column 2) cell?
A: A yellow circle

Class Activity—Option 2

- Demonstration using TM 24
  Q: Here we have a different type of overlapping-class matrix. Row 1 contains white things; row 2, black things. Column 1 contains squares; column 2, circles. A white square is already placed in the cell for row 1, column 1. It is correct because it is in the White row and the Square column.

TRANSPARENCY MASTER 24

OVERLAPPING CLASSES—MATRIX

Column 1 SQUARE | Column 2 CIRCLE | Row 1 WHITE | Row 2 BLACK

- Point to the cell just mentioned.
  Q: If row 1 is for white figures and column 2 contains circles, what figure should I draw in the cell for row 1, column 2?
- Point to the cell as you say it.
  A: A white circle
- Draw it on the transparency.
  Q: Why can I draw a black square in the row 2, column 1 cell?
  A: Row 2 is black figures; column 1 is squares.
- Draw the figure.
  Q: What figure should I draw in the remaining cell: row 2, column 2?
- Point to the cell.
  A: A black circle
- Draw and shade the circle.
  Q: Row 2 is black figures, and column 2 is circles. This figure is both black and a circle, so it fits into this cell.
  Q: You may look at page 139 in your book to see the completed matrix.

GUIDED PRACTICE

EXERCISE: **D-137**

INDEPENDENT PRACTICE

- Assign exercises **D-138** through **D-140**.

THINKING ABOUT THINKING

Q: What did you pay attention to when you classified the figures?

1. I looked carefully at the row and column headings.
2. I checked to be sure that any figure added to the matrix had the characteristics of both the row and the column.

PERSONAL APPLICATION

Q: When might you need to complete or read a matrix?

A: Examples include using or making charts and schedules, solving puzzles by filling in clues.

EXERCISES D-141 to D-145

WRITING DESCRIPTIONS OF CLASSES

ANSWERS D-141 through D-145 — Student book pages 142–44
Guided Practice: D-141 This is a group of gray triangles. They have nothing else in common. Some have two sides equal (isosceles), some have all sides equal (equilateral), and some have no sides equal (scalene).
Independent Practice: D-142 This is a group of five-sided figures (pentagons). They all have the same pattern. **D-143** This is a group of squares and rectangles (or a group of rectangles since a square is a special rectangle). They are striped in many ways. **D-144** This is a group of four-sided figures (quadrilaterals). They are all half black and half white. **D-145** This is a group of many-sided figures (polygons). They are striped in many ways.

LESSON PREPARATION

OBJECTIVE AND MATERIALS

OBJECTIVE: Students will look at a group of figures and decide what characteristics all the shapes have in common. Then they will write a description of the group.

MATERIALS: Transparency of page 142 of the student book • scratch paper

CURRICULUM APPLICATIONS

Language Arts: Describing the way things or people look, writing or giving directions for constructing something

Mathematics: Describing and reproducing geometric shapes

Science: Matching classification statements based on appearance with corresponding items, describing the results of an experiment or demonstration

Social Studies: Interpreting charts and graphs, drawing inferences from pictures

Enrichment Areas: Answering questions about a work of art or a dance

TEACHING SUGGESTIONS

Encourage students to use complete sentences.

MODEL LESSON

LESSON

Introduction

Q: We have been discussing many ways to classify figures. We also explained our thinking about classifying.

Explaining the Objective to the Students

Q: In this lesson, you will write a description of a class of figures.

Class Activity

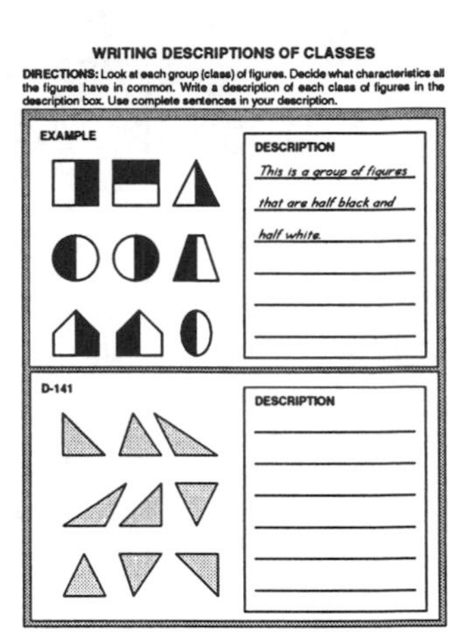

- Project the transparency of page 142, covering the answer.

  Q: Are all these figures the same shape?

  A: No, they are squares, triangles, circles, trapezoids, pentagons, and ovals.

  Q: What do all the shapes have in common?

  A: They all are half black and half white.

  Q: Write a complete sentence that expresses this idea on a piece of scratch paper.

- After sufficient reflection and writing time, expose the answer on the transparency.

GUIDED PRACTICE

EXERCISE: **D-141**

INDEPENDENT PRACTICE

- Assign exercises **D-142** through **D-145**.

THINKING ABOUT THINKING

Q: What did you pay attention to when you classified the figures?

1. I looked carefully at the details to decide how they are alike (size of the figure, length of the sides, size of the angle, pattern or color).
2. I checked to be sure that all the figures had the same characteristic(s).
3. I described the group using complete sentences.

PERSONAL APPLICATION

Q: When might you need to describe a group of things?

A: Examples include giving directions or instructions, or making requests.

Chapter 5

FIGURAL ANALOGIES
(Student book pages 145–55)

EXERCISES E-1 to E-18

ANALOGIES WITH SHAPES

ANSWERS E-1 through E-18 — Student book pages 146–51
Guided Practice: E-1 they are triangles; **E-2** the second triangle is larger than the first; **E-3** they are circles; **E-4** the second circle is larger than the first; **E-5** shape, color and size; **E-6** color and shape, size; **E-7** shape, size and color; **E-8** c, change in color; **E-9** c, flip up and down
Independent Practice: E-10 b, change in size; **E-11** a, change in color; **E-12** c, adding detail; **E-13** b, flip like a page of a book (flip about the vertical axis) **E-14** c, flip like a page of a book (flip about the vertical axis) **E-15** b, alternation of color, i.e., the part that is black in the first figure is gray in the second figure; **E-16** a, flip up and down like a calendar (flip about a horizontal axis) **E-17** d, tumble to the right (rotate clockwise) **E-18** c, tumble to the left (rotate counterclockwise)

LESSON PREPARATION

OBJECTIVE AND MATERIALS
OBJECTIVE: Students will recognize how two pairs of figures are related in a relationship called an analogy.
MATERIALS: Transparencies of student book pages 146 to 148 • washable transparency marker • attribute blocks or facsimiles (optional) • three 5" × 7" unlined cards marked as follows: 2 cards ":"; 1 card "::" (to mark analogies)

CURRICULUM APPLICATIONS
Language Arts: Using knowledge of parts to determine meaning of unfamiliar compound words, comparing and/or contrasting information gained from pictures
Mathematics: Changing numerical information to graphic information
Science: Recognizing the relationship of cloud formations and weather, recognizing analogous body parts or structural elements in different organisms
Social Studies: Recognizing parallel structures of governments (local to state to federal, monarchy to federalist); recognizing similar patterns in artifacts; comparing and/or contrasting information from maps, charts, or graphs
Enrichment Areas: Comparing/contrasting two types of music, art, or dance

TEACHING SUGGESTIONS
Encourage students to use precise vocabulary when describing shapes and relationships between shapes.

MODEL LESSON

LESSON

Introduction
Q: In the exercises you have done so far, you have practiced identifying relationships. You can examine and explain how figures both are alike and different. You can also put figures in order and separate them into classes.

Explaining the Objective to the Students

Q: In this lesson, you will learn to recognize how two pairs of figures are related.

Class Activity—Option 1

- Write *analogy* on the blackboard. Set up the following blocks on the chalk rail:

small red triangle	large red triangle	(Leave a space about a foot wide)	small red circle

- Point to the appropriate figures as you go.

  Q: How are the figures (red triangles) in this pair alike?

  A: They are both triangles and both red (same shape and color).

  Q: How are they different?

  A: The second shape is larger than the first (different size).

  Q: Now, look at the small red circle. Which block could we add to make this second pair of blocks like the first pair? Why?

  A: A large red circle because it would be the same shape (circle) and the same color (red), but the second figure would be larger than the first.

  Q: This relationship is called an analogy. This means that both pairs are related in the same way, in this example, small to large.

- Place dot cards into proper slots so that the analogy on the tray looks as shown below.

  Q: When an analogy is written down, it looks like this:

- Point to each section and write the words under each as you say it.

  Q: We read an analogy like this: *a* is to *b* as *c* is to *d.*

Class Activity—Option 2

- Lesson using transparencies of student book pages 146–48: Project the top section of the transparency of page 146, covering the bottom portion.

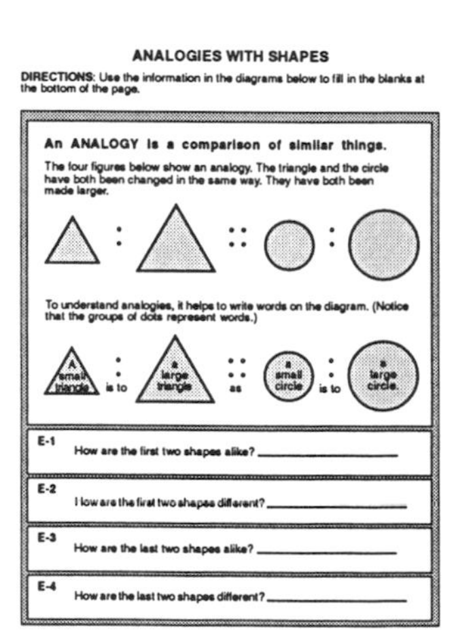

ANALOGIES WITH SHAPES

DIRECTIONS: Use the information in the diagrams below to fill in the blanks at the bottom of the page.

An ANALOGY is a comparison of similar things.

The four figures below show an analogy. The triangle and the circle have both been changed in the same way. They have both been made larger.

To understand analogies, it helps to write words on the diagram. (Notice that the groups of dots represent words.)

A small triangle is to a large triangle as a small circle is to a large circle.

E-1 How are the first two shapes alike? ____________

E-2 How are the first two shapes different? ____________

E-3 How are the last two shapes alike? ____________

E-4 How are the last two shapes different? ____________

  Q: Let's compare the first two figures. How are they alike?

  A: They are both gray triangles (same shape and color/pattern).

  Q: How are these triangles different?

  A: The second triangle is larger (different size).

  Q: What has been done to the first figure to make it into the second?

  A: It has been enlarged.

  Q: Now, look at the last two figures. How are they alike?

  A: They are both gray circles (same shape and color/pattern).

  Q: How are these circles different?

  A: The second circle is larger (different size).

Q: To make the first circle into the second circle, you would have to enlarge it. The circles have been changed in the same way as the triangles. These four figures, or two pairs of figures,...

- Point to the pair of triangles, then to the pair of circles.
  Q: ...are called a figural analogy. The groups of dots (**:** and **::**) represent words.
- Uncover the bottom half of the transparency.
  Q: In place of the two dots (**:**), you read "is to," and in place of the four dots (**::**), you read "as." So now the figural analogy reads: "A small gray triangle is to a large gray triangle as a small gray circle is to a large gray circle."
- Project the transparency of page 147.
  Q: This is the example from page 147 in your book. Use your attribute blocks to copy the first pattern on the page. What is the same about each pair of figures?
    A: The color and the shape
  Q: What is different?
    A: The size
  Q: Use your attribute blocks to copy each pattern on the page then answer the questions for each analogy.

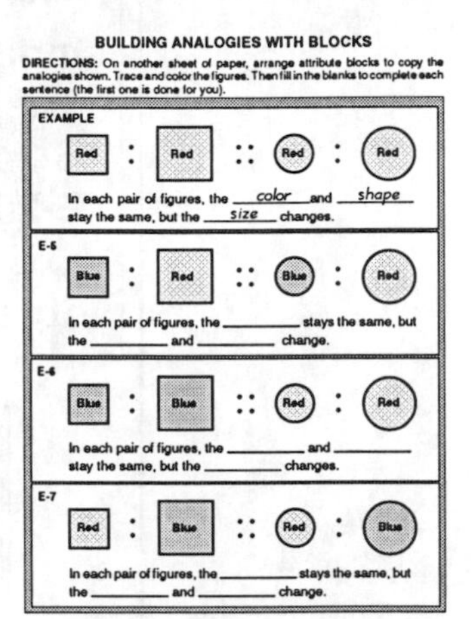
BUILDING ANALOGIES WITH BLOCKS
DIRECTIONS: On another sheet of paper, arrange attribute blocks to copy the analogies shown. Trace and color the figures. Then fill in the blanks to complete each sentence (the first one is done for you).
EXAMPLE
Red : Red :: Red : Red
In each pair of figures, the color and shape stay the same, but the size changes.
E-5
Blue : Red :: Blue : Red
In each pair of figures, the ______ stays the same, but the ______ and ______ change.
E-6
Blue : Blue :: Red : Red
In each pair of figures, the ______ and ______ stay the same, but the ______ changes.
E-7
Red : Blue :: Red : Blue
In each pair of figures, the ______ stays the same, but the ______ and ______ change.

- Project the transparency of page 148.
  Q: This is the example from page 148 in your book. Which shape will complete the analogy?
    A: Shape *c*

ANALOGIES WITH SHAPES—SELECT
DIRECTIONS: Circle the figure that completes the analogy.
EXAMPLE
E-8
E-9

GUIDED PRACTICE

EXERCISES: **E-1** through **E-9**

INDEPENDENT PRACTICE

- Assign exercises **E-10** through **E-18**. When students have had sufficient time to complete the exercises, check answers by discussion. Have them name the type of analogy for each problem and discuss why each answer was chosen or not chosen.

THINKING ABOUT THINKING

Q: What did you pay attention to when you found analogies between pairs of figures?

1. I looked carefully at the details (size of the figure, shape, pattern or color).
2. I looked at the first pair to see how the two figures were different (change of size, shape, pattern, or color).
3. I looked at the second pair to see if the two figures were different in the same way as the first pair.
4. I explained the relationship as an analogy: *a* is to *b* as *c* is to *d*.

PERSONAL APPLICATION

Q: When might you need to know or recognize how two pairs of things are related?

A: Examples include matching articles of clothing to wear; seeing consequences of moves in chess, checkers, or other games.

EXERCISES E-19 to E-30

ANALOGIES WITH SHAPES—COMPLETE

ANSWERS E-19 through E-30 — Student book pages 153–55
Guided Practice: E-19 color reversal (the part that was black is now white) **E-20** flip like the page of a book (flip about a vertical axis)
Independent Practice: E-21 Flip up and down like a calendar (horizontal axis) **E-22** tumble to the right (rotate clockwise) **E-23** color change; **E-24** flip about a vertical axis (like a book) **E-25** color reversal; **E-26** increase in size; **E-27** decrease in size; **E-28** flip about a horizontal axis (like a calendar) **E-29** flip about a vertical axis (like a book) **E-30** flip about a diagonal (or rotate twice) See below.

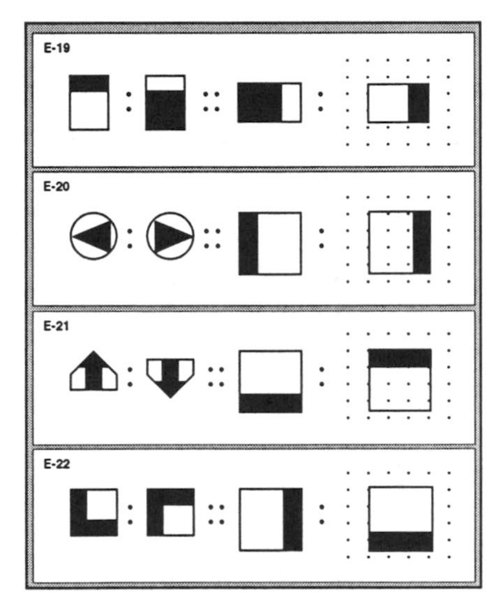

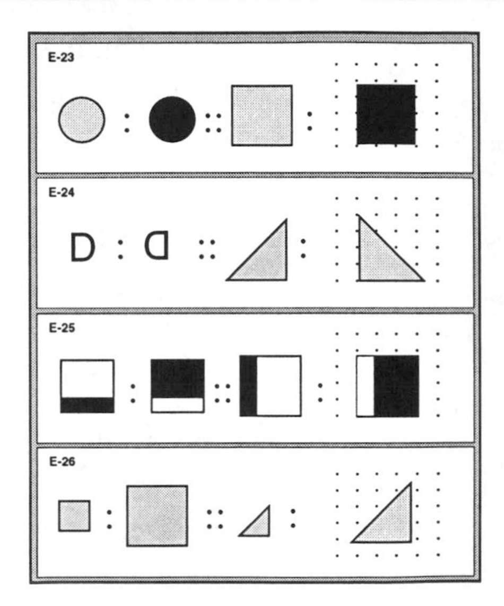

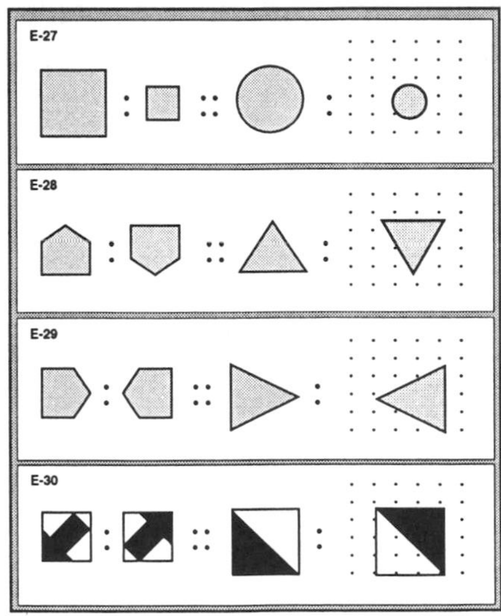

LESSON PREPARATION

OBJECTIVES AND MATERIALS

OBJECTIVE: Students will complete a figural analogy by drawing the missing figure.

MATERIALS: Transparency of TM 25 (p. 273) • transparency of student book page 152 • washable transparency marker

CURRICULUM APPLICATIONS

Language Arts: Recognizing correct pronunciation of unknown words by comparing letter patterns

Mathematics: Recognizing equivalent fractional parts, working with ratios, writing and recognizing arithmetic problems in pictorial form

Science: Seeing and stating relationships between different natural phenomena, animal, or plant classes, and/or minerals and rocks; naming analogous body parts of different organisms; comparing and/or contrasting organisms

Social Studies: Recognizing and stating the relationship between people and events, recognizing and stating causal relationships

Enrichment Areas: Learning new vocabulary in a foreign language, recognizing types of music by listening to the rhythm (march music, rumba music, waltz music, etc.), drawing things to scale

TEACHING SUGGESTIONS

Encourage students to describe the markings and the relationships using the terms learned in the section on figural sequences: *opposite*, *reflection* or *flip*, *rotation* or *turn*, *clockwise*, *counterclockwise*.

MODEL LESSON

LESSON

Introduction

Q: In the last exercise, you selected a figure which correctly completed an analogy. Now, you will draw the missing figure to complete the analogy.

Explaining the Objective to the Students

Q: In this lesson, you will draw a figure that completes an analogy.

Class Activity

- Project the first analogy from TM 25, covering the last two rows. Point to the two rectangles.

  Q: Let's compare these first two figures. How are they alike?

  A: They are both rectangles that are the same size and partly shaded.

  Q: How are these figures different?

  A: They are shaded opposite. The part that is black in the first rectangle is white in the second, and the part that is white in the first is black in the second.

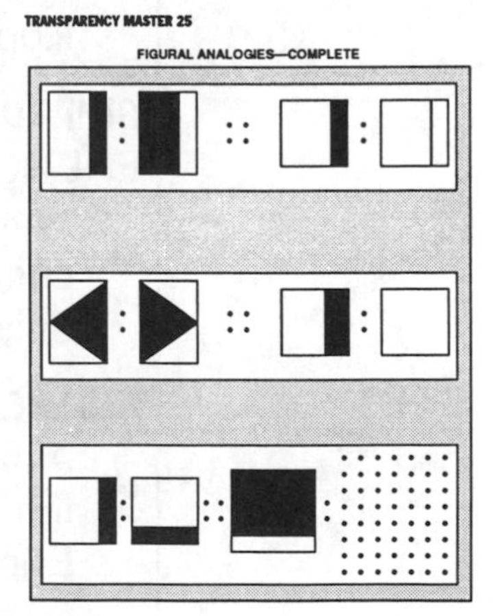

  Q: Look at the third figure. To correctly complete the analogy, how should the last figure be shaded?

  A: The small part on the right should remain white and the large part on the left should be shaded black.

- Shade the square properly then project the second example.

  Q: What has been done to the first rectangle to make it look like the second one? Look especially at the black triangle and the vertical line.

  A: The figure has been flipped about the vertical axis, like turning a page in a book.

  Q: What should the last figure look like if we want to show this same kind of relationship (a flip about a vertical axis) between the third and fourth figures?

  A: Draw a vertical line about a third of the way from the left edge. Darken the smaller (left) portion of the square.

- Fill in the figure then project the last example on the transparency.

  Q: So far, you have seen how to shade a given figure and how to provide the correct details to complete an analogy. Now we will see if we can provide (supply) the correct figure and the correct shading and details. What has been done to the first square to make it look like the second one?

  A: It has been rotated one position to the right (clockwise).

  Q: What type of figure will we have to draw on the grid to show this same relationship between the third and fourth figures?

  A: Draw a square, 4 blocks on each side, with a vertical line about one block in from the left-hand edge. The larger section should be shaded black, so the white (smaller) section is on the left.

- Project the transparency of page 152.

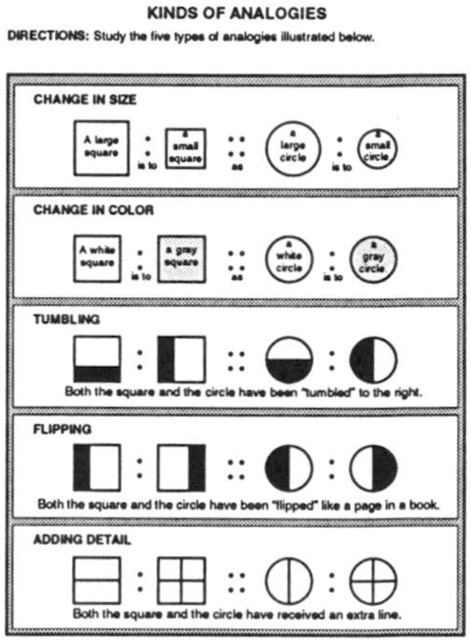

  Q: This is a transparency of page 152 in your workbook. How are the first two figures the same?

  A: They are the same shape, square.

  Q: How are they different?

  A: The second square is smaller than the first.

- Ask similar types of questions for the examples on page 152 illustrating color change, tumbling, flipping, details added.

KINDS OF ANALOGIES

DIRECTIONS: Study the five types of analogies illustrated below.

CHANGE IN SIZE

A large square : is to a small square :: as a large circle : is to a small circle

CHANGE IN COLOR

A white square : is to a gray square :: as a white circle : is to a gray circle

TUMBLING

Both the square and the circle have been "tumbled" to the right.

FLIPPING

Both the square and the circle have been "flipped" like a page in a book.

ADDING DETAIL

Both the square and the circle have received an extra line.

GUIDED PRACTICE

EXERCISES: **E-19**, **E-20**

- When students have had sufficient time to complete these exercises, check their answers by discussion. Encourage students to explain their thinking.

INDEPENDENT PRACTICE

- Assign exercises **E-21** through **E-30**. When students have had sufficient time to complete these exercises, check their answers by discussion. Encourage students to explain their thinking.

THINKING ABOUT THINKING

Q: What did you pay attention to when you found analogies between pairs of figures?

1. I looked carefully at the details (size of the figure, position of the figure, pattern or color).
2. I looked at the first pair to see how the two figures were different (change of the size, pattern, or color).
3. I looked at the second pair to see if the two figures were different in the same way as the first pair.
4. I explained the relationship as an analogy: *a* is to *b* as *c* is to *d*.

PERSONAL APPLICATION

Q: When could you use the ability to draw or supply something that has the same relationship as another pair or set?

A: Examples include repeating weaving or needlework patterns in different colors; matching clothing or linens; selecting proper hardware for construction projects; working with wood, metal, or leather.

Chapter 6

DESCRIBING THINGS

(Student book pages 157–74)

EXERCISES
F-1 to F-10

DESCRIBING THINGS—SELECT

ANSWERS F-1 through F-10 — Student book pages 158–62
Guided Practice: F-1A A–corn; **F-2A** F–orange; **F-2B** D–carrot
Independent Practice: F-3A I–owl; **F-3B** H–duck; **F-4A** J–zebra; **F-4B** L–giraffe; **F-5A** N–ambulance; **F-5B** O–truck; **F-6A** Q–airplane; **F-6B** R–helicopter; **F-7A** S–farm; **F-7B** U–restaurant; **F-8A** X–gas station; **F-8B** W–police station; **F-9A** BB–dentist; **F-9B** CC–teacher; **F-10A** EE–fire fighter; **F-10B** DD–police officer

LESSON
PREPARATION

OBJECTIVE AND MATERIALS

OBJECTIVE: Students will match the description of an item with its picture.
MATERIALS: Transparency of student book page 158 • washable transparency marker • optional—a red/purple onion • large pictures of the food, animals, vehicles, buildings, and occupations featured in the lesson

CURRICULUM APPLICATIONS

Language Arts: Dictionary activities, vocabulary development, using precise words, reading comprehension activities using illustrations and context
Mathematics: Matching terms with operations, recognizing sets
Science: Matching scientific terms with their descriptions; identifying types of plants, animals, etc.
Social Studies: Recognizing and matching topographic or geographic areas with their correct terms, identifying pictures of historical artifacts
Enrichment Areas: Identifying art styles, techniques, and artists' works; recognizing different types and periods in art and music

TEACHING SUGGESTIONS

Encourage students to discuss reasons for their choices and why the other items don't fit. The following cues model the discussion procedure demonstrated in the explanation and help students determine the correct answer.

1. Name the item in each picture.
2. Select the pictures that match each part of the description.
3. If the description has more than one part, select the picture that matches all parts of the description.
4. Verify that no other picture fits the description.

Note: This method simulates how to determine correct answers and eliminate incorrect answers on multiple choice tests.

MODEL LESSON

LESSON

Introduction

Q: In the figural exercises, you sometimes selected or drew figures which were opposite in color or position to another figure.

Explaining the Objective to the Students

Q: In this lesson, you will match the description of something to its picture.

Class Activity

- Project the directions and the example from the transparency of student book page 158.

  Q: This is a picture of page 158 in your book. Look at the three pictures.

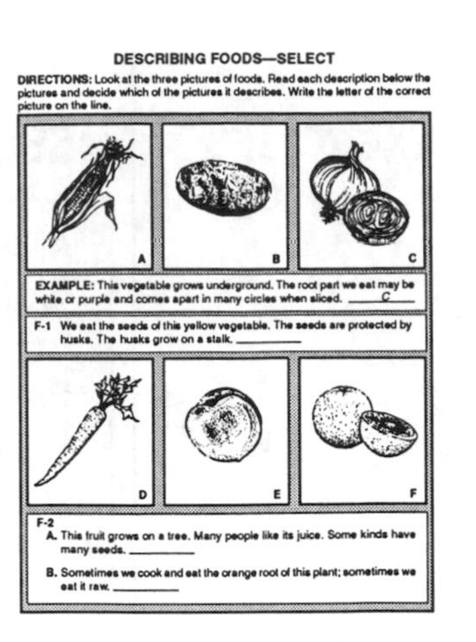

DESCRIBING FOODS—SELECT

DIRECTIONS: Look at the three pictures of foods. Read each description below the pictures and decide which of the pictures it describes. Write the letter of the correct picture on the line.

A B C

EXAMPLE: This vegetable grows underground. The root part we eat may be white or purple and comes apart in many circles when sliced. ___C___

F-1 We eat the seeds of this yellow vegetable. The seeds are protected by husks. The husks grow on a stalk. ________

D E F

F-2

A. This fruit grows on a tree. Many people like its juice. Some kinds have many seeds. ________

B. Sometimes we cook and eat the orange root of this plant; sometimes we eat it raw. ________

- Point to picture *A*.

  Q: What is the name of this vegetable?

  A: That's a picture of corn.

- Point to picture *B*.

  Q: What is the name of this vegetable?

  A: That's a potato.

- Point to picture *C*.

  Q: What is the name of this vegetable?

  A: That's an onion.

  Q: Now read the example. The first part of the sentence says this vegetable grows underground. Which vegetables grow underground?

  A: Corn does not, potatoes and onions do.

  Q: The rest of the description says the root part may be white or purple. Of the two plants that grow underground (potato and onion), which one may be purple?

  A: There are purple onions and purple potatoes.

- Show a purple onion (optional).

  Q: The last part of the description gives us a check on our answer, "and comes apart in many circles when sliced."

- Cut the onion and show the rings. (optional)

GUIDED PRACTICE

EXERCISES: **F-1** through **F-2B**

- When students have had sufficient time to complete these exercises, check answers by discussion to determine whether they have answered correctly. Discuss why the incorrect answers were eliminated.

INDEPENDENT PRACTICE

- Assign exercises **F-3** through **F-10**. Note: Depending on the reading level of your students, it may be necessary to read the descriptions to them.

ADDITIONAL EXERCISES USING PICTURES OR OBJECTS

Whenever possible, we recommend using actual objects; however, pictures usually offer sufficient detail even for students with limited backgrounds.

Supplement to DESCRIBING FOODS

The answers are underlined.

- Display an apple, strawberry, and tomato.

Q: Which of these fruits matches the following description? "This red fruit is often used for jelly and flavoring ice cream."

- Display butter, cheese, and milk.
  Q: Which of these dairy products matches the following description? "This dairy product is made by stirring milk until it becomes solid. People spread it on hot food."
- Display beans, peanuts, and peas.
  Q: Which of these foods matches the following description? "We eat the tiny root pods of this plant. They are roasted or made into a smooth spread for sandwiches."

Supplement to DESCRIBING ANIMALS

The answers are underlined.

- Display a fish, shark, and whale.
  Q: Which of these animals matches the following description? "This large, long sea animal eats other fish. There are many varieties of this animal, and they all have sharp teeth."
- Display a frog, lizard, and snake.
  Q: Which of these animals matches the following description? "This animal hatches from eggs and must live in the water when young. As it develops, it grows legs and loses a tail."

Supplement to DESCRIBING VEHICLES

The answers are underlined.

- Display an ambulance, bus, and car.
  Q: Which of these vehicles matches the following description? "Many people ride this vehicle to go to work or go shopping. They wait along its regular route and pay a fee to ride."
- Display an ambulance, police car, and truck.
  Q: Which of these vehicles matches the following description? "Many kinds of workers use this vehicle to haul supplies or equipment. It is also used for recreation."

Supplement to DESCRIBING PLACES

The answers are underlined.

- Display a barber shop, gas station, and supermarket.
  Q: Which of these places matches the following description? "People go there to buy a variety of foods. This business is large and has many workers."

Supplement to DESCRIBING OCCUPATIONS

The answers are underlined.

- Display a cook, farmer, and grocer.
  Q: Which of these workers matches the following description? "This worker prepares food for other people to eat. In this job, people bake, boil, fry, roast, and steam many foods."
- Display a barber, dentist, and nurse.

Q: Which of these workers matches the following description? "This worker makes people feel better by improving their appearance. This worker knows how to clean and trim hair."

THINKING ABOUT THINKING

Q: What did you pay attention to when you read the description?

1. I looked for the main characteristics (size, shape, color, etc.) of the item in the picture.
2. I made sure that characteristics in the description matched the item pictured.

PERSONAL APPLICATION

Q: When do you find it necessary in your life to match a picture to a description?

A: Examples include ordering food in a fast-food restaurant, finding the correct item in a store or catalog, solving puzzles that have picture clues, assembling objects or models from diagrams, explaining an object to someone who is not familiar with it.

EXERCISES F-11 to F-20

DESCRIBING THINGS—EXPLAIN

ANSWERS F-11 through F-20 — Student book pages 163–67

Guided Practice: F-11 celery: This green plant grows above the ground. It has a mild taste, is sometimes eaten raw, and is used in soups and salads. **F-12** cheese: This yellow-orange food is made from milk, comes in many varieties, and is used in making sandwiches and salads. Some people like a slice of it melted on a hamburger.

Independent Practice: F-13 hospital: People go to this place when they are sick or injured. Doctors and nurses treat illnesses or injuries here and have the equipment they need to help people get well. **F-14** mobile home: Some people live in mobile homes. Mobile homes can be pulled by large trucks and moved to a different location. Large numbers of these mobile homes are manufactured in factories. **F-15** desert: This dry part of the earth is covered with sand. Only plants and animals that can live on little water can survive here. **F-16** island: This piece of land has water on all sides. **F-17** spider: This animal has eight legs. It can spin a web to trap the insects that it eats. **F-18** lizard: This rough-skinned animal is a reptile. It lives in dry regions of the earth. It has legs and a tail and is hatched from an egg. **F-19** doctor: This person works in an office and a hospital. He or she treats illnesses and injuries. **F-20** dentist: This person works in an office. He or she takes cares of your teeth and gums.

LESSON PREPARATION

OBJECTIVE AND MATERIALS

OBJECTIVE: Students will look at a picture and write a description of what they see.

MATERIALS: Transparency of student book page 163 • washable transparency marker • (optional) large pictures of the items featured in the lesson

CURRICULUM APPLICATIONS

Language Arts: Writing descriptive sentences and paragraphs, doing vocabulary enrichment activities, using precise words
Mathematics: Writing and solving word problems, explaining steps in mathematical operations
Science: Defining scientific terms and processes; explaining the results of an experiment; describing and classifying animals, plants, etc.
Social Studies: Describing historical figures and events, identifying topographic and geographic areas
Enrichment Areas: Writing or giving directions for creating dance movements or art projects, describing types and periods of art and music

TEACHING SUGGESTIONS

Ask students to describe a family member and list the types of characteristics that students mention (age, gender, relationships to other members of the family, roles, feelings about them, interests or experiences that make them special). Students may use these characteristics to develop a story.

Ask students to describe a job and list the types of characteristics that students mention (specific tools or skills required, goods or services provided and what consumer seeks these goods or services, work location, training, and how the person spends his/her time). Emphasize the tasks that a person having this job carries out.

Encourage students to discuss their descriptions and why a description is or is not accurate. The suggested answers should be modified to fit the vocabulary level and needs of students. Remember the language that students use to describe food, animals, etc. Use the same words to remind students of the key characteristics of objects or people in this lesson and subsequent ones.

MODEL LESSON

LESSON

Introduction

Q: You have matched pictures with their descriptions.

Explaining the Objective to the Students

Q: In this lesson, you will look at a picture and describe what you see.

Class Activity

- Project the directions and exercise **F-11** from the transparency of student book page 163.
  Q: This is a picture of page 163 in your book.
- Point to exercise **F-11**.
  Q: What is the name of this food?
  A: That's a stalk of celery.
  Q: How can we describe "celery"?
  A: Class discussion will result in a number of responses: color, size, shape, uses, etc. Students may recognize that we slice and eat the long stems of this vegetable.

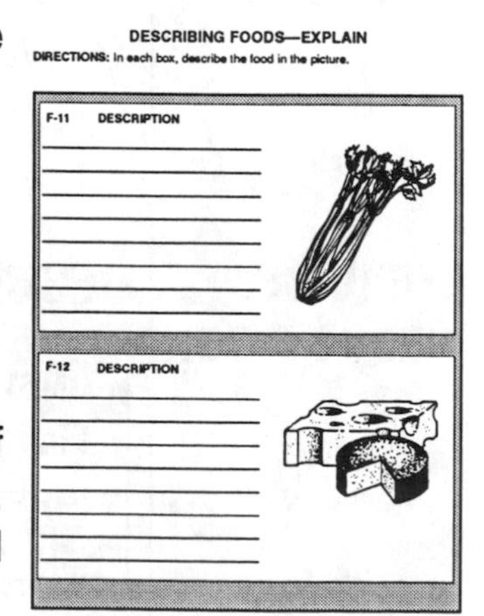
DESCRIBING FOODS—EXPLAIN
DIRECTIONS: In each box, describe the food in the picture.
F-11 DESCRIPTION
F-12 DESCRIPTION

- After class discussion, have a student write the consensus description on the chalkboard.

GUIDED PRACTICE
EXERCISE: **F-11** through **F-12**

INDEPENDENT PRACTICE
- Assign exercises **F-13** through **F-20**.

ADDITIONAL EXERCISES USING PICTURES OR OBJECTS
Whenever possible, we recommend using actual objects; however, pictures usually offer sufficient detail even for students with limited backgrounds. You may use the following activities to supplement exercises **F-11** through **F-20**.

Supplement to DESCRIBING FOODS
- Display a pineapple and ask students to write a description of it.
- Display lettuce and ask students to write a description of it.

Supplement to DESCRIBING PLACES
- Display a restaurant and ask students to write a description of it.
- Display an apartment building and ask students to write a description of it.

Supplement to DESCRIBING ANIMALS
- Display a butterfly and ask students to write a description of it.
- Display a turtle and ask students to write a description of it.

Supplement to DESCRIBING OCCUPATIONS
- Display a grocer and ask students to write a description of him or her.
- Display a teacher and ask students to write a description of him or her.

THINKING ABOUT THINKING
Q: What did you pay attention to when you wrote the description?
1. I looked for common characteristics to identify the item.
2. I looked for specific characteristics that were unique to that item.
3. I wrote as accurate a description as possible.

PERSONAL APPLICATION
Q: When do you need to describe something?
A: Examples include giving directions for going somewhere or making something, following directions that are given in pictures or diagrams; describing a gift, person, book, movie, or experience.

EXERCISES F-21 to F-26

DESCRIBING WORDS—SELECT

ANSWERS F-21 through F-26 — Student book pages 168–69
Guided Practice: F-21A photograph; **F-22A** lake; **F-22B** ocean
Independent Practice: F-23A country; **F-23B** continent; **F-24A** needle; **F-24B** blade; **F-25A** cloud; **F-25B** rainbow; **F-26A** yardstick; **F-26B** thermometer

LESSON PREPARATION

OBJECTIVE AND MATERIALS

OBJECTIVE: Students will select a word that matches a description.
MATERIALS: Transparency of student book page 168 • map of the United States • photograph of a landform • washable transparency marker

CURRICULUM APPLICATIONS

Language Arts: Doing vocabulary enrichment activities, using precise words, writing comparison paragraphs
Mathematics: Analyzing and solving word problems, recognizing sets
Science: Identifying processes, types of animals, plants, etc.
Social Studies: Comparing topographic or geographic areas
Enrichment Areas: Writing or giving directions for creating dance movements or art projects; comparing two pieces of music, two works of art, two dances, or two athletic activities

TEACHING SUGGESTIONS

Remember the language that students use in their descriptions. Use the same words to remind students of the key characteristics of objects in this lesson and subsequent ones.

When students have had sufficient time to complete the guided exercises, check answers by discussion to determine whether they have answered correctly. Discuss why the incorrect answers were eliminated. The following cues will encourage discussion and will help students model a method for determining the correct answer.

1. Recognize/pronounce word choices.
2. Define word choices.
3. Read the given definition.
4. Select a word that matches the given definition.
5. Eliminate detractors (check answer).

Note: This method also proves beneficial for determining correct answers (and eliminating incorrect answers) on any multiple choice test.

MODEL LESSON

LESSON

Introduction

Q: In the last exercises, you wrote a description of a picture.

Explaining the Objective to the Students

Q: In this lesson, you will select a word that matches a description.

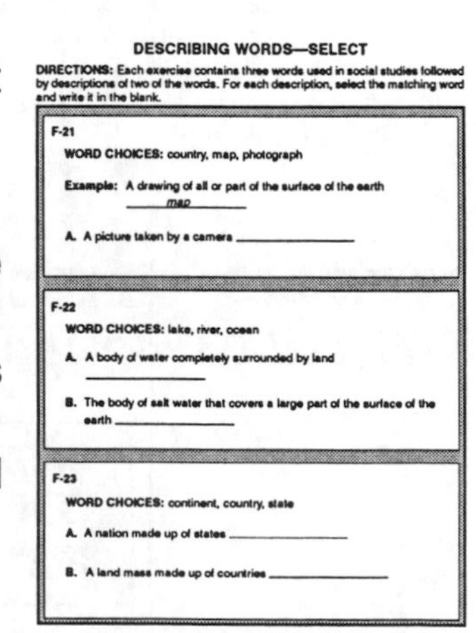
DESCRIBING WORDS—SELECT

DIRECTIONS: Each exercise contains three words used in social studies followed by descriptions of two of the words. For each description, select the matching word and write it in the blank.

F-21

WORD CHOICES: country, map, photograph

Example: A drawing of all or part of the surface of the earth *map*

A. A picture taken by a camera ____________

F-22

WORD CHOICES: lake, river, ocean

A. A body of water completely surrounded by land ____________

B. The body of salt water that covers a large part of the surface of the earth ____________

F-23

WORD CHOICES: continent, country, state

A. A nation made up of states ____________

B. A land mass made up of countries ____________

Class Activity

- Project the transparency of student book page 168, keeping the answer covered.

  Q: Look at the three word choices. The first is *country;* name our country.

  A: The United States of America, the United States, or America

  Q: The second word is *map.*

- Hold up a map of the United States.
  Q: What is this?
  A: A map (of the United States)
  Q: How is a map of the United States different from a country?
  A: The map is a drawing of the country and is much smaller than the country. The map is printed on paper; the country is made up of many things.
- Hold up a photograph of a landform (mountain, lake, ocean, canyon, glacier, etc.).
  Q: What is this?
  A: A picture (photograph) of a ________.
  Q: How is this photograph different from a country or a map?
  A: The photograph is a real picture of a part of the country. The map is a drawing, not a "real" picture.
  Q: Read the definition in the example to see which of the words it fits.
- Allow time for reflection.
  Q: Which of the words matches the definition "a drawing of all or part of the surface of the earth"? Explain your answer.
  A: Only a map is a drawing. A country is part of the surface of the earth. A photograph could be a picture of part of the surface of the earth, but is not a drawing.

GUIDED PRACTICE
EXERCISES: **F-21A** through **F-22B**

INDEPENDENT PRACTICE
- Assign exercises **F-23A** through **F-26B**.

THINKING ABOUT THINKING
Q: How did you match the word to the description?
1. I defined each of the words given.
2. I compared the differences in meaning between the words given.
3. I matched my definition of the word with the definition in the activity.

PERSONAL APPLICATION
Q: When do you need to match a word to a description?
A: Examples include solving word puzzles, reading directions, and understanding someone's description of an experience.

EXERCISES F-27 to F-53

DESCRIBING WORDS—SUPPLY/IDENTIFYING CHARACTERISTICS

ANSWERS F-27 through F-53 — Student book pages 170–74
Guided Practice: F-27 turkey; **F-28** lion; **F-29** tuna; **F-30** camel; **F-31** bee; **F-32** turtle

Independent Practice: **F-33** spinach, turnip or collard greens; **F-34** banana; **F-35** wheat, corn, or oats; **F-36** oak; **F-37** cactus; **F-38** pine; **F-39** tractor; **F-40** ambulance; **F-41** jet; **F-42** post office; **F-43** bank; **F-44** mall; **F-45** mail carrier; **F-46** barber or stylist; **F-47** dentist; **F-48** farmer; **F-49** grocer; **F-50** cook; **F-51** teacher; **F-52** doctor; **F-53** Students may list any four of the following characteristics: have backbones, hatch from eggs, change appearance as they grow, have moist skin, spend most of their time near water.

LESSON PREPARATION

OBJECTIVE AND MATERIALS

OBJECTIVE: Students will read a description and decide what is being described.
MATERIALS: None

CURRICULUM APPLICATIONS

Language Arts: Reading comprehension exercises, vocabulary enrichment activities
Mathematics: Matching terms and operations, analyzing and solving word problems, recognizing sets and set complements
Science: Recognizing scientific processes; identifying categories of plants, animals, etc.
Social Studies: Identifying topographic or geographic areas, and historical events and periods
Enrichment Areas: Identifying types of art or music, periods in art history, etc.

TEACHING SUGGESTIONS

Remember the language that students use in their descriptions. Use the same words to remind students of the key characteristics of objects in this lesson and subsequent ones.

MODEL LESSON

LESSON

Introduction

You have practiced selecting a word that matched a description.

Explaining the Objective to the Students

Q: In this lesson, you will read a description and decide what is being described.

Class Activity

- Read exercise **F-27**, page 170 of the student book, to your class and allow time for discussion.

  Q: This large bird is raised to be eaten. It is often purchased for holiday celebrations. Many meat products are made from this bird.

  A: Turkey

GUIDED PRACTICE

EXERCISES: **F-27** through **F-32**

- When students have had sufficient time to complete these exercises, check answers by discussion to determine whether they have answered correctly.

INDEPENDENT PRACTICE

- Assign exercises **F-33** through **F-53**. In exercise **F-53**, you are given the name of the animal being described and asked to write down the specific characteristics that identify the animal.

THINKING ABOUT THINKING

Q: What did you pay attention to when you read the description?

1. I looked for the main characteristics (size, shape, color, etc.) of the item in the picture.
2. I identified characteristics in the description that matched the picture.

PERSONAL APPLICATION

Q: When do you need to think of a word that fits a definition?

A: Examples include multiple choice tests, solving word puzzles, etc.

Chapter 7

VERBAL SIMILARITIES AND DIFFERENCES

(Student book pages 175–208)

EXERCISES
G-1 to G-24

OPPOSITES—SELECT (pictures and words)

ANSWERS G-1 through G-24 — Student book pages 176–80
Guided Practice: G-1 b; **G-2** b; **G-3** b; **G-4** a
Independent Practice: G-5 a; **G-6** c; **G-7** a; **G-8** b; **G-9** c; **G-10** b; **G-11** a; **G-12** c; **G-13** a; **G-14** a; **G-15** c; **G-16** b; **G-17** c; **G-18** c; **G-19** b; **G-20** b; **G-21** c; **G-22** b; **G-23** b; **G-24** a

DETAILED SOLUTIONS

Note: The key definition and its opposite are underlined in all answers for this section. The four words in each exercise are italicized.

G-1 b: *Up* indicates going in a rising direction. Look for a word which indicates going in a falling direction. *Down* is the direction of falling. *Above* is not an opposite because it means located in a higher position. *Right* is not an opposite because it indicates a position on one side.

G-2 b: *Left* refers to a position on or to one side of something. Look for a word which indicates a position on or to the opposite side of something. *Right* indicates the opposite side or direction from left. *Last* is not an opposite because it means a position after all others in time or order. *Up* is not an opposite because it describes a position above something.

G-3 b: *Top* means the highest part. Look for a word that means the lowest part. *Bottom* means the lowest part. *Between* means being somewhere in the middle. *Last* indicates a position at the end of a line or series of events.

G-4 a: *Above* means over or higher than something else. Look for a word that means under or lower than something. *Below* means under or beneath something. *Beside* means by or at the side of. *Between* describes a middle position.

G-5 a: A *stocking* is a piece of clothing for the foot. Look for a word which names a piece of clothing for the part of the body opposite the foot (hand). A *glove* is worn on the hand. *Shoe* is not an opposite because it is also worn on the foot. *Jacket* is not an opposite because it is worn over the middle of the body.

G-6 c: A *hat* is a piece of clothing for the top of the body (head). Look for a word which names a piece of clothing for the bottom of the body (foot). A *shoe* is worn on the foot. *Glove* is not an opposite because it is worn on the end of the arm at the side of the body. *Jacket* is not an opposite because it is worn over the middle of the body.

G-7 a: *Floor* names the bottom part of a room or house and is something that is under your feet. Look for a word that names the top part of a building and is over your head. *Roof* names the outside top of a building and is over your head. *Wall* is not an opposite because it names the sides of a building. *Window* is not an opposite because it names an opening in a wall.

G-8 b: A *nose* is usually found at the front of something. Look for a word which means something usually found at the back of something. A *tail* is found at the back of an airplane or an animal. *Airplane* is not an opposite because it refers to a whole object rather than any part. *Wing* is not an opposite because it refers to a side part.

G-9 c: *Black* describes the absence of color or light. Look for a word that describes the presence of color or light. *White* is the presence of all colors or light. *Night* is not an opposite because it describes a dark time of day. *Star* is not an opposite because it names an object that looks like a small point of light in the night sky.

G-10 b: A *stove* is an appliance used to heat or prepare food for eating. Look for a word that names an appliance used to cool or keep food. A *refrigerator* is an appliance used to cool food. Neither a *dryer* nor a *washer* is commonly used for food. They are appliances used to clean clothes.

G-11 a: A *washer* is an appliance used to clean clothes with water. Look for a word that names an appliance that is used to take the water out of wet clothes. A *dryer* is used to remove water from wet clothes. An *iron* is used to take the wrinkles out of dry clothes. A *sink* is a container in which things are washed.

G-12 c: A *chair* is a piece of furniture upon which a person sits. Look for a word that names a piece of furniture that people do not sit upon. A *table* is not used to sit upon. Both a *sofa* and a *stool* are furniture that people sit upon.

G-13 a: *Back* refers to the direction behind. Look for a word which means the direction ahead. *Front* means the direction ahead. *Side* means the direction to the left or right. *Top* means the direction above.

G-14 a: *Stop* means to halt or keep from going on. Look for a word that means to move or begin motion. *Go* means to begin or stay in motion. *Slow* is not an opposite because it means to reduce speed. *Turn* is not an opposite because it describes a change in direction.

G-15 c: *Winter* is the name of the coldest season of the year and the part of the year with the shortest daylight time. Look for a word that names the warmest season of the year or the part of the year with the longest daylight time. *Summer* names the warmest season and the part of the year with the longest days. Both *spring* and *fall* refer to mild seasons between summer and winter.

G-16 b; *Day* is the time during which the light of the sun is present. Look for a word that means a time when the light of the sun is not present. *Night* is a time when the light of the sun is not present. A *cloud* is mist formed by water in the air. The *sun* is what causes daytime, so it can't be opposite.

G-17 c: *Man* is a person who could become a father. Look for a word that names a person who could become a mother. A *woman* can become a mother. *Boy* is not an opposite because a boy becomes a man. *Father* is not an opposite because a father is a man.

G-18 c; *Arm* is an upper limb of a person. Look for a word that names a lower limb. *Leg* is a lower limb. *Foot* and *hand* are not opposites for they name the ends of the limbs.

G-19 b: *Hand* is the end of an upper limb. Look for a word that names the end of a lower limb. *Foot* names the end of a lower limb. *Arm* and *leg* are not opposites of hand because they name limbs and not the ends of limbs.
G-20 b: *Fire* means that something is hot or burning. Look for a word which means that something is cold or freezing. *Ice* is cold and frozen. *Cloud* is not an opposite because it names a visible form of water vapor. *Star* is not an opposite because it names an object that gives light.
G-21 c: *Hill* refers to a part of land higher than the land around it. Look for a word which means a part of land lower than the land around it. *Valley* refers to a lower part of the land. *Ocean* is not an opposite because it refers to a body of water. *Road* is not an opposite because it names a way for vehicles to travel from one place to another.
G-22 b: An *ocean* is a large body of water. Look for a word which means a large dry area. A *desert* is a large area that is almost without water. Neither *rainbow* nor *cloud* is an opposite because both are forms of moisture.
G-23 b: The *sun* is a hot object in the sky that gives off light. Look for a word which describes a cool object in the sky that does not give off light. The *moon* is a cool object in the sky that does not give off light. *Fire* is not a cool object. A *star* is an object like the sun but is very far away.
G-24 a: *Mountain* refers to a part of land much higher than the land around it. Look for a word which means a part of land much lower than the land around it. *Canyon* refers to a very low part of the land. *Cloud* is above the land not part of the land. *Road* is not an opposite because it names a way for vehicles to travel from one place to another.

LESSON PREPARATION

OBJECTIVE AND MATERIALS

OBJECTIVE: Using picture clues, students will select a word which has a meaning the opposite of a given word.
MATERIALS: Transparency of TM 25 (p. 273) • transparency of student book page 176 • washable transparency marker

CURRICULUM APPLICATIONS

Language Arts: Using precise words, doing vocabulary enrichment activities, working with antonym exercises, writing contrast paragraphs
Mathematics: Recognizing and using inverse operations, recognizing sets
Science: Recognizing reversed processes in simple experiments; describing differences between two objects, organisms, or concepts
Social Studies: Contrasting topographic or geographic areas
Enrichment Areas: Recognizing the differences between two pieces of music, two works of art, two dances, or two athletic activities

TEACHING SUGGESTIONS

Encourage students to discuss reasons for their choices and why the other given words are less correct. Remember the language that students use to describe their choices and use the same words to remind students of the key characteristics of objects or people in this lesson and subsequent ones. If appropriate for students' verbal proficiency, use the term *antonym* when discussing words with different meanings. These exercises can also be used to increase dictionary use as students check different meanings of the words. The following cues will help you to model for students a method for determining the correct answer:

1. Recognize/pronounce word choices.
2. Define word choices.
3. Define opposite word meaning.
4. Select opposites (answer).
5. Eliminate detractors (check answer).

Note: This method also proves beneficial for determining correct answers (and eliminating incorrect answers) on any multiple choice test.

MODEL LESSON

LESSON

Introduction

- Project the transparency of TM 25.

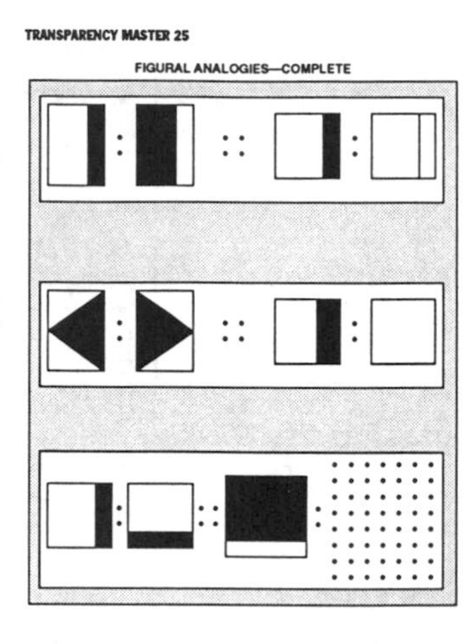

Q: You have sometimes selected or drawn figures which were opposite in color or position to another figure. Do you remember how the rectangles that you see here were different?

A: The section that is white on the first is black on the second. The section that is black on the first is white on the second.

Q: How should we shade the square to show that same relationship?

A: The large section on the left should be black, and the small section on the right should be white.

Q: We can say that the figures are colored opposite.

Explaining the Objective to the Students

Q: In this lesson, you will look at a word and then select a word which has the opposite meaning.

Class Activity

- Project the directions and the example from the transparency of student book page 176.

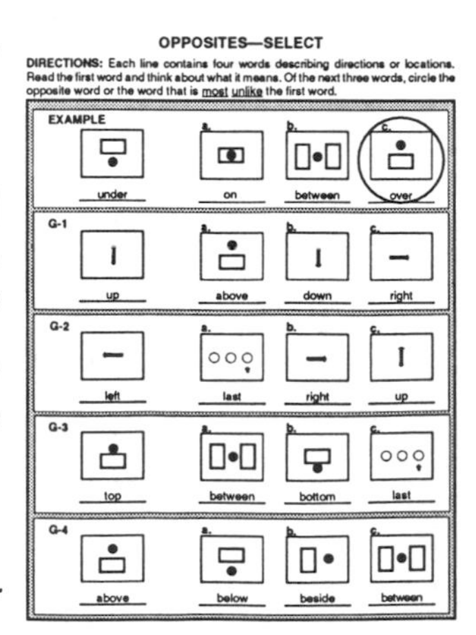

Q: On page 176, you will look for the word which is the opposite of the word on the left. The word *opposite* means that something is most unlike something else or that something is the reverse of a given thing or word. When you have chosen the opposite word, check the other choices to make sure that the other words are not also opposites.

- Point to the example.

Q: In the example, you are given the word *under* followed by three other words, one of which must be an opposite of *under*. These words are *on*, *between*, and *over*. The word *under* as illustrated by the picture...

- Point to the example.
  Q: ...means the dot is below the box. To do this exercise, we must look for a word which has the opposite meaning. *Over*..
- Point to the word and picture.
  Q: ...appears to be the opposite because *over* means above, and the dot is above the box. Before we can be certain, though, we must check the other two words to be sure that they are not more unlike *under*. Look at the word *on*...
- Point to the word and picture.
  Q: *On* means to be held up by or be in contact with a surface; the dot is touching the box. Now look at the word *between*.
- Point to the word and picture.
  Q: *Between* describes a middle position. Therefore, *over* is the word most unlike *under* and we can say it is the opposite. We'll draw a circle around the word and picture of *over* to show that it is the choice we made.

GUIDED PRACTICE

EXERCISES: **G-1** through **G-4**

- When students have had sufficient time to complete these exercises, check answers by discussion to determine whether they have answered correctly. Discuss why the incorrect answers were eliminated.

INDEPENDENT PRACTICE

- Assign exercises **G-5** through **G-24**.

THINKING ABOUT THINKING

Q: What did you pay attention to when you looked for a word with the opposite meaning?

1. I thought about what the first word meant.
2. I found another word that had the opposite meaning.
3. I substituted the second word for the first one to check whether the words had the opposite meaning.
4. I read the other words to see why they did not have the opposite meaning.

PERSONAL APPLICATION

Q: When do you need to know or find a word that means the opposite of another word?

A: Examples include solving crossword puzzles, giving directions for going somewhere or making something, following reverse directions, (e.g., turning left when you go to your friend's house and turning right when you go back home), reassembling models or appliances which have been taken apart.

EXERCISES G-25 to G-69

OPPOSITES—SELECT (words)

ANSWERS G-25 through G-69 — Student book pages 181–85
Guided Practice: **G-25** a; **G-26** a; **G-27** c; **G-28** a
Independent Practice: **G-29** a; **G-30** b; **G-31** b; **G-32** c; **G-33** b; **G-34** c; **G-35** b; **G-36** a; **G-37** b; **G-38** b; **G-39** c; **G-40** b; **G-41** a; **G-42** a; **G-43** c; **G-44** b; **G-45** a; **G-46** b; **G-47** c; **G-48** b; **G-49** c; **G-50** c; **G-51** b; **G-52** c; **G-53** a; **G-54** a; **G-55** c; **G-56** b; **G-57** c; **G-58** b; **G-59** b; **G-60** b; **G-61** c; **G-62** b; **G-63** a; **G-64** b; **G-65** b; **G-66** c; **G-67** a; **G-68** b; **G-69** b

DETAILED SOLUTIONS

Note: The key definition and its opposite are underlined in all answers for this section. The four words in each exercise are italicized.

G-25 a: *Clean* means the absence of stains or soil. Look for a word that means not clean. *Dirty* means stained or soiled. *New* means not used, and *old* means used, but new or old things could also be clean.

G-26 a: *Lost* means that something is missing. Look for a word that means something is located or not missing. *Found* means that something has been located or is no longer missing. *Gone* is not an opposite because it means the same as lost. *Look* is not an opposite because it is the action one takes to find lost objects.

G-27 c: *Catch* means to take hold of or grab an object with one's hands. Look for a word that means to let go of or release an object with one's hands. *Throw* means to release an object with one's hands. *Ball* is not an opposite because a ball is an object that can be caught or thrown. *Play* is not an opposite because play is recreation, some forms of which may involve catching or throwing objects.

G-28 a: *Die* means to stop life, motion, or action. Look for a word that means continuing life, motion, or action. *Live* means to have life, motion, or action. *Old* is not an opposite because it refers to the age of something. *Sick* is not an opposite because it is a condition of health.

G-29 a: *Enter* means to come or go into. Look for a word meaning to go out of. *Leave* means to go out of or away from. *Open* is not an opposite because it means to unfasten something. *Stay* is not an opposite because it means to remain in the same place.

G-30 b: *Hard* is sometimes used to describe a task that takes much effort. Look for a word that describes a task that takes little effort. *Easy* describes a task that takes little effort. *Difficult* and *problem* are words used in place of hard and are not opposites.

G-31 b: *Dull* is often used to describe something of little interest ("Her speech was dull"). Look for a word that describes something of great interest. *Exciting* describes something of great interest. *Boring* means the same as dull. *Long* is not an opposite because it is a measure of the time past, not a measure of interest.

G-32 c: *Hard* can also be used to mean uncomfortable ("This chair is too hard!") or difficult to scratch or break. Look for a word that means comfortable or easy to scratch or break. *Soft* can describe both comfortable and easy to scratch or break. *Brittle* is not an opposite because it describes a hard object that breaks easily. *Sharp* is not an opposite because it means capable of cutting.

G-33 b: *Dull* can also mean blunt or not capable of cutting (as a dull knife won't cut). Look for a word that means capable of cutting. *Sharp* means capable of cutting. *Blunt* means the same as dull. *Worn* has to do with the condition of something because of age or use.

G-34 c: *Back* refers to the part that is behind. Look for a word that means the part that is ahead. *Front* refers to the part that is ahead. *Bone* and *end* are not opposites to back, but rather are words often associated with back (backbone, back end).

G-35 b: *Upper* names a position on the top part. Look for a word that names a position on the bottom part. *Lower* means located on the bottom part. *Class* and *story* are not opposites of upper, but are words often associated with upper or lower (lower class, upper story).

G-36 a: *Begin* means to set into motion or the first action. Look for a word that means to stop a motion or the last action. *End* means the closing action. *Go* and *start* are words that mean nearly the same as begin.

G-37 b: *North* is the direction toward the top of a map. Look for a word that means the direction toward the bottom of a map. *South* is the direction toward the bottom of a map. *East* and *west* are directions toward the sides of a map.

G-38 b: *Give* means to pass or hand something over to another. Look for a word which means to get something from another. *Receive* means to get something. *Present* is not an opposite because a present is something that is given or received (or to present, meaning to give). *Rent* is not an opposite because rent is payment for the use of something. There is no payment for things that are given.

G-39 c: *Right* means a direction to the side. Look for a word that means the opposite of right. *Left* is the direction opposite to right. *Above* and *below* refer to positions up or down from an object.

G-40 b: *Head* refers to the top or front part of an object. Look for a word that refers to the bottom or back part of an object. *Foot* means the bottom part of an object. *First* might mean the same as head ("He's first in the class" or "He's at the head of the class"). Both *first* and *strong* are also words that are often associated with head (headfirst, headstrong).

G-41 a: *Leave* means to go out or go away. Look for a word that means to come in. *Arrive* means to come in or appear. *Exit* means the same as leave. *Run* means to move away swiftly.

G-42 a: *Go* means to leave. Look for a word that means to arrive. *Come* means to appear or arrive. *Exit* and *leave* mean the same as go.

G-43 c: *Dry* means absence of water. Look for a word that means presence of water. *Wet* indicates the presence of water. *Bright* is not an opposite because it means the presence of light. *Hot* is not an opposite because it means the presence of heat.

G-44 b: *Clear* means that all of something is visible. Look for a word that means that none or only part of something is visible. *Cloudy* means that only part of the sky is visible. *Bright* and *sunny* are not opposites since they mean nearly the same as clear.

G-45 a: *Hot* describes something that has a high temperature. Look for a word that describes something that has a low temperature. *Cold* means an object has a low temperature. *Heat* is not an opposite because it is what you use to make something hot. *Warm* is not an opposite because it describes a temperature between hot and cold.

G-46 b: *Sunny* means that the sun can be seen. Look for a word that means the sun cannot be seen. *Cloudy* describes a day when the sun cannot be seen. *Bright* describes the light that exists when it is sunny. *Hot* describes the temperature on a sunny summer day.

G-47 c: *Rainy* means that water is falling from the sky. Look for a word that means that water is not falling from the sky. *Sunny* describes a day when no rain is falling. *Cloudy* describes a day when it may be raining. *Cold* describes temperature and does not depend on whether it is rain.

G-48 b: *Big* means large in size. Look for a word which means small in size. *Little* means small in size. *Large* and *tall* are not opposites because they mean about the same as big.

G-49 c: *Warm* means not too hot. Look for a word that means not too cold. *Cool* is the opposite because it means not too cold. *Bright* is not an opposite because it means giving or having much light. *Clothes* is not an opposite because they are things you wear to keep you warm.

G-50 c: *Boiling* describes the rapid evaporation of very hot water. Look for a word that describes the action of cold water. *Freezing* describes the solidifying action of very cold water. *Baking* and *cooking* are two kinds of food preparation which are performed on a stove (just as one can boil water on a stove).

G-51 b: *Heavy* describes an object that has great weight. Look for a word that describes an object that has little weight. *Light* is a word describing something having little weight. *Firm* means an object is difficult to press in. *Weight* is not an opposite because it is the quality that heaviness measures.

G-52 c: *Often* means happening many times. Look for a word that means happening a few times. *Seldom* means happening a few times. *Again* and *repeat* mean to happen a second time.

G-53 a: *Always* means at all times. Look for a word that means at no time. *Never* means at no time. *Often* means frequently or at many times. *Some* means either occasionally or a few times.

G-54 a: *Tight* means fixed or fastened firmly in place. Look for a word that means not fastened. *Loose* means not fastened. *Close* is not an opposite because it means near in space or time. *Sharp* is not an opposite because it means having a fine point.

G-55 c: *None* means not any. Look for a word that means one or more. *Some* means one or more. *Empty* means containing nothing. *Nothing* means the same as none.

G-56 b: *All* means the whole quantity. Look for a word that means no quantity. *None* means no quantity. *Every* is similar to all and means each without exception. *Some* means a few things.

G-57 c: *Now* means at the present time. Look for a word that means at a past time. *Then* means at a past time. *Here* means in this place or at this time. *Often* means happening many times.

G-58 b: *Many* means a large number. Look for a word that means a small number. *Few* means a small number. *All* means the whole quantity. *None* means not any.

G-59 b: *Past* means in a time before now. Look for a word that means in a time after now. *Future* means in a time after now. *Before* means earlier than. *Then* means before now.

G-60 b: *Everything* means the whole quantity. Look for a word that means not any. *Nothing* means not any. *All* means the whole quantity. *Something* means part of the whole quantity.

G-61 c: *Dark* means not light in color. Look for a word that means light in color. So *light* is the opposite of dark. *Dim* means not well lighted. *Heavy* means thick or dark (in art).

G-62 b: *Small* means little. Look for a word that means big. *Large* means big. *Heavy* means thick or dark (in art). *Little* means the same as small.

G-63 a; *Dull* means not shiny. Look for a word that means shiny. *Bright* means shiny. *Dim* is similar to dull. *Light* is the opposite of dark or heavy.

G-64 b: *Smooth* means having an even surface. Look for a word that means having an uneven surface. *Rough* means having an uneven surface. *Glass* is an example of something that is smooth. *Slick* describes a surface upon which things can slide (smooth surfaces can be slick if they are wet).

G-65 b: *Tall* means having great height. Look for a word that means not having great height. *Short means* having little height, not tall. *High* means the same as tall. *Round* means shaped like a circle.

G-66 c: *Easy* means not hard or difficult. Look for a word that means difficult. *Hard* means difficult, not easy. *Soft* means easy to scratch or break. *Simple* means the same as easy.

G-67 a: *Fine* means delicate or ground into small particles. Look for a word that means not delicate or ground into large particles. *Coarse* means not delicate or ground into large particles. *Dim* means not well lighted. *Soft* can mean delicate in art.

G-68 b: *Thick* means wide. Look for a word that means narrow. *Thin* means narrow. *Fat* and *wide* are similar to thick.

G-69 b: *Zigzag* means a series of angles. Look for a word that means having no angles. *Straight* lines have no angles. *Bumpy* means a series of curves. *Up and down* can describe the angles in a zigzag path.

LESSON PREPARATION

OBJECTIVE AND MATERIALS

OBJECTIVE: Students will select an antonym of a given word.

MATERIALS: Transparency of student book page 181 • washable transparency marker

CURRICULUM APPLICATIONS

Language Arts: Doing vocabulary enrichment activities, using precise words, working with antonym exercises, writing contrast paragraphs

Mathematics: Recognizing and using inverse operations, using fractions and reciprocals, recognizing sets

Science: Recognizing reversed processes in simple experiments, accurately describing differences between two objects or concepts

Social Studies: Contrasting topographic or geographic areas, locating contrasting details between two or more object or artifacts
Enrichment Areas: Recognizing the differences between two pieces of music, works of art, dances, or athletic activities

TEACHING SUGGESTIONS

Encourage students to discuss reasons for their choices and why the other given words are less correct. Class discussion is a valuable technique for having children share their acquired knowledge. Remember the words that students use to describe their choices and use these same words to remind students of the key definitions of words in these lessons and subsequent lessons. These exercises can also be used to increase dictionary use as students check different meanings of the words, particularly those which are not pictured. The following cues model a method for determining the correct answer:

1. Recognize/pronounce word.
2. Define word.
3. Define opposite word.
4. Select opposites (answer).
5. Eliminate detractors (check answer).

Note: This method is also beneficial for determining correct answers (and eliminating incorrect answers) on any multiple choice test.

MODEL LESSON

LESSON

Introduction

Q: In the exercises which you just completed, you looked at pictures and words. In these exercises, there are no pictures.

Explaining the Objective to the Students

Q: In this lesson, you will read a word and then select a word that has the opposite meaning.

Class Activity

- Project the directions and **G-25** from the transparency of student book page 181.
  Q: On page 181, you will look for the word which is the opposite of the word on the left. The word *opposite* means that something is most unlike something else or that something is the reverse of a given thing or word. When you have chosen the opposite word, check the other choices to make sure that the other words are not also opposites.

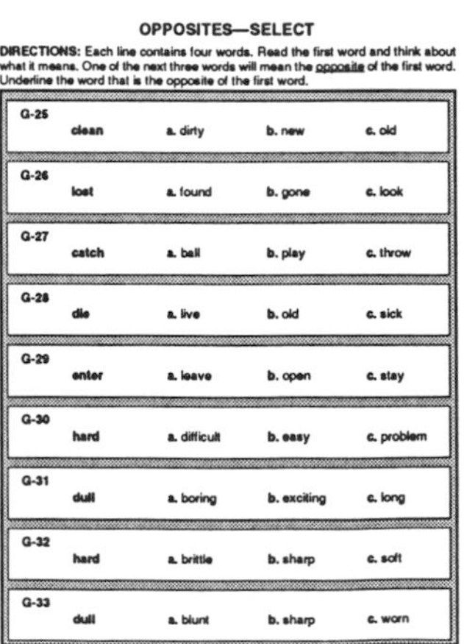
OPPOSITES—SELECT

DIRECTIONS: Each line contains four words. Read the first word and think about what it means. One of the next three words will mean the opposite of the first word. Underline the word that is the opposite of the first word.

G-25	clean	a. dirty	b. new	c. old
G-26	lost	a. found	b. gone	c. look
G-27	catch	a. ball	b. play	c. throw
G-28	die	a. live	b. old	c. sick
G-29	enter	a. leave	b. open	c. stay
G-30	hard	a. difficult	b. easy	c. problem
G-31	dull	a. boring	b. exciting	c. long
G-32	hard	a. brittle	b. sharp	c. soft
G-33	dull	a. blunt	b. sharp	c. worn

- Point to exercise **G-25**.
  Q: In this exercise, you are given the word *clean* followed by three other words, one of which must be an opposite of clean. These words are *dirty*, *new*, and *old*.

- Point to the word *clean.*
  Q: Clean means that something has no stains or soil.
- Point to the word *dirty.*
  Q: Dirty means stained or soiled.
- Point to the words *new* and *old.*
  Q: New means not used, and new things are usually clean. Old means used; used things can be either clean or dirty. Therefore, *dirty* is the word most unlike *clean* and we can say it is the opposite. We'll draw a line under *dirty* to show that it is the choice we made.

GUIDED PRACTICE

EXERCISES: **G-25** through **G-28**

- When students have had sufficient time to complete these exercises, check answers by discussion to determine whether they have answered correctly. Discuss why the incorrect answers were eliminated.

INDEPENDENT PRACTICE

- Assign exercises **G-29** through **G-69**. You may wish to divide these exercises into a number of lessons.

THINKING ABOUT THINKING

Q: What did you pay attention to in order to find a word with the opposite meaning?

1. I thought about what the first word meant.
2. I found another word that had the opposite meaning.
3. I substituted the second word for the first one to check whether the words had the opposite meaning.
4. I read the other words to see why they did not have the opposite meaning.

PERSONAL APPLICATION

Q: When do you find it necessary in your life to know or find a word that means the opposite of another word?

A: Examples include solving crossword puzzles, giving directions for going somewhere or making something, following reverse directions, (e.g., turning left when you go to your friend's house and turning right when you go back home), reassembling models or appliances which have been taken apart.

EXERCISES G-70 to G-96

OPPOSITES—SUPPLY

ANSWERS G-70 through G-96 — Student book pages 186–88

Guided Practice: G-70 slow; **G-71** hate; **G-72** follower; **G-73** dirty
Independent Practice: G-74 sad, down, depressed; **G-75** shout, yell, scream, cry, call, roar; **G-76** found, won, returned, regained; **G-77** play, recreation, sport; **G-78** awake; **G-79** far, distant, remote, removed; **G-80** finish, stop, complete, close, end, halt, wind up; **G-81** false, fake,

phony; **G-82** behind; **G-83** rear or back, tail, end; **G-84** bottom, base, floor, foot, or ground; **G-85** earlier, sooner, first, or before; **G-86** forget; **G-87** give, donate, present; **G-88** light; **G-89** stop, stay, halt; **G-90** walk; **G-91** few; **G-92** all, some, several, lots; **G-93** less; **G-94** before; **G-95** first; **G-96** always

DETAILED SOLUTIONS

Other antonyms or synonyms can be used to define words. Encourage students to arrive at as many acceptable antonyms as they can for each word. (Note: Possible definitions and their opposites are underlined throughout the answers to this section. The given word and its most likely opposite appear in italics.)

G-70 *Fast* means moving quickly. Think of a word that means not moving quickly. *Slow* means not moving quickly.

G-71 *Love* means a strong affection. Think of a word that means a strong dislike. *Hate* means a strong dislike.

G-72 *Leader* means one who goes ahead. Think of a word that means one who comes behind. *Follower* means one who comes after in the same direction.

G-73 *Clean* means free of stain or soil. Think of a word that means stained or soiled. *Dirty* means stained or soiled.

G-74 *Happy* means joyful. Think of a word that means lacking joy. *Sad* means full of sorrow or lacking joy. (Other possible opposites: down, depressed)

G-75 *Whisper* means to talk softly. Think of a word that means to talk loudly. *Shout* means to talk very loudly. (Other possible opposites: yell, scream, cry, call, roar)

G-76 *Lost* means that something is missing or you did not win. Think of a word that means you have something back. *Found* means something is located. (Other possible opposites: returned, regained, won)

G-77 *Work* means to do something for pay or purpose. Think of a word that means to do something for free or fun. *Play* means to do something for fun, not for money. (Other opposites: recreation, sport)

G-78 *Asleep* means to be at rest. Think of a word that means to come out of sleep or to be active and alert. *Awake* means to be alert and not asleep. (Other opposite: aware)

G-79 *Close* means near (as a direction word). Think of a word that means not near. *Far* means not near. (Other opposites: distant, remote, removed)

G-80 *Start* means to begin an action. Think of a word that means to end an action. *Finish* means to complete an action. (Other possible opposites: stop, complete, close, end, halt, wind up)

G-81 *True* means correct or real. Think of a word that means not correct or not real. *False* means incorrect. (Other possible opposites: fake, phony)

G-82 *Ahead* means at the front end of something. Think of a word that means at the rear end. *Behind* means to the rear of the leader.

G-83 *Front* means the part that faces forward. Think of a word that means the part that faces backward. *Rear* or *back* means the part that faces backward. (Other possible opposites: tail, end)

G-84 *Top* indicates the highest part of something. Think of a word that indicates the lowest part of something. *Bottom* is the lowest part. (Other possible opposites: base, floor, foot, or ground)
G-85 *Later* means coming after in time or space. Think of a word that means coming before. *Earlier* means coming before in time. (Other possible opposites: sooner, first, or before)
G-86 *Remember* means to bring back to your mind. Think of a word that means to take away from your mind. *Forget* means to lose from the mind.
G-87 *Receive* means to get something. Think of a word that means to make a present of something. The opposite of get is *give*. (Other possible opposites: donate, present)
G-88 *Dark* means without light. Think of a word that means not dark. *Light* is the opposite of dark.
G-89 *Go* means to start moving. Think of a word that means to not move. *Stop* means to cease moving. (Other possible opposites: stay, halt, close, turn off)
G-90 *Run* means to move quickly on foot. Think of a word that means to move slowly on foot. *Walk* means to move slowly on foot.
G-91 *Many* means a large number of things. Think of a word that means a small number of things. *Few* means a small number of things. (An extreme opposite would be none, and several may be considered less than many.)
G-92 *None* means not one. Think of a word that means every one. *All* means every thing or the entire extent. (Other possible opposites: some, several, many, lots)
G-93 *More* means a greater amount. Think of a word that means a smaller amount. *Less* indicates a smaller amount.
G-94 *After* means following in position or time. Think of a word that means ahead in position or time. *Before* means ahead in position or time.
G-95 *Last* means that which is after all others. Think of a word that means before all others. *First* means before all others.
G-96 *Never* means not at any time. Think of a word that means at all times. *Always* means at all times.

LESSON PREPARATION

OBJECTIVE AND MATERIALS

OBJECTIVE: Students must supply a word which means the opposite of the given word.
MATERIALS: Transparency of student book page 186 • washable transparency marker

CURRICULUM APPLICATIONS

Language Arts: Using antonyms, writing contrast paragraphs, expressing an opposite opinion, describing contrasting stories
Mathematics: Checking basic arithmetic problems by reversing processes (e.g., checking subtraction by addition)
Science: Describing differences between two objects or concepts; disassembling and reassembling motors, gears, or models
Social Studies: Locating and expressing contrasting details when comparing objects or maps

Enrichment Areas: Writing or giving directions for creating dance movements or art projects; expressing the differences between two pieces of music, two works of art, two dances, or two athletic activities

TEACHING SUGGESTIONS

Encourage students to discuss and define their antonyms. Remember the language that students use to describe their choices and use the same words to remind students of the key definitions of words in this lesson and subsequent ones. The following cues will encourage the discussion procedure demonstrated in the explanation:

1. Recognize/pronounce word.
2. Define word.
3. Define opposite word.
4. Supply opposite word.

Other antonyms or synonyms can be used to define words. For each word, encourage students to express as many acceptable antonyms as they can recall.

MODEL LESSON

LESSON

Introduction

Q: You have selected a word that has a meaning opposite to a given word.

Explaining the Objective to the Students

Q: In this lesson, you will read a word and then think of a word that has an opposite meaning.

Class Activity

- Project exercise **G-70** from the transparency of student book page 186.

  Q: This is exercise **G-70** on page 186 in your book. Before we say the answer, let's define the word *fast*. Most commonly, fast means moving quickly. Now, let's think of a word that means not moving quickly.

OPPOSITES—SUPPLY

DIRECTIONS: Each line contains a word. Read the given word and think about what it means. Think of a word that means the opposite. Write the word that you think of. (If you can think of other opposites, write them all down.)

G-70 fast ______

G-71 love ______

G-72 leader ______

G-73 clean ______

G-74 happy ______

G-75 whisper ______

G-76 lost ______

G-77 work ______

G-78 asleep ______

- Allow the students to generate several answers by discussion. Write students' answers on the transparency.

  Q: We'll write down all the words that we have found that mean the opposite of fast.

  A: *Slow* is the most probable answer.

GUIDED PRACTICE

EXERCISES: **G-70** through **G-73**

- When students have had sufficient time to complete these exercises, check by discussion to determine whether they have answered correctly. Discuss why students' answers are correct antonyms.

INDEPENDENT PRACTICE

- Assign exercises **G-74** through **G-96**. (You may wish to make this more than one lesson.)

THINKING ABOUT THINKING

Q: What did you pay attention to in order to find a word with the opposite meaning?

1. I defined the given word.
2. I thought about what the opposite meaning was.
3. I thought of words that had the opposite meaning.
4. I substituted the other word for the given one to check whether the words had the opposite meaning.

PERSONAL APPLICATION

Q: When do you need to think of a word that means the opposite of a certain word?

A: Examples include solving crossword puzzles, giving directions for going somewhere or making something, following reverse directions (e.g., turning left when you go to your friend's house and turning right when you go back home), reassembling models or appliances which have been taken apart.

EXERCISES G-97 to G-105

SIMILARITIES—SELECT (pictures and words)

ANSWERS G-97 through G-105 — Student book pages 189–90
Guided Practice: G-97 b; **G-98** c
Independent Practice: G-99 a; **G-100** c; **G-101** c; **G-102** c; **G-103** a; **G-104** b; **G-105** a

DETAILED SOLUTIONS

Note: The key definitions are underlined in all answers for this section. The four words used in each exercise are italicized.

G-97 b: A *shovel* is a tool used for digging. Look for a word that names another tool used for digging. A *hoe* is a tool used for digging. A *hammer* is a tool used for pounding or pulling nails, and a *paint brush* is a tool used for painting. The two words that are most alike are shovel and hoe.

G-98 c: An *automobile* is a motor vehicle which carries a few people and is driven along roads. A *jeep* fits this definition. An *airplane* is not driven on roads. A *bicycle* is not a motor vehicle.

G-99 a: A *garage* is a place used for keeping and protecting cars. Look for another protective shelter. A *barn* is a protective shelter for animals and farm equipment. Protection is not the main purpose of either a *school* or a *store.*

G-100 c: A *grocery store* is a store that sells food and household supplies. Look for another kind of store that sells the same things. A *supermarket* is a large food store where shoppers serve themselves. An *apartment* is a building divided into several residences. A *hospital* is a place used to care for the sick and injured.

G-101 c: *Pine* is the name of a kind of tree. Look for a word that names another kind of tree. *Oak* names a kind of tree. A *carrot* is a kind of vegetable, and a *daisy* is a kind of flower. Oak is most similar to pine.

G-102 c: A *chicken* is a bird raised for its food value. Look for another edible bird. A *turkey* is a bird which is frequently eaten. Neither *dogs* nor *cats* are birds, nor are they usually raised for food.

G-103 a: *Cheese* is a food made from milk. Look for another food made from milk. *Butter* is another food made from milk. *Eggs* are a food produced by chickens, and *milk* is a food produced by cows and used to make cheese and butter.

G-104 b: A *snake* is a reptile usually found on land. Look for another reptile that lives on the land. A *lizard* is another reptile that lives on the land. *Fish* and *sharks* live in the water, and neither of them is a reptile.

G-105 a: A *pea* is a small green vegetable that grows in a pod on a vine. Look for another small vegetable that grows in a pod on a vine. A *bean* is also a small vegetable that grows in a pod on a vine. *Carrots* and *potatoes* are larger root vegetables that grow underground.

LESSON PREPARATION

OBJECTIVE AND MATERIALS

OBJECTIVE: Using picture clues, students will select a word which has a meaning similar to a given word.

MATERIALS: Transparency of student book page 189 • washable transparency marker

CURRICULUM APPLICATIONS

Language Arts: Using context clues and illustrations in reading comprehension exercises, using synonyms, using knowledge of word parts to determine meaning of compound words

Mathematics: Recognizing and using equivalent values of money, time, or measurement; understanding directions for solving mathematics problems

Science: Following directions in performing basic experiments, inferring meanings of unfamiliar words by using definition or synonym clues

Social Studies: Paraphrasing or summarizing key concepts, building vocabulary using definition or synonym clues from text, identifying parallel or similar functions of different governmental levels

Enrichment Areas: Recognizing foreign language vocabulary by similarity to English (or native tongue) words

TEACHING SUGGESTIONS

Encourage students to discuss reasons for choosing or rejecting an answer. Remember the language that students use to describe the items, and use the same words to remind students of the key characteristics of items in this lesson and subsequent ones. If appropriate for the students' verbal proficiency, use the term *synonym* when discussing words with similar meanings. Use the following cues to model for students a method for determining the correct answer:

1. Recognize/pronounce word.
2. Define word.

3. Select similar word (answer).
4. Define similar word (confirmation).
5. Eliminate detractors (check answer).

MODEL LESSON

LESSON

Introduction

Q: You have selected or drawn figures which had similar patterns or shapes.

Explaining the Objective to the Students

Q: In this lesson, you will select words with similar meanings.

Class Activity

- Project the transparency of page 189.
  Q: You will look for a word which means most nearly the same as the given word. When you have selected the word that seems most like the given word, check the other word choices to see how they are different. In the example on page 186, the word you are given is *sneaker.*

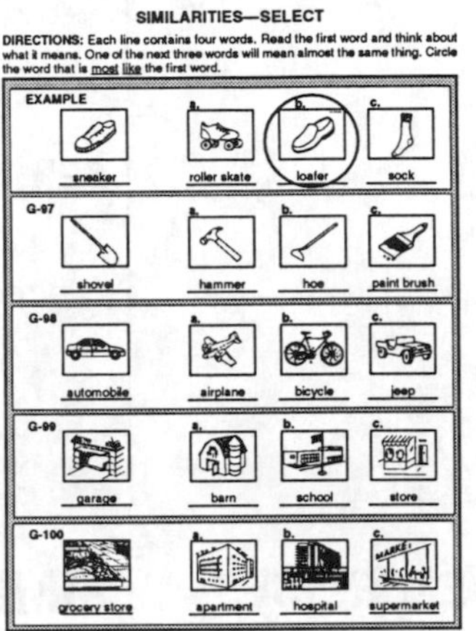

- Point to the word and picture.
  Q: Three other words,...
- Point as you say the words.
  Q: ...*roller skate*, *loafer*, and *sock,* follow the given word. One of these words is similar in meaning to *sneaker.* Before we can find a word that is similar, we must make sure that we can define and identify the characteristics of a sneaker.
- Ask students to brainstorm as many characteristics of a sneaker as they can identify. List the characteristics on the transparency or on the board.
  Q: Next, we must choose a word which has a similar meaning.
- Point to the word and picture of loafer.
  Q: *Loafer* seems to be the word closest in meaning to *sneaker.* A loafer is a type of shoe and a sneaker is a kind of sports shoe. Before we finally decide, though, we must check the other two words to be sure that *loafer* is really the best choice as a word most like *sneaker.*
- Point to each word and picture as you talk about it.
  Q: A roller skate also covers the feet but is used for skating rather than for protection. A skate would not be very useful for everyday wear. A sock also covers one's foot but is a cloth cover worn inside shoes. A sock does not offer the protection given by a shoe. We have checked out all of the words, and *loafer* is the word closest in meaning to *sneaker.* We'll circle *loafer* to indicate our choice.

GUIDED PRACTICE

EXERCISES: **G-97, G-98**

- When students have had sufficient time to complete these exercises, check by discussion to determine whether they have answered correctly. Alternate answers are acceptable if the student can explain the similarity. Discuss why the incorrect answers were eliminated; do not accept an answer without an explanation.

INDEPENDENT PRACTICE

- Assign exercises **G-99** through **G-105**.

THINKING ABOUT THINKING

Q: What did you pay attention to when you looked for a word with a similar meaning?

1. I thought about what the first word meant.
2. I found another word that meant almost the same thing.
3. I substituted the second word for the first one to check whether the words had almost the same meaning.

PERSONAL APPLICATION

Q: When might you need to choose or recognize another word that means the same thing as a given word?

A: Examples include following written or oral directions, understanding what others are saying to you, doing crossword puzzles or word games.

EXERCISES G-106 to G-141

SIMILARITIES—SELECT (words)

ANSWERS G-106 through G-141 — Student book pages 191–94
Guided Practice: G-106 a; **G-107** b; **G-108** b
Independent Practice: G-109 a; **G-110** a; **G-111** a; **G-112** c; **G-113** b; **G-114** b; **G-115** c; **G-116** a; **G-117** b; **G-118** a; **G-119** a; **G-120** a; **G-121** c; **G-122** b; **G-123** c; **G-124** b; **G-125** a; **G-126** c; **G-127** c **G-128** b; **G-129** a; **G-130** a; **G-131** b; **G-132** a; **G-133** a; **G-134** c; **G-135** c; **G-136** c; **G-137** a; **G-138** a; **G-139** b; **G-140** c; **G-141** b

DETAILED SOLUTIONS

Note: The key definitions are underlined in all answers for this section. The four words used in each exercise are italicized.

G-106 a: *Listen* means to use one's ears to detect sound. Look for another word that means to detect sound. *Hear* has the same meaning as listen. Both *speak* and *talk* mean the opposite of *listen.*

G-107 b: *Talk* means to use one's ability to speak. Look for a word that means to speak. *Say* means to speak. *Read* means to understand something written or printed. *Sing* means to utter words or sounds musically.

G-108 b: *Draw* means to make a picture with a pen or pencil. Look for a word that also means to make a picture. *Color* means to draw or fill in a picture with color using crayons or colored pencils. *Around* is a word that means to go in a circle. *Work* is the opposite of play.

G-109 a: *Study* describes the mental effort of understanding. Look for a word that means understanding. *Learn* means to get understanding. *Lesson* is that which is studied. *Talk* means to use one's ability to speak, which may be an aid to study.

G-110 a: *Act* means to perform. *Do* also means to perform. *Relax* and *rest* describe lack of action and are opposites of act.

G-111 a: *Learn* means to get understanding. *Discover* means to gain knowledge of something unknown. *Reply* means to answer. *Show* means to exhibit or explain.

G-112 c: *Ask* means to try to get an answer. To *question* means to ask. *About* is a word associated with ask as in "ask about." *Receive* means to get.

G-113 b: *Answer* means to make a response or reply. *Reply* means to respond or answer. *Hear* means to perceive by the ear. *Question* is the statement which requires an answer; question is the opposite of answer.

G-114 b: *Solve* means to find the answer. *Figure out* is a slang term for solve. *Ask* means to try to get an answer. *Problem* is the statement which is to be solved.

G-115 c: A *newspaper* contains news and is published daily or weekly. Look for another news publication that comes out regularly. A *magazine* contains news and is published either weekly or monthly. A *book* is a publication that tells a story or gives information. A *dictionary* is a publication which defines words.

G-116 a: *Cook* means to prepare food by heating. *Bake* has the same meaning. *Dinner* and *soup* are things that are cooked.

G-117 b: *Fix* means to put into working condition. *Repair* has the same meaning. *Build* means to construct from parts or materials. *Wreck* means the opposite of fix.

G-118 a: A *door* is a passageway for people through a wall. A *gate* is a passageway for people through a fence. A *roof* is the top covering of a building. A *window* is a passageway for light into a room. (Note that door and window may be seen as the synonyms since both are openings or passageways through a wall.)

G-119 a: A *rug* is a decorative woven floor covering. A *carpet* has the same definition. A *floor* is what the rug covers. A *sweeper* is a device used to clean rugs.

G-120 a: *Curtains* are thin window coverings. *Drapes* are thick window coverings. A *door* is a passageway through a wall. A *window* is a passageway for light into a room.

G-121 c: *Clean* means to remove soil or stains. *Wash* has a similar meaning. *Clothes* are wearing apparel (which must be cleaned). A *house* is a residence (which must be cleaned).

G-122 b: A *meal* is a variety of food. *Dinner* is one of the three daily meals. *Cook* means to prepare food by heating. *Vegetable* is one of the parts of some meals.

G-123 c: A *broom* is a hand-operated sweeper used to clean dry surfaces. A *sweeper* is an appliance used to clean dry surfaces. A *sponge* is used with water to clean wet surfaces. *Handle* is part of a broom.

G-124 b: *Pay* means to give out money. Look for a word with a similar meaning. *Spend* means to give out money. *Owe* means to have a debt (the need to give money, not yet given). *Take* means to seize or grab something.

G-125 a: *Run* means to move rapidly on foot. Look for a word meaning to move quickly on foot. *Jog* means to move moderately fast on foot and is most like run in meaning. *Walk* means to move relatively slowly on foot. *Stop* means to not move at all.

G-126 c: *Look* means to use one's sense of sight. Find another word pertaining to sight. *Watch* means to observe for a long time. *Hear* means to use one's ears (sense of hearing). *Talk* means to use one's ability to speak, which is not one of the five senses.

G-127 c: *Stay* means to continue in the same place. Look for a word meaning to continue one's position. *Remain* means to be left behind or not to leave. *Go* and *leave* mean to change one's location.

G-128 b: Call means to request or speak in a loud voice. Look for a word that means to say loudly. *Shout* means to talk in a loud voice. *Say* and *tell* mean to speak and neither need be in a loud voice. (Call and shout are higher degrees of saying or telling.)

G-129 a: *Take* means to move from one place to another. Look for a word meaning to move from one place to another. *Carry* has that meaning. *Give* means to release something to someone else. *Keep* means to hold on to something and not let go.

G-130 a: *Throw* means to hurl through the air. Look for another word meaning to propel something through the air. *Toss* means to throw lightly. *Catch* means the opposite of throw. *Hit* means to strike something.

G-131 b: *Build* means to construct from parts or materials. Look for a word which means to construct. *Make* means to construct. *Repair* and *fix* mean to put a device in working order.

G-132 a: *Hide* means to keep out of sight. Look for a word that has a similar meaning. *Cover* means to place something over an object to protect it or keep it from being seen. *Seek* means to look for an object. *Show* means to make something available for viewing.

G-133 a: *Sound* is a vibration that can be detected by the ear. *Noise* is also detected by the ear and is a form of sound. *Quiet* and *silence* represent the absence of sound.

G-134 c: *Heat* is energy that causes the temperature of objects to increase. Look for a word related to increase in temperature. *Warmth* indicates the presence of heat. *Light* is energy used for illumination (for seeing). A *stove* is a device used to warm food, but it is not always hot. (Note: Alternate synonyms might be heat and light if the student sees them both as forms of energy.)

G-135 c: The *sun* is the star at the center of our solar system. Since the sun is a star, *star* is most like sun. *Beam* and *shine* are words associated with sunlight (sunbeam and sunshine).

G-136 c: A *rock* is a large piece of hard mineral matter found in or on the earth. Look for a word meaning a hard piece of earth. A *stone* has the same definition, but is usually smaller and lighter than a rock. A *mountain* is a large formation of rock and soil that often extends a mile or more above the earth's

surface. An *ocean* is a large body of water covering a major portion of the earth's surface.
G-137 a: *Fog* is a thick mist near the earth's surface that is made up of a large mass of tiny drops of water. Look for another kind of mist. A *cloud* is a mass of fine drops of water or ice floating in the air above the earth. *Ice* is water frozen solid by cold. *Snow* is soft white flakes formed from drops of frozen water that fall to the earth.
G-138 a: A *beak* is the hard part of a bird's mouth. *Bill* is another name for beak. A *claw* is the handlike part of a bird. *Wing* is the armlike part of a bird.
G-139 b: A *baby* is a very young human being. Look for a word meaning young. A *child* is a young human being. (A baby is a very young child.) *Brothers* or *sisters* are children of your parents, but they may or may not be young. Some children have adult brothers or sisters.
G-140 c: The *earth* is the planet upon which we live. *World* is another name for the earth. *Continents* and *countries* are parts of the earth. Continents are made up of countries.
G-141 b: A *map* is a drawing of part of the surface of the earth. *Drawing* is most like map. A *country* can be represented on a map. A *photograph* of the surface of the earth is not a map because the names of geographic features are not printed on it.

LESSON PREPARATION

OBJECTIVE AND MATERIALS

OBJECTIVE: Students will select a word which has a meaning similar to a given word.
MATERIALS: Transparency of student book page 191 • washable transparency marker

CURRICULUM APPLICATIONS

Language Arts: Doing synonym and vocabulary enrichment exercises, avoiding overused words and trite expressions in compositions, using knowledge of word parts to decode compound words, using knowledge of prefixes and suffixes to build or comprehend new words, recognizing denotative and connotative meanings
Mathematics: Recognizing key words that indicate proper function or order in word problems; recognizing face and place value of numbers; recognizing and using equivalent values of money, time, or measurement
Science: Following directions in performing basic experiments, gaining meanings of unfamiliar words by using definitions or synonym clues
Social Studies: Using definition or synonym clues from text, identifying similar functions of different governmental levels
Enrichment Areas: Increasing vocabulary from pleasure reading (clues from context), recognizing foreign language vocabulary by similarity to English (or native tongue) words, following directions in creating a work of art or in performing a dance or musical work

TEACHING SUGGESTIONS

Encourage students to discuss reasons for choosing or rejecting an answer. Remember the language that students use to describe the items and use the same words to remind students of the key characteristics of items in this lesson and subsequent ones. If appropriate to the students' verbal proficiency, use the term *synonym* when discussing words with similar meanings.

This lesson may require more than one session. You may wish to assign these exercises a page at a time since each page is devoted to a single topic. The following cues encourage the discussion procedure demonstrated in the explanation and answers:

1. Recognize/ pronounce word.
2. Define word.
3. Select similar word (answer).
4. Define similar word (confirm answer).
5. Eliminate detractors (check answer).

MODEL LESSON

LESSON

Introduction

Q: You have selected or drawn figures which had similar patterns or shapes.

Explaining the Objective to the Students

Q: In this lesson, you will read a word and then select a word that has a similar meaning.

Class Activity

- Project transparency of page 191.
  Q: In the first item on page 191, the word you are given is *listen*. Look for a word which means most nearly the same as *listen*. When you have selected the word that seems most like *listen,* check the other word choices to be sure that they are different.

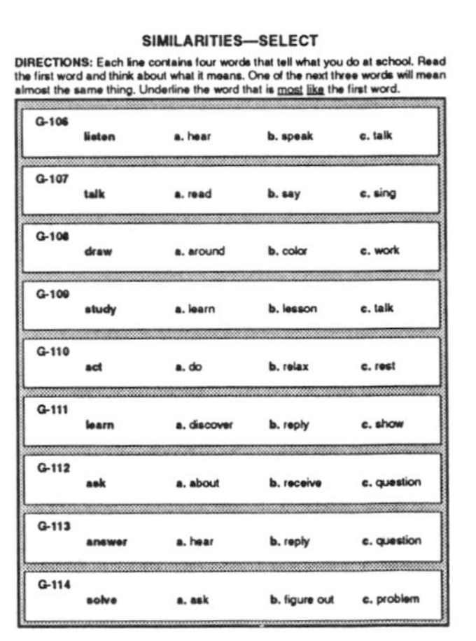

SIMILARITIES—SELECT

DIRECTIONS: Each line contains four words that tell what you do at school. Read the first word and think about what it means. One of the next three words will mean almost the same thing. Underline the word that is most like the first word.

G-106	listen	a. hear	b. speak	c. talk
G-107	talk	a. read	b. say	c. sing
G-108	draw	a. around	b. color	c. work
G-109	study	a. learn	b. lesson	c. talk
G-110	act	a. do	b. relax	c. rest
G-111	learn	a. discover	b. reply	c. show
G-112	ask	a. about	b. receive	c. question
G-113	answer	a. hear	b. reply	c. question
G-114	solve	a. ask	b. figure out	c. problem

- Point to the word.
  Q: Three other words,...
- Point as you say the words.
  Q: ...*hear*, *speak*, and *talk*, follow the given word. Select the word that is most like the word *listen* in meaning. Before we can find a word that is similar, we must make sure that we can give a definition of and identify the characteristics of the first word.
- Have students brainstorm as many characteristics and definitions of *listen* as they can.
  Q: Next, we must choose a word which has a similar meaning. *Hear* has the same meaning as *listen*, but *speak* and *talk* mean the opposite of *listen*. We have checked out all of the words, and *hear* is the word closest in meaning to *listen*. We'll underline the word *listen* to indicate our choice.

GUIDED PRACTICE

EXERCISES: **G-106** through **G-108**

- When students have had sufficient time to complete these exercises, check by discussion to determine whether they have answered correctly. Alternate answers are acceptable if the student can explain the similarity. Discuss why the incorrect answers were eliminated; do not accept an answer without an explanation.

INDEPENDENT PRACTICE

- Assign exercises **G-109** through **G-141**.

THINKING ABOUT THINKING

Q: What did you pay attention to in order to find a word with a similar meaning?

1. I thought about what the first word meant.
2. I found another word that meant almost the same thing.
3. I substituted the second word for the first one to check whether the words had almost the same meaning.

PERSONAL APPLICATION

Q: When might you need to choose or recognize another word that means the same thing as a given word?

A: Examples include understanding and following written or oral directions, playing crossword puzzles or word games.

EXERCISES G-142 to G-168

SIMILARITIES—SUPPLY

ANSWERS G-142 through G-168 — Student book pages 195–97

Note: Other acceptable synonyms may be used to define the given word. Many acceptable answers that are not mentioned may come up in discussion.

Guided Practice: G-142 not long, not tall, brief, rude, sudden; **G-143** heavy, tall; **G-144** helpful, loving; **G-145** massive, fat, stout (weight); firm, serious, weighty (seriousness); forceful, excessive, severe (force) **G-146** joyful, glad, pleased, satisfied, contented; **G-147** worthy, helpful, proper, pleasant

Independent Practice: G-148 pleasant, sugary, good, fresh; **G-149** attractive, good-looking, handsome, pretty; **G-150** secure, unhurt, healthy, unharmed; **G-151** feel, handle, stroke, pet, rub, pat; **G-152** behold, look, note, notice, observe, peep, peer, view; **G-153** catch on, find out, learn, listen; **G-154** converse, lecture, talk, utter, verbalize; **G-155** employment, job, occupation, chore, drudgery, effort, exertion, grind, labor; **G-156** consider, examine, inspect, ponder, think about, think over; **G-157** assemble, construct, erect, make, manufacture, put up; **G-158** mend, overhaul, rebuild, recondition, repair; **G-159** demolish, dismantle, dissolve, pull down, ruin, tear down, wreck; **G-160** go, exit, quit, depart; **G-161** aloft, higher, over, overhead; **G-162** admit, come in, enlist, enroll, go in, join; **G-163** beneath, under, underneath; **G-164** carry on, carry through, endure, last, persist, remain, stay; **G-165** close up, block, clog, close, plug, cease, discontinue, halt, quit; **G-166** desire, elect, pick, select, take, want; **G-167** catch on, find out, hear, learn; **G-168** adhere, anchor, attach, fix, join, link, unite

LESSON PREPARATION

OBJECTIVE AND MATERIALS

OBJECTIVE: Students will supply a word which has a meaning similar to a given word.

MATERIALS: Transparency of student workbook page 195 • washable transparency marker

CURRICULUM APPLICATIONS

Language Arts: Paraphrasing written or spoken sentences or paragraphs, avoiding overused words and trite expressions in compositions, using knowledge of word parts to determine meaning of compound words, using knowledge of prefixes and suffixes to build or comprehend new words

Mathematics: Writing solutions to word problems; reading and understanding directions for solving mathematics problems; recognizing and using equivalent values of money, time, or measurement

Science: Writing directions and solutions for basic experiments, inferring meanings of unfamiliar words by using definition or synonym clues

Social Studies: Paraphrasing key concepts, identifying similar functions of different governmental levels

Enrichment Areas: Describing similarities in forms of art and music, increasing vocabulary from pleasure reading (clues from context), recognizing foreign language vocabulary by similarity to English (or native tongue) words

TEACHING SUGGESTIONS

Encourage students to discuss and define their synonyms. Remember the language that students use to describe their choices and use the same words to remind students of the key definitions of items in the lesson.

Discussion should include as many synonyms as possible, especially when words may have more than one meaning. The following cues will encourage discussion procedures like those demonstrated in the explanation and answers:

1. Recognize/pronounce word.
2. Define word.
3. Supply similar word (answer).
4. Define similar word (confirm answer).
5. Recognize/use other synonyms.

MODEL LESSON

LESSON

Introduction

Q: You have selected a word which had a meaning similar to a given word.

Explaining the Objective to the Students

Q: In this lesson, you will read a word and then think of a word that has a similar meaning.

Class Activity

- Project exercise **G-142**, covering remaining exercises.

  Q: You will search your memory for a word which means almost the same as the given word. Look at page 195, exercise **G-142**. Before you give an answer, think of the definition of *short*. The word *short* can mean any of the following: (1) not long, (2) not tall, (3) brief or rude, (4) having less than enough, or (5) sudden.

SIMILARITIES—SUPPLY

DIRECTIONS: Each line contains a word used to describe something. Read the word and think about what it means. Think of a word that means almost the same as the word and write it down. Write as many similar words as you can think of.

G-142 short ______

G-143 big ______

G-144 friendly ______

G-145 heavy ______

G-146 happy ______

G-147 good ______

G-148 sweet ______

G-149 beautiful ______

G-150 safe ______

- Before a word is eliminated as a possible synonym, the students should state their definitions of the given word.

  Q: Now, let's think of other words that mean not long or not tall.

- Discussion should focus on generating synonyms. Write them on the transparency.

GUIDED PRACTICE

EXERCISES: **G-142** through **G-147**

- When students have had sufficient time to complete these exercises, check answers by discussion to determine whether they have answered correctly.

INDEPENDENT PRACTICE

- Assign exercises **G-148** through **G-168**. You may wish to make this two lessons.

THINKING ABOUT THINKING

Q: What did you pay attention to when you thought of a word similar to a given word?

1. I thought about what the first word meant.
2. I found another word that meant almost the same thing.
3. I substituted the second word for the first one to check whether the words had almost the same meaning.

PERSONAL APPLICATION

Q: When might it be necessary to provide a word that means the same thing as a given word?

A: Examples include giving clear directions (either written or oral), understanding what others are saying to you, crossword puzzles or word games, test-taking skills.

EXERCISES G-169 to G-175

HOW ALIKE?—SELECT

ANSWERS G-169 through G-175 — Student book pages 198–99

Guided Practice: G-169 b, c, d (Sentence *a* is not true because a candle is not electric.) **G-170** a, c, d (Sentence *b* is not true of both because an ear cannot close itself.)

Independent Practice: G-171 b, d; **G-172** c, d; **G-173** a, b, c; **G-174** a, c; **G-175** a, b

LESSON PREPARATION

OBJECTIVE AND MATERIALS

OBJECTIVE: Students will select the sentences which are true of two words.
MATERIALS: Transparency of student book page 198 • washable transparency marker

CURRICULUM APPLICATIONS

Language Arts: Organizing and writing compare and contrast statements or paragraphs, recognizing denotative and connotative words or phrases
Mathematics: Evaluating geometric shapes for type and congruence, interpreting word problems, interpreting different forms of statistical presentations (e.g., charts, graphs, pictures, schedules)
Science: Identifying classes of plants or animals by similar characteristics
Social Studies: Expressing similarities and differences between historical events, eras, people, or artifacts
Enrichment Areas: Contrasting and/or comparing different works of art, music, or dance

TEACHING SUGGESTIONS

Encourage students to discuss and define their synonyms. Remember the language that students use to describe their choices and use the same words to remind students of the key characteristics of items in the lesson.

MODEL LESSON

LESSON

Introduction

Q: You have selected a word that had a meaning similar to a given word.

Explaining the Objective to the Students

Q: In this lesson, you will select the sentences that are true of two words.

Class Activity

- Project the transparency of the example from page 198, covering the right side so students cannot see the sentences they are to evaluate.
  Q: In these exercises, you are given two items, such as a bus and a truck. How are a bus and a truck alike?

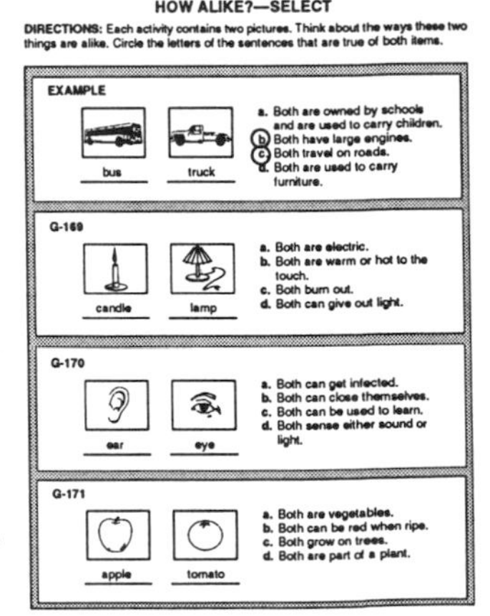
HOW ALIKE?—SELECT

DIRECTIONS: Each activity contains two pictures. Think about the ways these two things are alike. Circle the letters of the sentences that are true of both items.

EXAMPLE
bus truck
a. Both are owned by schools and are used to carry children.
(b) Both have large engines.
(c) Both travel on roads.
d. Both are used to carry furniture.

G-169
candle lamp
a. Both are electric.
b. Both are warm or hot to the touch.
c. Both burn out.
d. Both can give out light.

G-170
ear eye
a. Both can get infected.
b. Both can close themselves.
c. Both can be used to learn.
d. Both sense either sound or light.

G-171
apple tomato
a. Both are vegetables.
b. Both can be red when ripe.
c. Both grow on trees.
d. Both are part of a plant.

- Encourage discussion then expose responses *a* through *d*.
  Q: Is sentence *a* true of both a bus and a truck?
  A: No, both can be owned by schools, but a truck is not usually used to carry children.

  Q: Is sentence *b* true of both a bus and a truck?
  A: Yes, both have large engines.

  Q: Is sentence *c* true of both a bus and a truck?
  A: Yes, both travel on roads.

Q: Is sentence *d* true of both a bus and a truck?
A: No, a bus is not used to carry furniture.

Q: Sentences *b* and *c* are true and have been circled.

GUIDED PRACTICE

EXERCISES: **G-169**, **G-170**

- When students have had sufficient time to complete these exercises, check answers by discussion to determine whether they have answered correctly. Students should follow the above format when discussing answers and should always give their underlying reasons.

INDEPENDENT PRACTICE

- Assign exercises **G-171** through **G-175**.

THINKING ABOUT THINKING

Q: What did you pay attention to when you decided whether the same statement was true of two different things?

1. I thought about the meaning of each word.
2. I read each given statement to see if it was true of each word.

PERSONAL APPLICATION

Q: When in your everyday life might you need to decide whether the same statement is true of two different things?
A: Examples include making value judgements (good/bad, right/wrong) in different situations, playing word games, taking tests (especially true-false or multiple-choice objective tests), giving or following directions.

EXERCISES G-176 to G-187

HOW ALIKE AND HOW DIFFERENT?

ANSWERS G-176 through G-187 — Student book pages 200–203

Guided Practice: G-176 ALIKE: Both have wheels, can be turned to go different directions (steered), are used for transportation, and are commonly used vehicles. DIFFERENT: Cars have engines, bicycles don't. Cars can carry more passengers than an ordinary bicycle. A bicycle is lighter and easier to store, is less expensive, and offers exercise. Cars offer more safety and protection from the weather. Bicycles are driven by both children and adults, but only adults drive cars; you have to take a test and have a license to drive a car. **G-177** ALIKE: Both float, can carry people or things, and were early forms of transportation that are still used today. Both can be used for pleasure or business but are slower than cars, trains, or airplanes. DIFFERENT: Ships are larger and more expensive than boats. Families can own a boat, but ships are usually owned by companies. A person can row a boat, but a ship is too big to be rowed. **G-178** ALIKE: Both are called in emergencies. Both have sirens and go to fires. DIFFERENT: Police cars are used to chase law breakers, and they carry weapons. Fire trucks carry fire fighting equipment.
Independent Practice: G-179 ALIKE: Both provide information and

entertainment, can deliver fact or fiction to the user, are common in most households, can be purchased or "borrowed" (books can be borrowed from a library or a friend, televisions can be rented from a company or sometimes borrowed from a friend), and involve the sense of sight. DIFFERENT: Television is electronic, but a book is printed. You "read" a book, but "see" and "hear" a television. Books are an older form of communication. It takes skill to read a book, but no skill is required to watch television. Books do not have commercials. Television programs (especially news reports) are usually more "timely" than books. You can read a book anytime and anywhere, but usually you can only watch a television program once (unless you wait for reruns or have a VCR!). Books don't break down or require repairmen! **G-180** ALIKE: Both are types of balls, are used in team sports, are used by both professionals and amateurs, and can be thrown and caught. DIFFERENT: Baseballs are round and not hollow in the middle; footballs are oblong and hollow in the middle. You can add air to a football but not to a baseball. A baseball doesn't go flat. Footballs can be kicked; baseballs can be hit with a bat. **G-181** ALIKE: Both care for people who are usually younger than they, teach skills, and are responsible for others. DIFFERENT: Mothers are women, but a teacher can be a man or a woman. Teachers usually have different students (children) every year. Mothers and children have a family relationship; teachers and students have a professional relationship. Teachers are paid a salary and are hired by others to teach; mothers are not (usually) paid a salary nor can they normally be hired and fired! **G-182** ALIKE: Both are long and thin, extend from a "trunk," can grow longer and thicker, and contain smaller, flexible growths at their ends. DIFFERENT: An arm is jointed and can be moved in many directions; a branch moves only when the wind blows or someone pulls it. Arms can be used for lifting objects. Arms belong to people (animals) and branches belong to trees (plants). **G-183** ALIKE: Both are common covered containers having the same name. Both can be made of metal. Some people claim that both contain "junk"! DIFFERENT: A trash can holds waste; a beverage can becomes waste. A trash can is used many times and can be opened and closed; a beverage can is used once (although it can be recycled) and cannot be closed once it is opened. Trash cans can be made of plastic and beverage cans are not. Trash cans can contain both solids and liquids; beverage cans contain only liquid. Trash cans are larger and more expensive than beverage cans. Trash cans are purchased empty; beverage cans are purchased full. You can drink out of a beverage can. **G-184** ALIKE: Both are used to support something, are longer and thinner than what they support, can get scratched or damaged, and have the same name. DIFFERENT: Human legs bend and move (walk); table legs do not normally bend or move. Human legs are alive and have feeling; table legs are not alive and do not have feeling. **G-185** ALIKE: Both are living animals (reptiles). Both lay soft-shelled eggs and have dry, scaly skin. DIFFERENT: A snake does not have legs and moves by slithering; a lizard has legs. **G-186** ALIKE: Both are living

four-footed, hoofed animals of the mammal group, are sometimes used as "characters" in children's literature (Black Beauty, Bambi), and have hides that can be used to make articles of clothing (deerskin or horsehide gloves or shoes). DIFFERENT: Horses are usually domesticated; deer are usually wild. Deer have horns, horses do not. Horses are larger and stronger than deer. Some types of deer are associated with a holiday.
G-187 ALIKE: Both are large living sea animals. Both have fins and tails. DIFFERENT: Whales are larger than sharks. Whales are mammals (they produce milk for their young); sharks are fish. Whales have bones; sharks have cartilage.

LESSON PREPARATION

OBJECTIVE AND MATERIALS

OBJECTIVE: Students will write sentences that tell how two words are alike and how they are different.
MATERIALS: Transparency of student book page 200 • washable transparency marker

CURRICULUM APPLICATIONS

Language Arts: Writing compare and contrast statements, paragraphs, or papers; recognizing denotative and connotative words
Mathematics: Identifying geometric shapes for type and congruence, interpreting word problems, interpreting different forms of statistical presentations (e.g., charts, graphs, pictures, and schedules relating to the same statistics)
Science: Identifying classes of plants or animals by similar characteristics
Social Studies: Identifying similarities and differences between historical events, eras, people, or artifacts
Enrichment Areas: Contrasting and/or comparing different works of art, music, or dance

TEACHING SUGGESTIONS

Encourage students to discuss their answers. Remember the language that students use to describe their choices and use the same words to remind students of the key characteristics of items in the lesson. The brief answers provided should be modified and tailored to the vocabulary and level of your students.

MODEL LESSON

LESSON

Introduction

Q: You have selected sentences which were true for two different words.

Explaining the Objective to the Students

Q: In this lesson, you will write sentences that tell how two words are alike and how they are different.

Class Activity

- Project the transparency of page 200, exposing only exercise **G-176**.

HOW ALIKE AND HOW DIFFERENT?

G-176 bicycle car How alike? How different?

G-177 boat ship How alike? How different?

G-178 fire truck police car How alike? How different?

Q: Here you have two words: bicycle and car. How are a bicycle and a car alike?

A: Both have wheels, can be turned to go different directions (steered), are used for transportation, and are commonly used vehicles.

- Allow time for discussion, writing answers on the transparency.

Q: How are a bicycle and a car different?

A: Cars have engines, bicycles don't; cars can carry more (people or cargo) than ordinary bicycles can; bicycles are driven by both children and adults, but only adults drive cars; a bicycle is lighter and easier to store, is less expensive, and offers exercise; cars offer more safety and protection from the weather; you have to take a test and have a license to drive a car.

GUIDED PRACTICE

EXERCISES: **G-176** through **G-178**

- When students have had sufficient time to complete these exercises, check answers by discussion to determine whether they have answered correctly. Students should be asked to explain their answers.

INDEPENDENT PRACTICE

- Assign exercises **G-179** through **G-187**.

THINKING ABOUT THINKING

Q: What did you think about when you decided how two words were alike and how they were different?

1. I thought about the meaning of each word.
2. I listed the things that were true of each word.
3. I listed the things that were different about each word.

PERSONAL APPLICATION

Q: When might you need to decide how two things are alike and how they are different?

A: Examples include making value judgements (good/bad, right/wrong) in different situations, playing word games, taking tests (either essay or objective tests), making purchases.

EXERCISES G-188 to G-191

COMPARE AND CONTRAST—GRAPHIC ORGANIZER

ANSWERS G-188 through G-191 — Student book pages 204–8

Guided Practice: G-188 ALIKE: Both are kinds of words that appear in sentences; both may change spelling to describe more than one. DIFFERENT: Nouns name people, places, or things; verbs name actions. Nouns don't tell when. Verbs tell whether something happens in the past, present, or future.

Independent Practice: G-189 ALIKE: Both are kinds of sentences and are whole thoughts. Both contain a subject and a verb and end with a punctuation mark. Both are used to exchange information. DIFFERENT:

Statements give information; questions ask for information. Statements end with a period. Questions end with a question mark. The whole verb usually follows the subject in statements. The subject usually comes between parts of the verb in questions. **G-190** ALIKE: Both can be found on maps (they are geographic areas). Both have governments which pass laws. DIFFERENT: The U.S.A. is large; it is made up of 50 states. (Your state) is part of the U.S.A., so it is smaller. The president is the chief U.S. official; the governor is the chief state official. Federal lawmakers meet in Washington, D.C.; state lawmakers meet at the state capital (your state capital). Federal lawmakers meet in the U.S. Capitol building; state lawmakers meet in the state capitol building. **G-191** ALIKE: Both have weight, both take up space, and both are many sizes and shapes. DIFFERENT: Living things move by themselves; nonliving things do not. Living things breathe and need nourishment; nonliving things do not. Living things can reproduce themselves; nonliving things cannot.

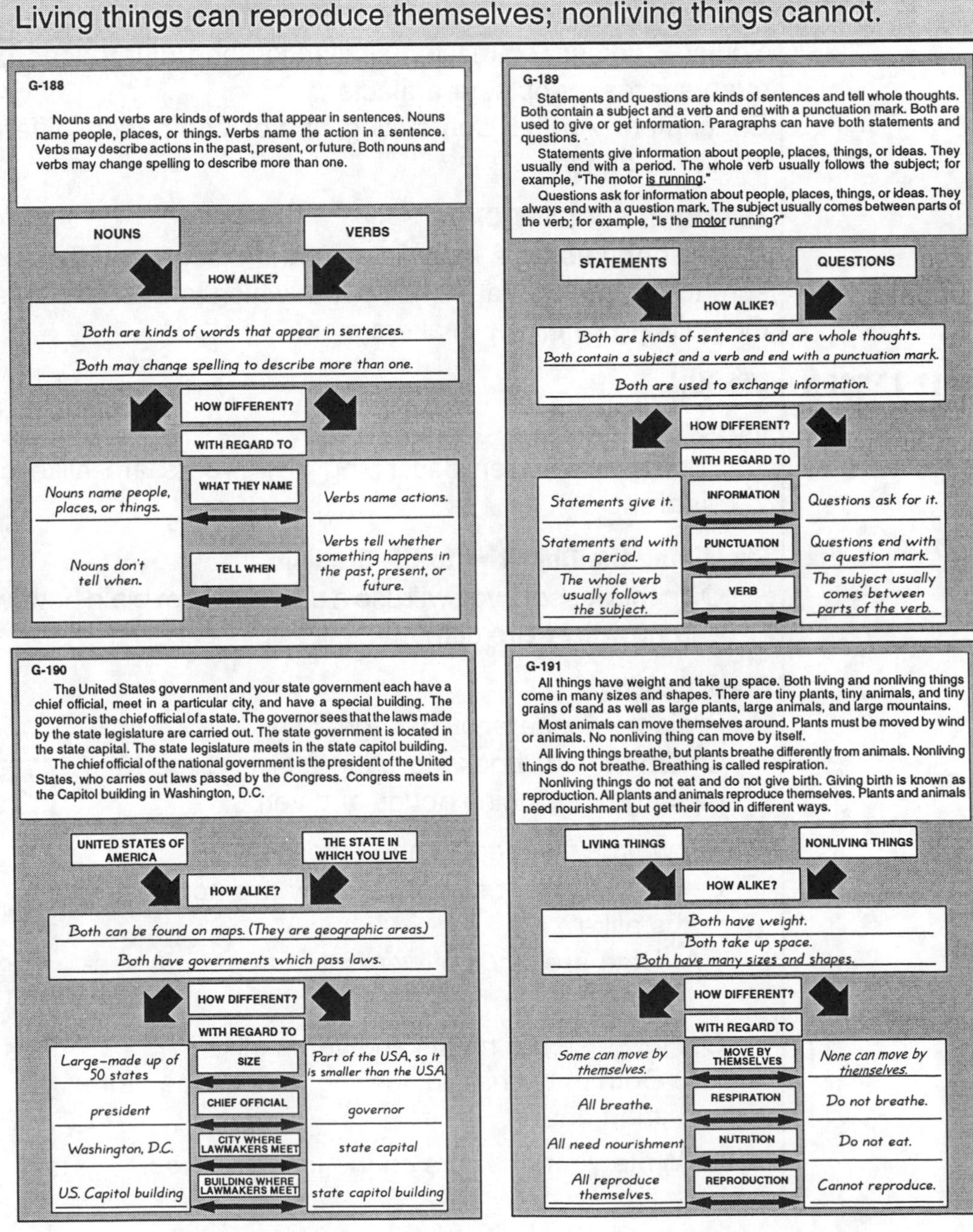

G-188

Nouns and verbs are kinds of words that appear in sentences. Nouns name people, places, or things. Verbs name the action in a sentence. Verbs may describe actions in the past, present, or future. Both nouns and verbs may change spelling to describe more than one.

NOUNS — VERBS

HOW ALIKE?

Both are kinds of words that appear in sentences.

Both may change spelling to describe more than one.

HOW DIFFERENT? WITH REGARD TO

NOUNS	WITH REGARD TO	VERBS
Nouns name people, places, or things.	WHAT THEY NAME	Verbs name actions.
Nouns don't tell when.	TELL WHEN	Verbs tell whether something happens in the past, present, or future.

G-189

Statements and questions are kinds of sentences and tell whole thoughts. Both contain a subject and a verb and end with a punctuation mark. Both are used to give or get information. Paragraphs can have both statements and questions.

Statements give information about people, places, things, or ideas. They usually end with a period. The whole verb usually follows the subject; for example, "The motor is running."

Questions ask for information about people, places, things, or ideas. They always end with a question mark. The subject usually comes between parts of the verb; for example, "Is the motor running?"

STATEMENTS — QUESTIONS

HOW ALIKE?

Both are kinds of sentences and are whole thoughts.

Both contain a subject and a verb and end with a punctuation mark.

Both are used to exchange information.

HOW DIFFERENT? WITH REGARD TO

STATEMENTS	WITH REGARD TO	QUESTIONS
Statements give it.	INFORMATION	Questions ask for it.
Statements end with a period.	PUNCTUATION	Questions end with a question mark.
The whole verb usually follows the subject.	VERB	The subject usually comes between parts of the verb.

G-190

The United States government and your state government each have a chief official, meet in a particular city, and have a special building. The governor is the chief official of a state. The governor sees that the laws made by the state legislature are carried out. The state government is located in the state capital. The state legislature meets in the state capitol building.

The chief official of the national government is the president of the United States, who carries out laws passed by the Congress. Congress meets in the Capitol building in Washington, D.C.

UNITED STATES OF AMERICA — THE STATE IN WHICH YOU LIVE

HOW ALIKE?

Both can be found on maps. (They are geographic areas.)

Both have governments which pass laws.

HOW DIFFERENT? WITH REGARD TO

UNITED STATES OF AMERICA	WITH REGARD TO	THE STATE IN WHICH YOU LIVE
Large—made up of 50 states	SIZE	Part of the USA, so it is smaller than the USA
president	CHIEF OFFICIAL	governor
Washington, D.C.	CITY WHERE LAWMAKERS MEET	state capital
U.S. Capitol building	BUILDING WHERE LAWMAKERS MEET	state capitol building

G-191

All things have weight and take up space. Both living and nonliving things come in many sizes and shapes. There are tiny plants, tiny animals, and tiny grains of sand as well as large plants, large animals, and large mountains.

Most animals can move themselves around. Plants must be moved by wind or animals. No nonliving thing can move by itself.

All living things breathe, but plants breathe differently from animals. Nonliving things do not breathe. Breathing is called respiration.

Nonliving things do not eat and do not give birth. Giving birth is known as reproduction. All plants and animals reproduce themselves. Plants and animals need nourishment but get their food in different ways.

LIVING THINGS — NONLIVING THINGS

HOW ALIKE?

Both have weight.

Both take up space.

Both have many sizes and shapes.

HOW DIFFERENT? WITH REGARD TO

LIVING THINGS	WITH REGARD TO	NONLIVING THINGS
Some can move by themselves.	MOVE BY THEMSELVES	None can move by themselves.
All breathe.	RESPIRATION	Do not breathe.
All need nourishment	NUTRITION	Do not eat.
All reproduce themselves.	REPRODUCTION	Cannot reproduce.

LESSON PREPARATION

OBJECTIVE AND MATERIALS

OBJECTIVE: Students will use a "graphic organizer" to compare and contrast two concepts, objects, or organisms.

MATERIALS: Transparency of TM 26 (p. 274) • transparency of page 204 in student book

CURRICULUM APPLICATIONS

Language Arts: Using mind-mapping techniques for reading comprehension, organizing compare and contrast statements or paragraphs, differentiating between parts of speech or types of stories

Mathematics: Identifying geometric shapes for type and congruence, interpreting word problems, writing and interpreting different forms of statistical presentations (e.g., charts, graphs, pictures, and schedules relating to the same statistics)

Science: Identifying and organizing classes of plants or animals by similar characteristics

Social Studies: Expressing similarities and differences between historical events, eras, people, or artifacts

Enrichment Areas: Contrasting and/or comparing different works of art, music, or dance

TEACHING SUGGESTIONS

Allow adequate time for reflection and discussion. For exercise **G-190**, be sure to define "capital" as the city in which lawmakers meet and "capitol" as the building where lawmakers meet.

MODEL LESSON

LESSON

Introduction

Q: We have been describing how words are alike and how they are different.

Explaining the Objective to the Students

Q: In this lesson, you will use a diagram to explain how two things are alike and how they are different.

Class Activity

- Project transparency of page 204.

  Q: Open your books to page 204 and read the passage describing nouns and verbs.

- Allow time for reading and reflection.

  Q: Read the first sentence. How are nouns and verbs alike?

  A: Both are kinds of words that appear in sentences.

  Q: How else are nouns and verbs alike?

  A: Both may change spelling to describe more than one.

  Q: Write your answers on the first two lines.

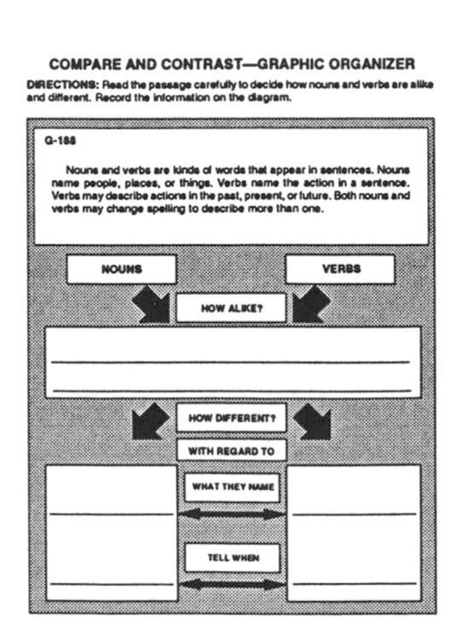

COMPARE AND CONTRAST—GRAPHIC ORGANIZER

DIRECTIONS: Read the passage carefully to decide how nouns and verbs are alike and different. Record the information on the diagram.

G-188

Nouns and verbs are kinds of words that appear in sentences. Nouns name people, places, or things. Verbs name the action in a sentence. Verbs may describe actions in the past, present, or future. Both nouns and verbs may change spelling to describe more than one.

NOUNS
VERBS
HOW ALIKE?
HOW DIFFERENT?
WITH REGARD TO
WHAT THEY NAME
TELL WHEN

- Allow time for writing.

  Q: Read the passage again to learn how nouns and verbs are different with regard to "what they name." Write your answers about nouns on the left side and verbs on the right.

  A: Nouns name people, places, or things. Verbs name actions.

  Q: Read the passage again to learn how nouns and verbs are different with regard to how they "tell when." Write your answers about nouns on the left side and verbs on the right.

  A: Nouns don't tell when. Verbs tell whether something happens in the past, present, or future.

GUIDED PRACTICE

EXERCISES: Finish the **G-188** graphic organizer "Nouns and Verbs."

INDEPENDENT PRACTICE

- Assign exercises **G-189** through **G-191**.

ADDITIONAL EXERCISES USING PICTURES AND A GRAPHIC ORGANIZER

Whenever possible, we recommend using actual objects; however, pictures usually offer sufficient detail even for students with limited backgrounds. You may use the following activities to supplement exercises **G-188** through **G-191**, using a transparency of TM 26:

- Display a bank and a post office; ask students to compare and contrast these places.
- Display a mobile home and an apartment building; ask students to compare and contrast these places.
- Display a fire station and a hospital; ask students to compare and contrast these places.
- Display a butterfly and a spider; ask students to compare and contrast these animals.
- Display ham and bacon; ask students to compare and contrast these foods.

THINKING ABOUT THINKING

Q: What did you pay attention to when you compared words or concepts?

1. I carefully read the passage.
2. I looked for the ways the two concepts were alike.
3. I recorded the similarities.
4. I looked for the ways the two concepts were different.
5. I recorded the differences.

PERSONAL APPLICATION

Q: When in your everyday life might you need to describe how two words or ideas are alike and different?

A: Examples include making value judgements (good/bad, right/wrong) in different situations, playing word games, taking tests (either essay or objective tests).

Chapter 8

VERBAL SEQUENCES

(Student book pages 209–52)

EXERCISES H-1 to H-5

FOLLOWING YES-NO RULES—A

ANSWERS H-1 through H-5 — Student book pages 210–11
Guided Practice: H-1 to **H-2** See below.
Independent Practice: H-3 through **H-5** See below.

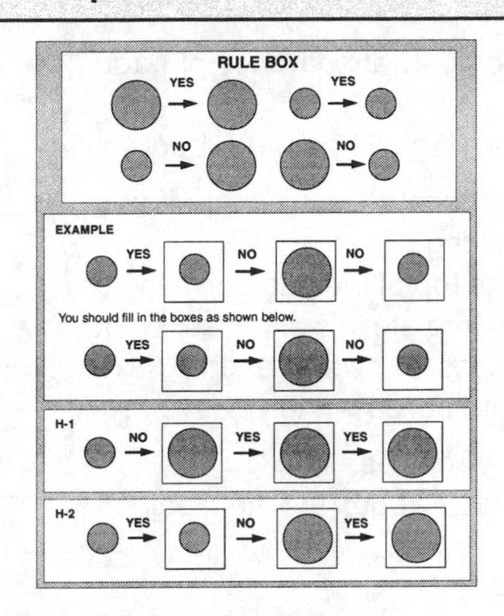

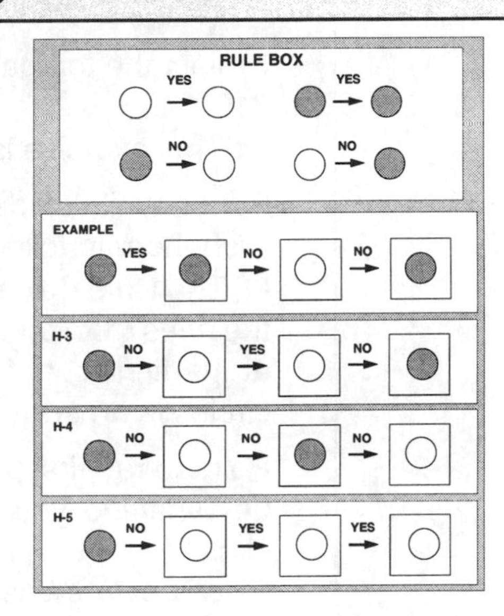

LESSON PREPARATION

OBJECTIVE AND MATERIALS

OBJECTIVE: Students will practice following rules which tell when to change or not change the size or color of a circle.
MATERIALS: Transparency of student book page 210 • washable transparency marker

CURRICULUM APPLICATIONS

Language Arts: Recognizing the effect of "no" or "not" on meaning, interpreting double negatives
Mathematics: Following multistep operations in solving problems, recognizing similarity and congruence in geometric figures
Science: Recognizing the change of a single variable in demonstrations or experiments, tracing the path of an electrical circuit
Social Studies: Following or creating a chart or map route
Enrichment Areas: Creating simple computer programs; learning or duplicating dance steps; duplicating plays in organized sports or games; following instructions for art, needlework, or craft projects

TEACHING SUGGESTIONS

Remember the language that students use to describe their choices and use the same words to remind students of the key concepts in this lesson and subsequent ones. Note: For students who speak Spanish or African-American dialects, double negatives mean "emphatically not."

MODEL LESSON

LESSON

Introduction

Q: One of the things you learn in language exercises is when to use capital letters and when to use lower case letters. After you learn the rules, you are often given sentences which contain errors in capitalization. You then have to decide which words do not follow the rules and correct the errors.

Explaining the Objective to the Students

Q: In this lesson, you will follow rules which tell whether to change the size or color of a circle.

Class Activity

- Project the top part of the transparency of page 210.
  Q: This exercise is on page 210 of your book.

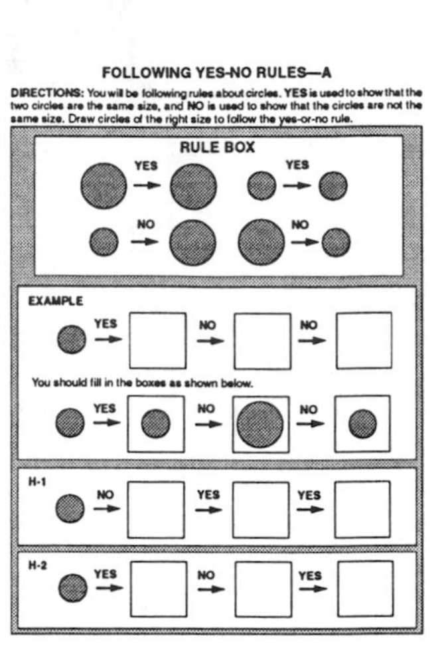
FOLLOWING YES-NO RULES—A

DIRECTIONS: You will be following rules about circles. YES is used to show that the two circles are the same size, and NO is used to show that the circles are not the same size. Draw circles of the right size to follow the yes-or-no rule.

- Pause while students locate the exercise if you wish them to follow along.
  Q: The Rule Box at the top of the page gives you the rules to follow to complete this exercise. Look at the first row of circles. Between each pair of circles is an arrow with a Yes written above it.
- Point to the first pair of circles.
  Q: What do you notice about the first pair of circles?
  A: They are both large.

  Q: Is the first circle the same size as the second circle?
  A: Yes
- Point to the second pair of circles.
  Q: Are the circles in the second pair the same size?
  A: Yes, they are both small.
- Point to the second row of circles.
  Q: Look at the next row of circles. Each pair of circles has an arrow with a No written above it.
- Point to the arrow.
  Q: Look at the first pair of circles.
- Point to the first pair of circles.
  Q: Is the first circle the same size as the second circle?
  A: No, the second circle is larger.
- Point to the second pair of circles.
  Q: Are the circles in the second pair the same size?
  A: No, the second circle is smaller than the first circle.

  Q: In the Rule Box, when the arrow above a pair of circles says Yes, what does Yes tell about the circles?
  A: They are the same size.

Q: And when the arrow says No, what does that tell you about the circles?
A: They are not the same size.

Q: Now, look at the example. You are going to draw a circle in the empty box.

- Point to the arrow.
  Q: The starting circle is small, and the first arrow has a Yes written above it. What size circle should we draw in the box?
  A: A small circle

  Q: According to the Rule Box, the Yes means that the next circle should be the same size as the first circle. Draw a small circle in the box.
  Q: The second arrow...
- Point to the No arrow.
  Q: ...is marked No. This means that the next circle should not be the same size. What size circle should we draw in this blank?
- Point to the second blank.
  A: A larger circle
- Draw a large circle in the box.
  Q: The third arrow...
- Point to the second No arrow.
  Q: ...is marked No. This means that the next circle should not be the same size. What size circle should we draw in this blank?
- Point to the second blank.
  A: A small circle
- Draw a small circle in the box.

GUIDED PRACTICE

EXERCISES: **H-1**, **H-2**

- Remind students to read carefully. When they have had sufficient time to complete the exercises, check answers by discussion to determine whether students have answered correctly. They should be able to justify their answers and tell why the other possible answer is incorrect.

INDEPENDENT PRACTICE

- Assign exercises **H-3** through **H-5**. Note: For exercises **H-3** through **H-5**, the rule refers to a change in color. Yes means the color is the same; No means the color is different.

THINKING ABOUT THINKING

Q: What did you pay attention to when you followed a series of steps?
1. I read the instructions carefully.
2. I looked at the diagrams and drew in the correct answer.

PERSONAL APPLICATION

Q: When might you be asked to follow a series of steps or rules?

A: Examples include assembling games or models involving multistep directions; following directions involving switches on equipment such as tape recorders, calculators, computers, projectors; following the rules in games or sports.

EXERCISES H-6 to H-16

WRITING YES-NO RULES

ANSWERS H-6 through H-16 — Student book pages 212–13
Guided Practice: H-6 Yes, Yes; **H-7** Yes, No
Independent Practice: H-8 No, No; **H-9** No, No; **H-10** Yes, No; **H-11** Yes, Yes; **H-12** No, Yes, No, No; **H-13** Yes, No, Yes, No; **H-14** No, Yes, No, No; **H-15** No, No, No, No; **H-16** No, No, No, No

LESSON PREPARATION

OBJECTIVE AND MATERIALS

OBJECTIVE: Students will write the rules which describe the changes.
MATERIALS: Transparency of student book page 212 • washable transparency marker

CURRICULUM APPLICATIONS

Language Arts: Determining whether examples of usage, spelling, grammar, or paragraph construction follow a given set of rules
Mathematics: Recognizing whether multistep operations have been correctly followed, recognizing similarity and congruence in geometric shapes
Science: Explaining basic experimental reactions and their causes
Social Studies: Placing artifacts according to usage, era, or culture; determining specifics from a map (type of road, size of city, elevations, points of interest, etc.)
Enrichment Areas: Matching the time signature to a given piece of music, learning rules of a game by playing the game rather than by reading the instructions, playing hidden-rule games on a computer or video

TEACHING SUGGESTIONS

Remember the language that students use to describe their choices and use the same words to remind students of the key concepts in this lesson and subsequent ones.

MODEL LESSON

LESSON

Introduction

Q: In the previous exercises, you drew circles following rules about changing sizes or colors.

Explaining the Objective to the Students

Q: In this lesson, you will write the rules which describe changes in the size or color of a circle.

Class Activity

- Project the top part of the example from the transparency of page 212.
  Q: In this exercise, the circles are drawn, and you are to write "Yes" or "No" above the arrows between the circles. If the circles are the same color or

the same size, you will write "Yes" on the arrow. If the circles are different colors or different sizes, you will write "No" above the arrow. The Rule Box...

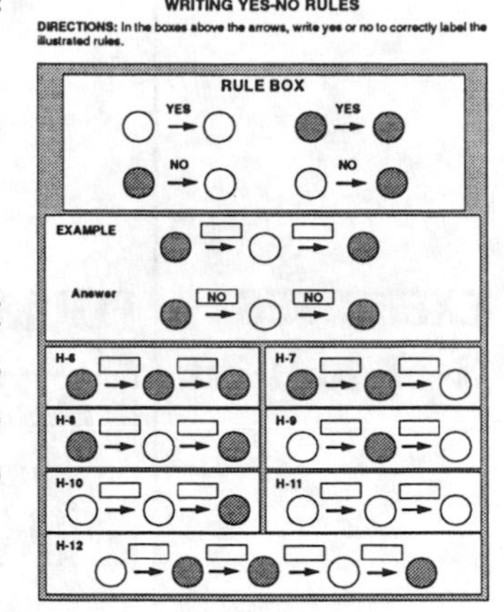

- Point to the Rule Box on the transparency.

  Q: ...at the top of each page tells you the rules that apply to the set of exercises on that page. This is the example from page 212. By looking at the Rule Box, we know that the rules on this page apply to color.

- Point to the first two circles in the example.

  Q: Are these two circles the same color?

  A: No

  Q: Since they are not the same, we should write a No above the arrow.

- Write "No" above the first arrow.

  Q: Look at the second arrow. It points from a white circle to a gray circle. What should we write above this arrow?

  A: No

- Write "No" above the second arrow.

  Q: Why are both the first arrow and the second arrow marked No?

  A: The first two circles are not the same color. The last two circles are not the same color.

  Q: Remember to look carefully at the circles each arrow connects before deciding what to write above the arrow.

GUIDED PRACTICE

EXERCISES: **H-6**, **H-7**

- Remind students to pay careful attention to the Rule Box on each page. When they have had sufficient time to complete these exercises, check answers by discussion to determine whether students have answered correctly.

INDEPENDENT PRACTICE

- Assign exercises **H-8** through **H-16**. Note: For exercises **H-13** through **H-16**, the rules refer to a change in size. Yes means the size is the same; No means the size is not the same.

THINKING ABOUT THINKING

Q: What did you pay attention to when you determined the rule?

1. I looked carefully at the circles each arrow connects.
2. I decided what to write above the arrow. ("Yes" if the circles were the same color or size and "No" if they were not.)

PERSONAL APPLICATION

Q: At what other times might you have to use a given number of examples to determine the rule?

A: Examples include assembling games or models involving pictorial directions; following directions involving on-off switches on equipment such as tape recorders, calculators, computers, or projectors; watching a game or sport with which you are unfamiliar.

EXERCISES H-17 to H-20

FOLLOWING YES-NO RULES—B

ANSWERS H-17 through H-20 — Student book pages 214–15
Guided Practice: H-17 to **H-18** See below.
Independent Practice: H-19 to **H-20** See below.

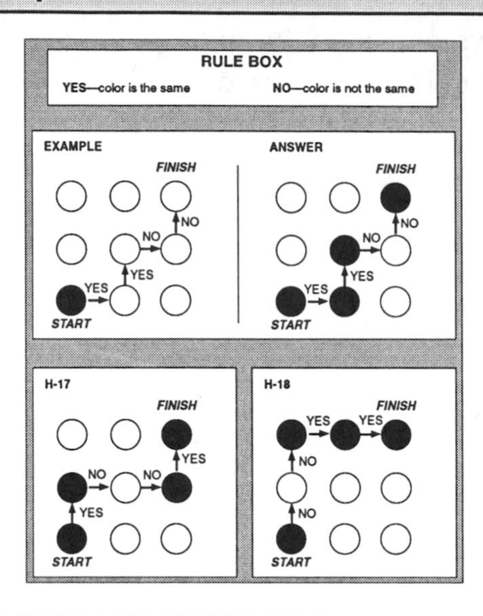

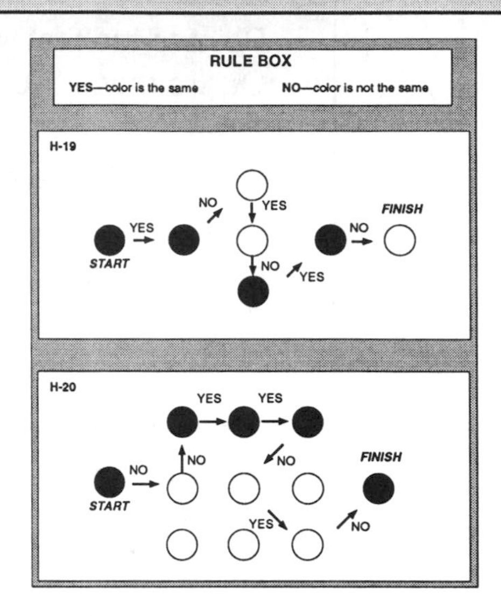

LESSON PREPARATION

OBJECTIVE AND MATERIALS

OBJECTIVE: Students will complete a path having a sequence of four or more color changes by following the given rule.
MATERIALS: Transparency of student book page 214 • washable transparency marker

CURRICULUM APPLICATIONS

Language Arts: Determining whether examples of usage, spelling, or grammar follow a given set of rules
Mathematics: Recognizing whether multistep operations have been correctly followed, recognizing similarity and congruence in geometric shapes
Science: Explaining the outcomes of science activities
Social Studies: Placing artifacts according to usage, era, or culture; applying a map key (type of road, size of city, elevations, points of interest, etc.)
Enrichment Areas: Matching the time signature to a given piece of music from written samples of the music, learning rules of a game by playing the game rather than by reading the instructions, playing hidden-rule games on a computer or video

TEACHING SUGGESTIONS

Remember the language that students use to describe their choices and use the same words to remind students of the key concepts in this lesson.

MODEL LESSON

LESSON

Introduction

Q: In the previous exercise, you identified the rules for changing sizes and colors of circles.

Explaining the Objective to the Students

Q: In this lesson, you will complete a sequence of changes in color by using a rule.

Class Activity

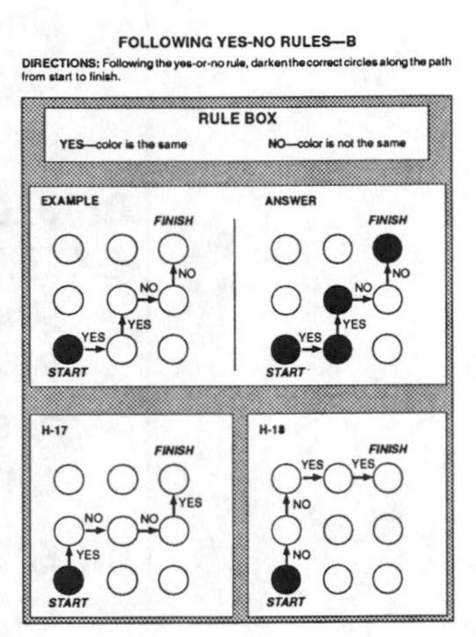

- Project the top left portion of the example from the transparency of page 214.

  Q: In the example on page 214, you will darken the correct circles according to the Rule Box that you see at the top of the page. If Yes is written above the arrow, then the next circle is the same color. If No is written above the arrow, then the next circle is the opposite color. The first arrow from the Start circle has Yes above it.

- Point to the Start circle and the arrow.

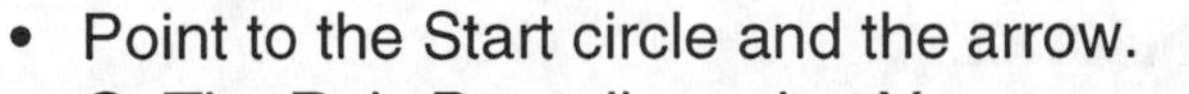

  Q: The Rule Box tells us that Yes means that the color is the same. According to this rule, what should we do to the circle on the right?

- Point to the appropriate circle.

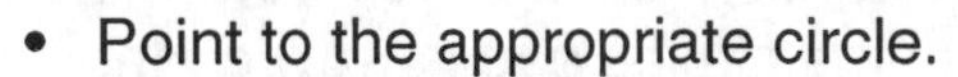

  A: Darken it or shade it in.

- Darken the circle on the transparency.

  Q: The second arrow on the indicated path is also marked Yes.

- Point to the arrow.

  Q: According to the Rule Box, should these circles be the same color?

  A: Yes

  Q: What should we do to the next circle above the arrow to make the rule true?

  A: Darken it.

- Darken the center circle.

  Q: The third arrow on the path is marked No.

- Point to the third arrow.

  Q: Should these circles be the same color?

- Point to the third and fourth circles.

  A: No

  Q: According to the Rule Box, we should leave the circle to the right of the center circle white. Why?

  A: The Rule Box says that a No above the arrow means that the color is not the same. Since the last circle was dark, the next circle should be white.

Q: Look at the last arrow near the top. It is marked No.

- Point to the last arrow.

  Q: Should these two circles be the same color?

- Point to the fourth circle and the Finish circle.

  A: No

  Q: What should be done to the Finish circle?

  A: Darken it.

- Darken the last circle on the transparency.

  Q: Remember to look carefully at the arrows and the circles before you decide your answers.

GUIDED PRACTICE

EXERCISES: **H-17**, **H-18**

- Remind students to pay careful attention to the Rule Box on each page. When they have had sufficient time to complete these exercises, check answers by discussion to determine whether students have answered correctly.

INDEPENDENT PRACTICE

- Assign exercises **H-19**, **H-20**.

THINKING ABOUT THINKING

Q: What did you pay attention to when you determined the rule?

1. I looked carefully at the circles each arrow connects.
2. I decided what to write on the arrow. ("Yes" if the circles were the same color or size and "No" if they were not.)

PERSONAL APPLICATION

Q: When might you have to use several examples to determine the rule?

A: Examples include assembling games or models involving pictorial directions; following directions involving on-off switches on equipment such as tape recorders, calculators, computers, or projectors; watching a game or sport with which you are unfamiliar.

EXERCISES H-21 to H-25

COMPLETING TRUE-FALSE TABLES

ANSWERS H-21 through H-25 — Student book pages 216–18
Guided Practice: H-21 It is black: True, False, True; It is white: False, True, False; It is not white: True, False, True
Independent Practice: H-22 It is white: False, True, True; It is black: True, False, False; It is not black: False, True, True; **H-23** It is a square: True, False, False; It is a circle: False, True, True; It is not a circle: True, False, False; **H-24** It is all black: True, False, False; It is all white: False, True, False; It is not all white: True, False, True; **H-25** It is half black: False, True, True; It is half white: False, True, True; It is not a circle: False, True, False

LESSON PREPARATION

OBJECTIVE AND MATERIALS

OBJECTIVE: Students will complete a matrix by deciding whether statements are true or false.

MATERIALS: Transparency of student book page 139 • transparency of student book page 216 (with answers removed) • washable transparency marker

CURRICULUM APPLICATIONS

Language Arts: Finding errors in spelling, grammar, or punctuation; answering true/false questions about a story or passage

Mathematics: Checking computations to determine if correct procedures were followed, choosing correct answer sets to a problem, checking computations by estimation

Science: Verifying animal or plant characteristics in laboratory demonstrations

Social Studies: Finding facts to support or negate statements, determining correct time in different time zones

Enrichment Areas: Determining whether instructions and rules have been followed in sports or recreation activities

TEACHING SUGGESTIONS

Some students may experience difficulty with the truth or falsity of negative statements. Encourage students to discuss reasons for their choices.

MODEL LESSON

LESSON

Introduction

Q: You have identified or followed yes-no rules.

Explaining the Objective to the Students

Q: In this lesson, you will fill in a chart (matrix) by deciding if the statement in each heading is true or false for each shape.

Class Activity

- Project the transparency of page 139.
  Q: Remember this exercise? This is a matrix. A matrix has rows...

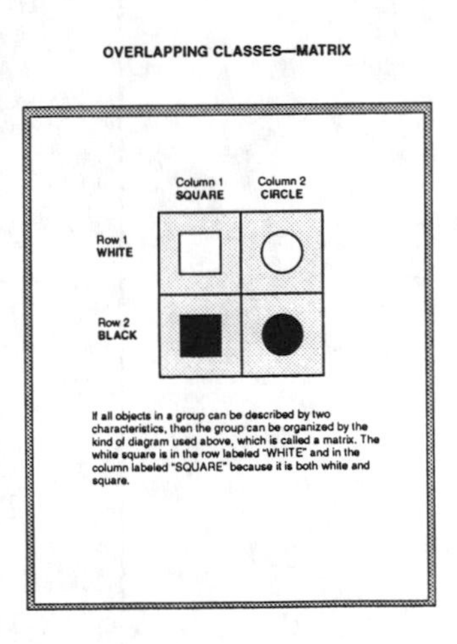

If all objects in a group can be described by two characteristics, then the group can be organized by the kind of diagram used above, which is called a matrix. The white square is in the row labeled "WHITE" and in the column labeled "SQUARE" because it is both white and square.

- Indicate rows on the transparency.
  Q: ...and columns.
- Indicate columns. Replace transparency of page 139 with transparency of page 216.
  Q: This is the example from page 216. It is another kind of matrix called a True-False Table. What is the heading for column one?
- Indicate column one.
  A: It Is Black.

  Q: What is the heading for column two?
  A: It Is White.

  Q: What is the shape in the first row?

A: It is a white circle.

COMPLETING TRUE-FALSE TABLES

DIRECTIONS: Study the given true false table and then complete the blank table.

The shapes below will be used in a special matrix called a true-false table.

○ △ □ ● ▲ ■

EXAMPLE

	IT IS BLACK	IT IS WHITE
○	*FALSE*	*TRUE*
▲	*TRUE*	*FALSE*
□	*FALSE*	*TRUE*

H-21

	IT IS BLACK	IT IS WHITE	IT IS NOT WHITE
●			
△			
■			

Q: As you move across the row with the white circle, you will write *true* if the statement at the head of that column is true for a white circle. You will write *false* if the statement at the head of that column is not true for a white circle. The first column heading says, "It Is Black." Is this statement true for a white circle?

A: No

Q: Since the statement is not true for a white circle, what should be written in the first row-first column cell?

A: False

- Write "false" in the proper cell on the transparency.

  Q: Now we will move to the next column in the first row. This column heading says, "It Is White." Is this statement true for a white circle?

  A: Yes

  Q: Since the statement is true for a white circle, what should we write in the first row-second column cell?

  A: True

- Write "true" in the proper cell on the transparency.

  Q: Now look at the second row. What is the shape at the beginning of this row?

  A: It shows a black triangle.

  Q: Look at the cell in row two, column one. To complete this cell, we must ask ourselves the question: "Is the statement 'It Is Black' true for the black triangle in row two?"

  A: Yes

  Q: What should we do to show this on the matrix?

  A: Write "true" in the row two-column one cell.

  Q: Following the same rules, what should go in the row two-column two cell on this matrix and why?

  A: False should go in the cell because the statement "It is white" is not true of the black triangle. (Repeat a similar line of questioning until all cells in the True-False Table have been filled.)

  Q: Remember to read and think carefully as you complete these exercises.

GUIDED PRACTICE

EXERCISE: **H-21**

- When students have had sufficient time to complete this exercise, check answers by discussion to determine whether students have answered correctly. Students should always be asked to explain their answers.

INDEPENDENT PRACTICE

- Assign exercises **H-22** through **H-25**.

THINKING ABOUT THINKING

Q: What did you pay attention to when you decided whether a statement was true or false?

1. I looked at the shape being tested.
2. I read the heading of the column.
3. If the shape matched the heading, I wrote true in the box.
4. If the shape did not match the heading, I wrote false in the box.

PERSONAL APPLICATION

Q: At what other times might you need to decide whether a statement was true or false or whether a rule is being correctly followed?

A: Examples include answering true/false questions on a test, recognizing whether instructions for operating equipment or directions for construction have been correctly followed, recognizing correct directions and procedures for preparing recipes, observing when directions for taking medicine are being correctly followed, recognizing that negative statements can be either true or false.

EXERCISES H-26 to H-37

FINDING LOCATIONS ON MAPS

ANSWERS H-26 through H-37— Student book pages 219–20
Guided Practice: H-26 through **H-30** See below.
Independent Practice: H-31 through **H-33** See below. **H-34** R; **H-35** P; **H-36** Q; **H-37** "A" Street

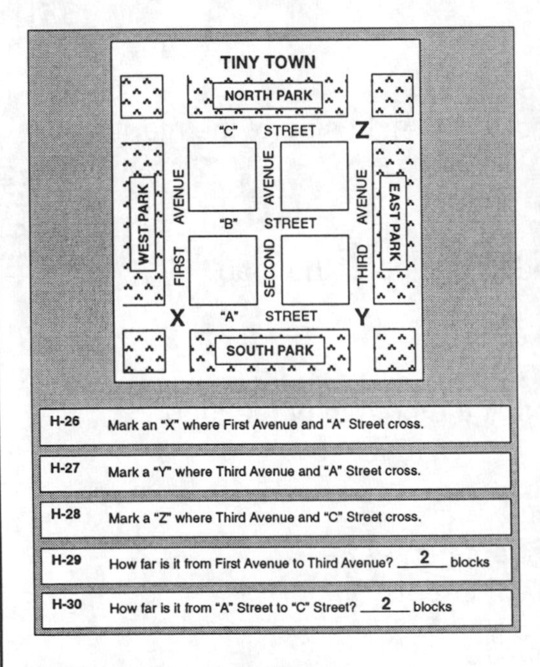

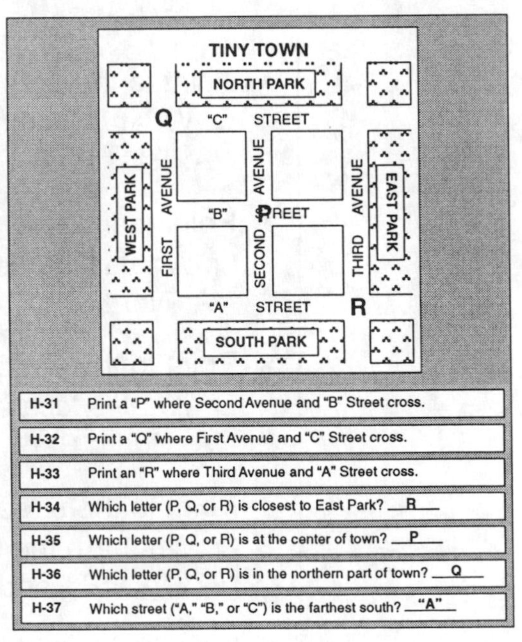

LESSON PREPARATION

OBJECTIVE AND MATERIALS

OBJECTIVE: Students will locate street intersections on a map.
MATERIALS: Transparency of TM 27 (p. 275) • washable transparency marker

CURRICULUM APPLICATIONS

Language Arts: Finding facts by using book sections (Table of Contents, Index), using reference materials (atlas, encyclopedia index, Readers' Guide)
Mathematics: Locating and charting data on a graph, locating coordinates on a grid, geoboard activities
Science: Constructing a motor or gear following written directions, using text book sections to locate facts
Social Studies: Reading maps, graphs, or charts
Enrichment Areas: Creating or following pattern plays in team sports

TEACHING SUGGESTIONS

Encourage students to discuss their answers. They should be specific when describing locations, using the terms *right, left, top, bottom*, and *center*.

MODEL LESSON

LESSON

Introduction

Q: You have followed instructions regarding such positions as right or left, upper or lower, and center or corner.

Explaining the Objective to the Students

Q: In this lesson, you will locate street crossings (intersections) on a map.

Class Activity

- Project TM 27.
  Q: This is the map from page 219.

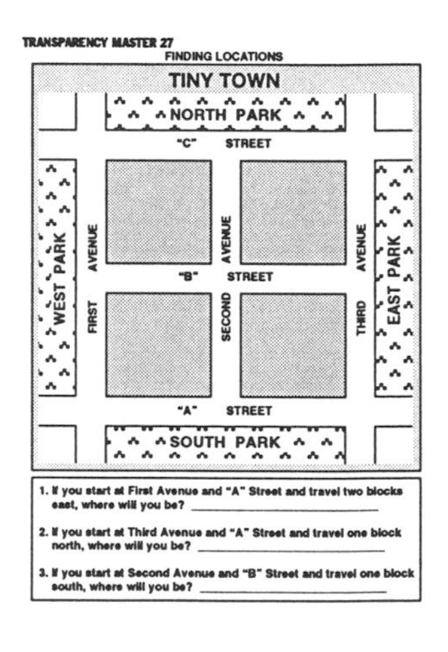

- Pause. Point to the left side of the map.
  Q: What is located on the left side of the map?
  A: West Park
- Point to the right side.
  Q: What is located on the right side of the map?
  A: East Park
- Point to the top.
  Q: What is located at the top of the map?
  A: North Park
- Point to the bottom.
  Q: What is located at the bottom of the map?
  A: South Park
- Point to the streets.
  Q: What are the street names?
  A: "A," "B," and "C" streets
- Indicate the avenues.
  Q: What are the names of the avenues?
  A: First, Second, and Third avenues

  Q: In what direction do the streets in Tiny Town run?
  A: East and west

Q: In what direction do the avenues in Tiny Town run?

A: North and south

Q: Now that you are familiar with the map, let's indicate some crossings. Let's make an *L* where Second Avenue and "A" Street cross. First, we have to find one of the roads then follow it to find where it meets the other one.

- Have a student indicate the avenue and street then mark the intersection with an *L* on the transparency. Ask other students for verification.

  Q: Let's make an *M* where Third Avenue and "B" Street cross.

- Have a student indicate the given route and mark the intersection with an *M* on the transparency. Ask other students for verification.

  Q: Let's make an *N* where First Avenue and "C" Street cross.

- Have a student indicate the location and mark an *N* on the transparency. Ask other students for verification.

  Q: Now you have seen how to locate and mark crossing points.

GUIDED PRACTICE

EXERCISES: **H-26** through **H-30**

- When students have had sufficient time to complete these exercises, check answers by discussion to determine whether students have answered correctly.

INDEPENDENT PRACTICE

- Assign exercises **H-31** through **H-37**.

THINKING ABOUT THINKING

Q: What did you pay attention to when you were finding locations on maps?

1. I looked at the map and read the directions.
2. I found the streets and avenues described.
3. I marked the map or counted blocks.

PERSONAL APPLICATION

Q: Where else might you use this ability to find locations by following directions?

A: Examples include following or giving street directions, following a map for a treasure hunt game, constructing something by following written directions.

EXERCISES H-38 to H-45

DESCRIBING LOCATIONS—A

ANSWERS H-38 through H-45 — Student book pages 221–22

Guided Practice: H-38 Third Avenue; **H-39** "B" Street; **H-40** 1 block; **H-41** 2 blocks

Independent Practice: H-42 "C" Street; **H-43** Second Avenue; **H-44** 1 block; **H-45** 2 blocks

LESSON PREPARATION

OBJECTIVE AND MATERIALS

OBJECTIVE: Students will find the locations of map features that are near other features.

MATERIALS: Transparencies of TM 27 (optional, p. 275) and TM 28 (p. 276) • washable transparency marker

CURRICULUM APPLICATIONS

Language Arts: Writing a set of instructions or giving a demonstration speech

Mathematics: Identifying particular points on a graph, describing an arrangement of geometric shapes

Science: Stating or writing directions for construction of motors or gears, detailing the position of a star or planet

Social Studies: Reading maps, graphs, or charts; describing a geographic location

Enrichment Areas: Creating or following pattern plays in team sports, describing the performance of a dance step or the creation of an art project, directing or writing stage directions for a play

TEACHING SUGGESTIONS

Encourage students to discuss reasons for their choices and why the other explanations don't fit. Students' explanations of their thinking are often more meaningful than an adult's explanation.

MODEL LESSON

LESSON

Introduction

Q: You have located positions on a map.

Explaining the Objective to the Students

Q: In this lesson, you use information describing a location on a map to locate a place nearby.

Class Activity—Optional Review Lesson

Note: If it has been a few days since the Finding Locations lesson, use the following review lesson before proceeding with the lesson on describing locations.

- Project TM 27.

  Q: This is the map of Tiny Town which we used in the last lesson.

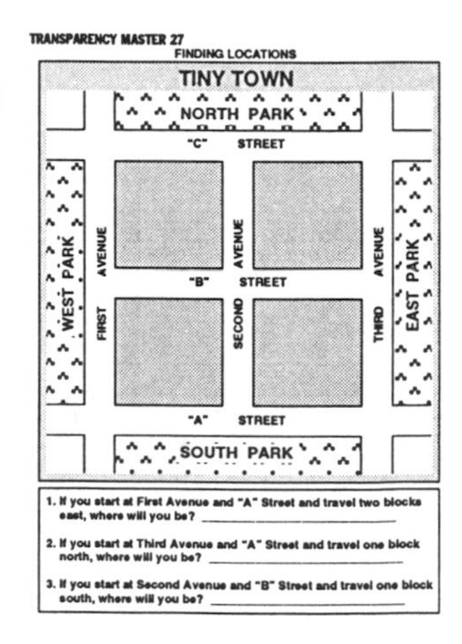

- Point to the left side of the map.

  Q: What is located on the left side of the map?

  A: West Park

- Point to the right side.

  Q: What is located on the right side of the map?

  A: East Park

- Point to the top.

  Q: What is located at the top of the map?

  A: North Park

- Point to the bottom.

  Q: What is located at the bottom of the map?

  A: South Park

- Point to the streets.

  Q: What are the street names?

  A: "A," "B," and "C" streets

- Indicate the avenues.

  Q: What are the names of the avenues?

  A: First, Second, and Third Avenues

  Q: In what direction do the streets in Tiny Town run?

  A: East and west

  Q: In what direction do the avenues in Tiny Town run?

  A: North and south

Class Activity

- Project the transparency of TM 28.

  Q: This map shows a path in Tiny Town. Look at the part of the path from point *S* to point *T*.

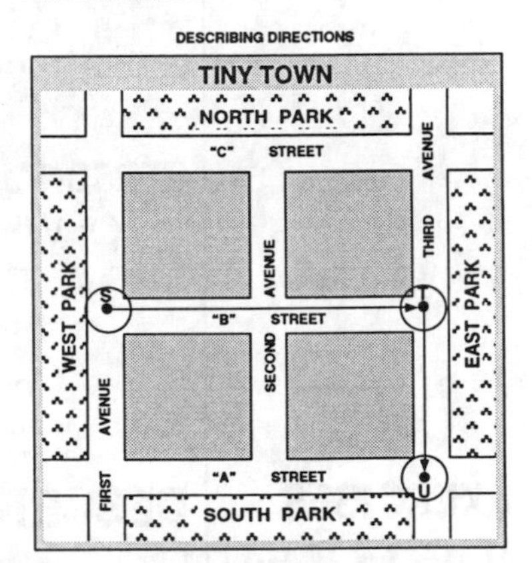

- Indicate the path.

  Q: Which of the town's four parks is closest to point *S*?

  A: West Park

  Q: Which park is closest to point *T*?

  A: East Park

  Q: When describing directions, it is useful to say that the path proceeds from one direction to another. How can we describe the direction we would have to go to get from point *S* near West Park to point *T* near East Park?

  A: From west to east

  Q: Now look at the path from point *T* to point *U*. Toward which park does the arrow point?

  A: Toward South Park

  Q: Which park is located behind point *T*?

  A: North Park

  Q: Describe the direction of the path that goes from point *T* to point *U*.

  A: From north to south

  Q: Now let's look at the lengths of the paths. How far is it from point *S* to point *T*?

- Point to the indicated path, putting a finger under each of the two blocks.

  A: Two blocks

  Q: How far is it from point *T* to point *U*?

  A: One block

Q: Now you have seen how to describe directions using the directional words north, south, east, and west.

GUIDED PRACTICE

EXERCISES: **H-38** through **H-41**

- When students have had sufficient time to complete these exercises, check answers by discussion to determine whether they have answered correctly.

INDEPENDENT PRACTICE

- Assign exercises **H-42** through **H-45**.

THINKING ABOUT THINKING

Q: What did you pay attention to when you described specific locations or directions on maps?

1. I located my starting position.
2. I carefully read the question.
3. I checked the street and avenue names.
4. I visualized the move from the original location to the new one.
5. I located my new position and wrote the location.

PERSONAL APPLICATION

Q: When might you need to describe specific locations or directions?

A: Examples include following or giving street directions, telling someone where something is located in your house (how to find the hidden cookie jar, for example).

EXERCISES H-46 to H-54

DESCRIBING DIRECTIONS—A/LOCATIONS—B

ANSWERS H-46 through H-54 — Student book pages 223–25
Guided Practice: H-46 north, 2 blocks; **H-47** east, 1 block
Independent Practice: H-48 south, 2 blocks; **H-49** west, 2 blocks; **H-50** north, 1 block; **H-51** "A" Street and Third Avenue; **H-52** "C" Street and Third Avenue; **H-53** "B" Street and First Avenue; **H-54** "A" Street and Second Avenue

LESSON PREPARATION

OBJECTIVE AND MATERIALS

OBJECTIVE: Students will use the terms *north*, *south*, *east*, and *west* to describe directions on the map.

MATERIALS: Transparency of TM 28 (p. 276) • washable transparency marker

CURRICULUM APPLICATIONS

Language Arts: Matching a set of instructions with an actual demonstration, describing activities

Mathematics: Describing particular points on a graph or grid, describing an arrangement of geometric shapes

Science: Describing a circuit; following the path of blood vessels, digestive

system, etc. in the human body; detailing the position of a star or planet
Social Studies: Reading maps, graphs, or charts; describing a geographic location; tracing a family history
Enrichment Areas: Following pattern plays in team sports, directing or writing stage directions for a play

TEACHING SUGGESTIONS
Encourage students to discuss reasons for their choices and why the other items don't fit. Students' explanations of their thinking are often more meaningful than an adult's explanation.

MODEL LESSON

LESSON

Introduction

Q: You have described locations or positions on a map.

Explaining the Objective to the Students

Q: In this lesson, you will use the terms *north*, *south*, *east*, and *west* to describe directions on a map.

Class Activity

Note: If it has been a few days since the lessons on Finding and Describing Locations, the review lesson on page 174 of the manual may be repeated.

- Project the transparency of TM 28.

  Q: This map shows a path in Tiny Town. Look at the part of the path from point *S* to point *T*.

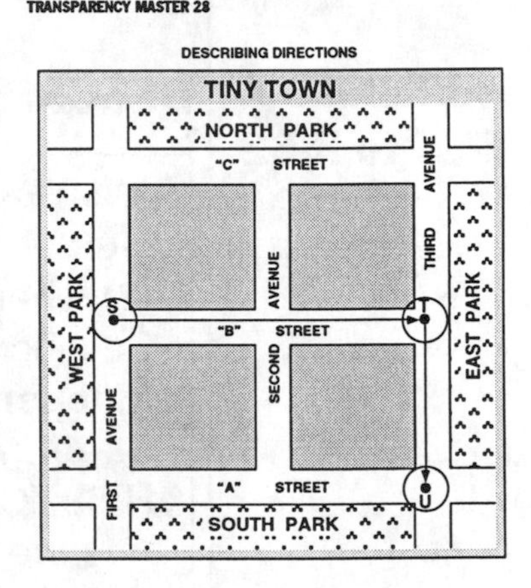

- Indicate the path.

  Q: When describing directions, it is useful to say that the path proceeds in one direction from where you start. How can we describe the direction we would have to go to get from point *S* near West Park to point *T* near East Park?

  A: East

  Q: Now, look at the path from point *T* to point *U*. Toward which park does the arrow point?

  A: Toward South Park

  Q: Describe the direction we need to go to get from point *T* to point *U*.

  A: South

  Q: Now, let's look at the lengths of the paths. How far is it from point *S* to point *T?*

- Point to the indicated path, putting a finger under each of the two blocks.

  A: Two blocks

  Q: How far is it from point *T* to point *U?*

  A: One block

GUIDED PRACTICE
EXERCISES: **H-46, H-47**

- When students have had sufficient time to complete these exercises, check answers by discussion to determine whether they have answered correctly.

INDEPENDENT PRACTICE

- Assign exercises **H-48** through **H-54**.

THINKING ABOUT THINKING

Q: What did you pay attention to when you described specific locations or directions on maps?

1. I located my starting position.
2. I carefully read the question.
3. I checked the street and avenue names.
4. I visualized the move from the original location to the new one.
5. I located my new position and wrote the location.

PERSONAL APPLICATION

Q: When might you need to describe specific locations or directions?

A: Examples include following or giving street directions, reading a map or atlas, telling someone where something is located in your house (how to find the hidden cookie jar, for example).

EXERCISES H-55 to H-62

DESCRIBING DIRECTIONS—B

ANSWERS H-55 through H-62 — Student book pages 226–28
Guided Practice: TM 27 (1) where "A" Street meets Third Avenue, (2) where "B" Street meets Third Avenue, (3) where "A" Street meets Second Avenue
Independent Practice: H-55 "B" Street meets Third Avenue; **H-56** "C" Street meets Third Avenue; **H-57** "B" Street meets Second Avenue; **H-58** "C" Street meets Third Avenue; **H-59** "B" Street meets Second Avenue; **H-60** "A" Street meets Second Avenue; **H-61** "A" Street meets First Avenue; **H-62** "C" Street meets Second Avenue See below.

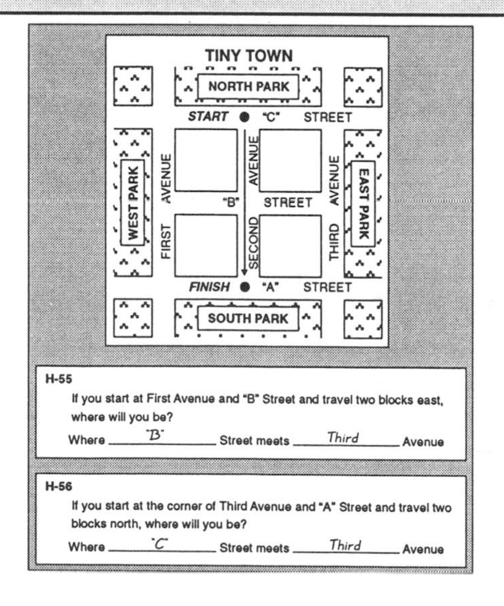

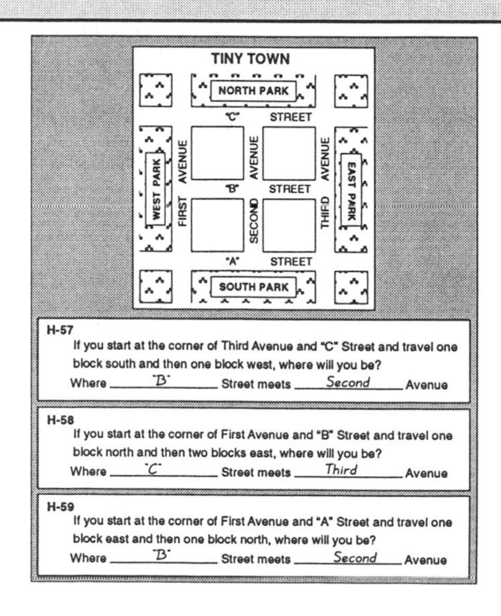

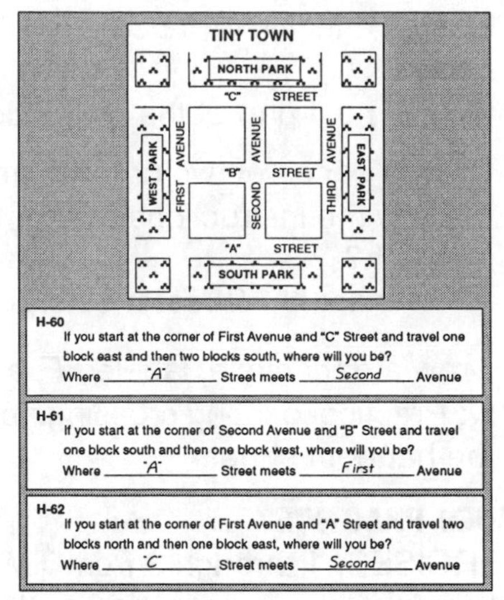

LESSON PREPARATION

OBJECTIVE AND MATERIALS

OBJECTIVE: Students will locate intersections by following written instructions for a route.

MATERIALS: Transparency of TM 27 (p. 275) • photocopies of TM 27 (one per student) • washable transparency marker

CURRICULUM APPLICATIONS

Language Arts: Using references and cross-references in research materials, doing crossword puzzles and similar word games

Mathematics: Locating or charting consecutive points on a graph, doing follow-the-dots activities

Science: Using a celestial map to locate stars or planets

Social Studies: Creating a time line; reading maps, graphs, or charts

Enrichment Areas: Putting a series of dance steps into a sequence, creating or following pattern plays in team sports

TEACHING SUGGESTIONS

Encourage students to discuss reasons for their answers using accurate terms for street directions.

MODEL LESSON

LESSON

Introduction

Q: You have learned to describe a path marked on a street map.

Explaining the Objective to the Students

Q: In this lesson, you will locate street crossings (intersections) by marking a path following written instructions.

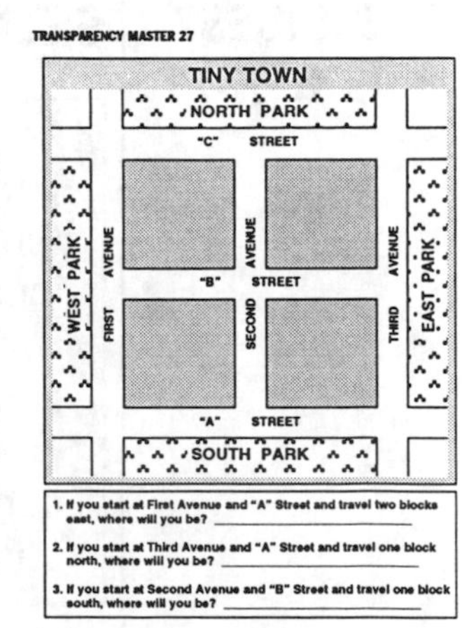

Class Activity

- Project the transparency of TM 27.

  Q: Here is a blank map of Tiny Town. The instructions for question 1 at the bottom of the page say to start at First Avenue and "A" Street. We need to mark this starting point on the page.

- Draw a dark circle, labeled *S*, at the intersections of First Avenue and "A" Street.

  Q: The next part of the instructions says to travel two blocks east.
- Draw a series of two or three arrows to indicate the path.

  Q: Now the instructions ask for the final location. Where have we ended our path?

  A: Where "A" Street meets Third Avenue
- Draw a dark circle, labeled *F*, at the ending location.

  Q: Practice locating crossings by completing the other two questions on the handout sheet.

GUIDED PRACTICE

EXERCISES: 1 through 3 from TM 27

- When students have had sufficient time to complete these exercises, check answers by discussion to determine whether students have answered correctly.

INDEPENDENT PRACTICE

- Assign exercises **H-55** through **H-62**.

THINKING ABOUT THINKING

Q: What did you pay attention to when you described specific locations or directions on maps?

1. I located my starting position.
2. I carefully read the question.
3. I checked the street and avenue names.
4. I visualized the move from the original location to the new one.
5. I located my new position and wrote the location.

PERSONAL APPLICATION

Q: When do you want or need to find a location?

A: Examples include following or giving street directions, reading a road map to find out how to get from one place to another (sometimes by more than one route), writing or following clues in a "treasure hunt" game.

EXERCISES H-63 to H-77

SELECT THE WORD THAT CONTINUES THE SEQUENCE

ANSWERS H-63 through H-77 — Student book pages 229–31

Guided Practice: H-63 after (time order, increasing) **H-64** Wednesday (time order, increasing) **H-65** year (degree, increasing) **H-66** third (order, increasing) **H-67** Christmas (order, increasing)

Independent Practice: H-68 gallon (volume [rank], increasing) **H-69** less (degree or amount, decreasing) **H-70** large (size, increasing) **H-71** many (degree or amount, increasing) Note: Most is also an acceptable answer; if all five words are arranged in an increasing sequence then many would be the next word after more.) **H-72** yard (size or length, increasing) **H-73** below (order [space], decreasing) **H-74** worst (degree, decreasing) **H-75** man (rank [or possibly size], increasing) **H-76** leave (time order, decreasing) **H-77** best (degree of worth or goodness, increasing)

DETAILED SOLUTIONS

Note: The detailed solutions follow the model shown in the Teaching Suggestions on page 182.

H-63 before, during, **after**

1. The relationship between the first two words is time order: *during* follows (or comes after) *before*
2. The sequence is increasing.
3. Look for a word that comes after *during*.
4. *After* would occur later than *during*.
5. *Whole* refers to entire actions or things, not a section; *beside* is a space indicator, not a time indicator.

H-64 Monday, Tuesday, **Wednesday**

1. represent time order: days of the week
2. increasing
3. Look for the day that comes after Tuesday.
4. Wednesday is next.
5. The other days are either before or after Tuesday.

H-65 day, month, **year**

1. represent degree: a month is a longer measure of time than a day
2. increasing
3. Look for a word that names a longer measure of time than a month.
4. A year is longer than a month.
5. Both hour and week are shorter periods of time than day or month and do not continue the increasing sequence. Note the five-word sequence: *hour, week, day, month, year.*

H-66 first, second, **third**

1. represent order: second comes next after first
2. increasing
3. Look for a word meaning coming next after second.
4. Third comes next after second.
5. *Beginning* has a meaning similar to first; *last* is less definite, it does not convey the concept of number.

H-67 Halloween, Thanksgiving, **Christmas**

1. represent order: Thanksgiving is the next holiday after Halloween
2. increasing
3. Look for the next holiday after Thanksgiving.

4. Christmas is the next holiday after Thanksgiving.
5. Both April Fools' Day and Labor Day come before Halloween and Thanksgiving in the calendar year.

H-68 pint, quart, **gallon**

1. measures of volume (rank): the quart is larger than the pint
2. increasing
3. Look for a word that represents a volume larger than a quart.
4. A gallon is larger than a quart.
5. Both a cup and a teaspoon are smaller volumes than a pint.

H-69 more, same, **less**

1. represent degree or amount: *same* indicates a smaller amount than *more*
2. decreasing
3. Look for a word that would indicate a smaller amount than *same*.
4. *Less* indicates a smaller amount than *same*.
5. *Most* is the largest amount, not a smaller amount; *equal* is a synonym for *same*.

H-70 small, medium, **large**

1. represent size: medium is bigger than smaller
2. increasing
3. Look for a word that indicates a size bigger than medium.
4. Large is bigger than medium.
5. *Tiny* and *regular* can be used as synonyms for the given words; they do not continue the increasing sequence.

H-71 some, more, **most**

1. represent degree (amount): *more* is a greater amount than *some*
2. increasing
3. Look for a word that indicates a greater amount than *more*.
4. *Most* means the greatest amount.
5. *Few* indicates a lesser amount than *more* and does not indicate an increasing sequence; *many* could indicate a greater amount than *some,* but *more* could be a greater amount than *many.*

H-72 inch, foot, **yard**

1. represent size or length: inch is the smallest length
2. increasing
3. Look for a word that indicates a size longer than a foot.
4. A yard represents 3 feet.
5. A minute measures time, not length; a pound measures weight, not length.

H-73 above, beside, **below**

1. represent order (space): *beside* indicates a position in space lower than *above*
2. decreasing
3. Look for a word that indicates a space lower than *beside*.
4. *Below* indicates a position lower in space than *beside*.
5. *Over* means the same as *above*, and *between* indicates a spatial relationship similar to *beside*.

H-74 bad, worse, **worst**

1. represent degree: *worse* is less desirable than *bad*
2. decreasing
3. Look for a word that indicates a degree less desirable than *worse*.
4. *Worst* indicates a degree lower than *worse*.
5. *Better* is more desirable than *bad* and does not indicate decreasing sequence; *less* is a measure of amount, rather than degree.

H-75 boy, teen, **man**

1. represent rank (or possibly size): a teen is older (bigger) than a boy
2. increasing
3. Look for a word that names someone older (bigger) than a teen.
4. A man is older (bigger) than a teen.
5. Both a baby and a child are younger (smaller) than a teen and do not indicate increasing sequence. Note again that all five words can be sequenced: *baby, child, boy, teen, man.*

H-76 enter, stay, **leave**

1. represent time order: after you enter, you stay
2. decreasing
3. Look for a word that comes after *stay*.
4. *Leave* comes later in time than *stay*; after you stay, you leave.
5. *Arrive* means the same as *enter*, and *remain* means the same as *stay*. They are synonyms for the given words.

H-77 good, better, **best**

1. represents the degree of worth or goodness
2. increasing
3. Look for a word that means of more or higher worth than *better*.
4. *Best* is the highest degree of quality.
5. *Worse* expresses the lower degree of worth; *well* expresses a degree of worth similar to *good*.

LESSON PREPARATION

OBJECTIVE AND MATERIALS

OBJECTIVE: Students will select the word that best continues a sequence of rank, degree, size, or order.

MATERIALS: Transparency of student book page 229 • washable transparency marker

CURRICULUM APPLICATIONS

Language Arts: Expressing comparative and superlative rank of adjectives or adverbs; exercises involving chronological order (e.g., writing narratives or story plots); speech preparation; selecting nouns and verbs to express degree, rank, or order

Mathematics: Solving word problems involving transitivity or inequality, describing geometric proportions in angle or size

Science: Identifying stages of development, recognizing and predicting size or frequency, writing reports of science demonstrations

Social Studies: Recognizing and using chronological order for historical events or people, recognizing divisions and subdivisions of governmental or political structure

Enrichment Areas: Describing gradations of color or size in art, describing gradations of pitch or volume in music

TEACHING SUGGESTIONS

Encourage students to discuss reasons for their choices and why the other items don't fit. Each time students answer, conduct the following dialogue:

1. What is the relationship between the first two words?
2. Is the sequence increasing or decreasing?
3. What should be the relationship between the next two words?
4. Select the answer and confirm that the relationship exists.
5. Explain why other possible choices were eliminated.

While this procedure may seem somewhat lengthy, time-consuming, or obvious, it models the precision and thoroughness with which capable thinkers rapidly—and often unconsciously—examine a problem. By insisting on this procedure, you are establishing habits which will become speedy and routine. Promote language acquisition by encouraging learners to explain their answers using additional words implying rank, degree, size, or order.

MODEL LESSON

LESSON

Introduction

Q: Sometimes you get back an assignment with a B grade, and you are really relieved that you didn't get a C! Sometimes when you get back an assignment, your teacher tells you that you did good work. You really hoped she would say that the paper was your best work. In describing your grades, you are comparing your estimate of the degree of the quality of your work with your actual grades.

Explaining the Objective to the Students

Q: In this lesson, you will select a word that best continues a sequence of rank, degree, size, or order. Knowing the small differences in a sequence of words helps us to understand what we read and hear and to choose the best words to tell someone else what we mean.

Class Activity

Q: We will be thinking about word sequences. There are two kinds of word sequences...

- Write the types on the chalkboard.
  Q: ..."increasing sequences" and "decreasing sequences."
- Write the numerals "1," "2," and "3" on the board.
  Q: Is this number sequence increasing or decreasing?
  A: Increasing
- Write the letters "A," "B," and "C" on the board.
  Q: What kind of letter sequence is this, increasing or decreasing?
  A: Increasing: each letter has a later position in the alphabet.
- Write the numerals "10," "9," and "8" on the board.
  Q: Astronauts listen to this type of sequence. What kind of number sequence is this?
  A: Decreasing

  Q: Let's think of some words describing increasing size. I'll say a word and you think of two words that will continue the sequence by increasing size. The first word is *large*.
  A: Larger, largest; huge, monstrous; or various advertising slogans such as queen size, king size

  Q: Now, let's make a three-word sequence that shows increasing position starting with the word *back*.
  A: Middle, front

  Q: Now let's make up some decreasing word sequences. Think of words that show decreasing size starting with the word *small*.
  A: Smaller, smallest

  Q: Starting with the word *top*, think of two words that show a decrease in position.
  A: Middle, bottom
- Project the left side of the transparency of page 229, covering the choice boxes with a strip of paper.
  Q: In **H-63** on page 229, you are given the words *before* and *during*. What kind of sequence is this?
  A: It is an increasing time sequence because *before* happens earlier than *during*.

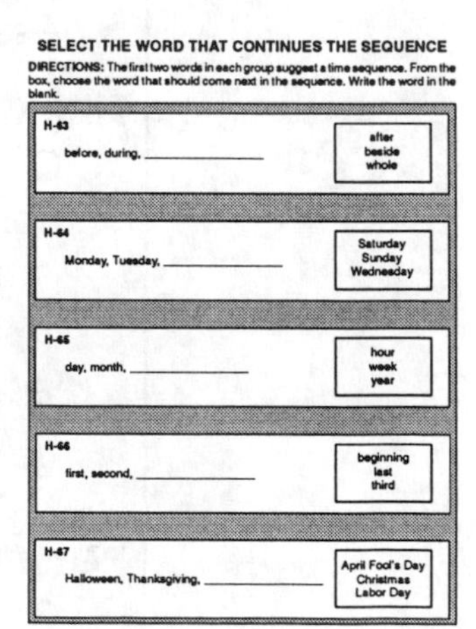

SELECT THE WORD THAT CONTINUES THE SEQUENCE

DIRECTIONS: The first two words in each group suggest a time sequence. From the box, choose the word that should come next in the sequence. Write the word in the blank.

H-63
before, during, ____________ (after / beside / whole)

H-64
Monday, Tuesday, ____________ (Saturday / Sunday / Wednesday)

H-65
day, month, ____________ (hour / week / year)

H-66
first, second, ____________ (beginning / last / third)

H-67
Halloween, Thanksgiving, ____________ (April Fool's Day / Christmas / Labor Day)

- Project the Choice Box for exercise **H-63**.
  Q: To continue this time sequence, I must find a word that means "happens later." Which of these words...
- Indicate the choices.
  Q: ...means "something that happens later"?
  A: After

- Write "after" on the answer line.

  Q: To make sure we have the best answer, we should always check the other words. Why is *beside* not a good choice?

  A: *Beside* refers to position; it does not indicate "happens later."

  Q: Why is *whole* not a good choice?

  A: *Whole* represents the entire amount; it does not indicate "happens later."

  Q: Let's look at **H-64**. What is the relationship between Monday and Tuesday?

  A: Tuesday is one day after Monday.

  Q: To continue this sequence, then, we must look for a word that means "the next day" or "one day after Tuesday."

- Expose the Choice Box for **H-64**.

  Q: Which of these words names one day after Tuesday?

  A: Wednesday

- Write "Wednesday" on the answer line.

  Q: What kind of sequence is this?

  A: Increasing

  Q: Why is Saturday not a good choice?

  A: It is the end of the week; several days come between Tuesday and Saturday.

  Q: Why is Sunday not a good choice?

  A: Because Sunday is the day before Monday (decreasing series) and would be the first item in the series.

GUIDED PRACTICE

EXERCISES: **H-63** through **H-67**

- When students have had sufficient time to complete these exercises, check by discussion to determine whether they have answered correctly.

INDEPENDENT PRACTICE

- Assign exercises **H-68** through **H-77**.

THINKING ABOUT THINKING

Q: What did you pay attention to when you decided which word came next?

1. I looked for the relationship between the words.
2. I determined if the sequence was increasing or decreasing.
3. I decided which word would come next in the sequence.
4. I checked that the other words didn't fit the pattern.

PERSONAL APPLICATION

Q: When do you want or need to know what comes next in a given order or rank?

A: Examples include understanding and describing test or game results; telling and understanding jokes, puns, or stories; distinguishing the size or worth of objects; playing word puzzles and games; understanding consumer product terms describing size or volume.

EXERCISES H-78 to H-94

RANKING

ANSWERS H-78 through H-94 — Student book pages 232–34
Guided Practice: H-78 morning, afternoon, night; **H-79** never, sometimes, always; **H-80** regular, large, giant; **H-81** many, more, most; **H-82** first, middle, last
Independent Practice: H-83 dime, quarter, dollar; **H-84** sidewalk, street, highway; **H-85** pond, lake, ocean; **H-86** city, state, nation; **H-87** month, year, century; **H-88** penny, nickel, dime; **H-89** tail, body, head; **H-90** thread, string, rope; **H-91** out, inning, game; **H-92** start, continue, finish; **H-93** word, page, book; **H-94** Ping-Pong ball, baseball, basketball

DETAILED SOLUTIONS

Note: The detailed solutions follow the model shown in the Teaching Suggestions on page 189.

H-78

1. The given words are morning, afternoon, and night. They are related as times of day.
2. The ranking is from earliest to latest.
3. Synonyms arranged in order are the earliest time, the middle of the day, the latest time.
4. The given words in order are *morning, afternoon, night.*
5. Morning would be the earliest (first), afternoon would be second, and night would be latest (third).

H-79

1. always, never, sometimes: frequency of occurrence
2. increasing frequency
3. happens at no time, happens now and then (occasionally), happens all the time
4. *never, sometimes, always*
5. Something that never happens is less likely to occur than something that happens sometimes or always. Something that sometimes happens occurs more often than never but less often than always. Something that always happens occurs most often.

H-80

1. giant, large, regular: measurement of size
2. increasing size
3. average or normal size, bigger than average or normal size, largest size
4. *regular, large, giant*

5. Regular indicates that something is of average or normal size. Large is bigger than average, but smaller than giant. Giant is bigger than both regular and large (think of containers or shirt sizes).

H-81

1. many, more, most: measurement of amount, number, or quantity
2. increasing amount, number, or quantity
3. a large amount or number, a larger amount or number, the largest amount or number
4. *many, more, most*
5. Many indicates a large number or amount of something. More indicates a larger number or amount than many but a smaller number or amount than most. Most indicates the greatest percentage, larger than either many or more.

H-82

1. first, last, middle: order of position
2. increasing order
3. to be in the beginning position, to be halfway between the beginning and the end, to be at the end of something
4. *first, middle, last*
5. To be first indicates something at the beginning position, in front of last and middle. Middle names the position halfway between first and last. Last is the name of the end position, behind first and middle.

H-83

1. dime, dollar, quarter: amounts or value of money
2. increasing value
3. 10¢, 25¢, 100¢
4. *dime, quarter, dollar*
5. A dime has a value of 10¢, which is less than the 25¢ value of a quarter. A dollar is worth 100¢, which is greater than the value of a dime or a quarter.

H-84

1. highway, sidewalk, street: types of passageways
2. increasing size (width)
3. smallest (narrowest) passageway, middle-sized passageway, largest (widest) passageway
4. *sidewalk, street, highway*
5. A sidewalk is smaller and narrower than both a street and a highway. A street is larger than a sidewalk and smaller (narrower) than a highway. A highway is both longer and wider than both a street and a sidewalk. Note: Some students may rank these passageways according to how fast you can travel on them; the final order, however, will be the same.

H-85

1. lake, ocean, pond: bodies of water
2. increasing size
3. smallest body of water, middle-sized body of water, largest body of water
4. *pond, lake, ocean*
5. Ponds are very small lakes. Lakes vary in size from extra large ponds to lakes the size of the Great Lakes. All the oceans are larger than any of the Great Lakes.

H-86

1 city, nation, state: geographic region
2. increasing size
3. smallest geographic region, middle-sized geographic region, largest geographic region
4. *city, state, nation*
5. Cities are contained within states and states are contained within nations.

H-87

1. century, month, year: amount of time
2. increasing length of time
3. shortest time given, medium length of time given, longest length of time given
4. *month, year, century*
5. A month is one-twelfth of a year; a year is 12 months; a century is 100 years.

H-88

1. dime, nickel, penny: values of money (Alternate: dimensions of coins)
2. increasing value (size)
3. least valuable (smallest) coin, medium-valued (middle-sized) coin, most valuable (largest) coin
4. *penny, nickel, dime* (Alternate: dime, penny, nickel)
5. A penny has a value of 1 cent, a nickel is worth 5 cents, and a dime has the highest value of the given words at 10 cents. (Alternate: A dime is smaller in size than both a penny and a nickel; a penny is larger than a dime and smaller than a nickel in size; a nickel is larger than both a penny and a dime in dimension.)

H-89

1. body, head, tail: parts of an animal's body
2. increasing position (order)
3. the back part of an animal's body, the middle part of an animal's body, the front part of an animal's body
4. *tail, body, head*

5. The tail is located at the back of an animal. The body is in the middle, between the head and the tail. The head is located at the front.

H-90

1. rope, string, thread: things used for tying
2. increasing thickness (strength) (weight)
3. a thin (weak) type of tying material, a medium-thick (medium-strength) type of tying material, a thick (strong) type of tying material
4. *thread, string, rope*
5. A thread is a thin, light, and usually weak cord used for tying things together. A string is thicker, heavier, and stronger than a thread but thinner, lighter, and usually weaker than a rope. It is made up of several threads. A rope is a thick, heavy, strong cord made up of several strings of thread.

H-91

1. game, inning, out: parts of a baseball or softball contest
2. increasing order (part of)
3. part of an inning, part of a game, entire contest
4. *out, inning, game*
5. An out is a part of an inning (6 outs to an inning). An inning is part of a game (usually 9 innings to each game). A game refers to the entire contest (54 outs and 9 innings).

H-92

1. continue, finish, start: parts of an action
2. increasing order
3. to begin an action, to be in the middle of an action, to end an action
4. *start, continue, finish*
5. To start something means to begin to do it. To continue something means to be in the middle of it (already started but not yet finished). To finish something means to be done with it or to end it.

H-93

1. book, page, word: forms (units) of printed communication
2. increasing size (part of)
3. smallest unit of printed communication, middle-sized unit of printed communication, a large unit of printed communication
4. *word, page, book*
5. A word is the smallest unit of communication; it is part of a sentence. A page is a medium-sized unit of communication; it is made up of many words and sentences but is smaller than a book. A book is larger than both a word and a page; it is made up of many words and many pages.

H-94

1. baseball, basketball, Ping-Pong ball: balls used in sports
2. increasing size

3. smallest ball, medium-sized ball, largest ball
4. *Ping-Pong ball, baseball, basketball*
5. A Ping-Pong ball is smaller than a baseball. A baseball is larger than a Ping-Pong ball and smaller than a basketball. A basketball is larger than both a baseball and a Ping-Pong ball.

LESSON PREPARATION

OBJECTIVE AND MATERIALS

OBJECTIVE: Students will rearrange three words to form an increasing sequence of words.

MATERIALS: Transparency of student page 232 (with answers removed) • washable transparency marker

CURRICULUM APPLICATIONS

Language Arts: Sequencing events and ideas in a story; outlining exercises; expressing comparative and superlative rank of adjectives or adverbs; exercises involving chronological order (e.g., writing narratives or letters, relating story plots); selecting nouns and verbs to express degree, rank, or order

Mathematics: Ordering numbers by size; measurement activities involving weight, volume, etc.; solving word problems involving transitivity or inequality; describing geometric proportions in angle or size

Science: Classifying physical phenomena, recognizing and predicting size or frequency, writing reports of science demonstrations

Social Studies: Recognizing chronological order and using it to place historical events, eras, artifacts, cultures, and people into proper time relationships; recognizing divisions and subdivisions of governmental or political structures

Enrichment Areas: Describing gradations of color or size in art; describing gradations of pitch, rhythm, or volume in music; recognizing and describing degrees of expertise in any area

TEACHING SUGGESTIONS

Encourage students to discuss their answers using the model provided and to classify each sequence (size, degree, rank, order, time, length, etc.). Remember the language that students use to describe their choices and use the same words to remind students of the key concepts in this lesson. The following cues model the discussion:

1. Pronounce given words and state their area of commonality.
2. State characteristic of sequence, i.e., what is being ranked.
3. Arrange synonyms of given words into desired order.
4. Arrange given words into desired order (answer).
5. Confirm arranged order.

MODEL LESSON

LESSON

Introduction

Q: You have selected a word to continue a sequence.

Explaining the Objective to the Students

Q: In this lesson, you will put the given words in increasing order from lowest or smallest to highest or largest based on size, degree, rank, or order.

Class Activity

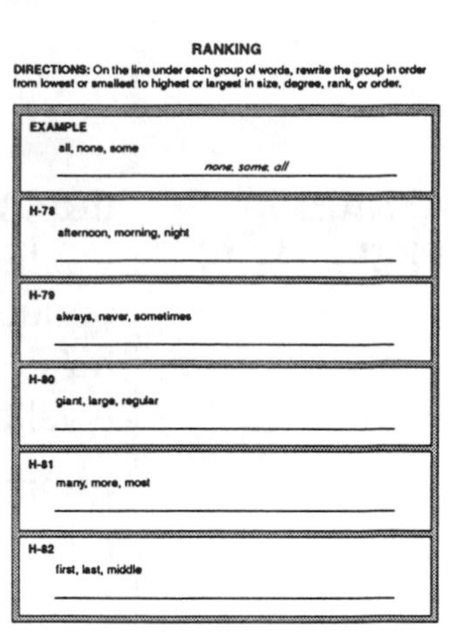

RANKING

DIRECTIONS: On the line under each group of words, rewrite the group in order from lowest or smallest to highest or largest in size, degree, rank, or order.

EXAMPLE
all, none, some
none, some, all

H-78
afternoon, morning, night

H-79
always, never, sometimes

H-80
giant, large, regular

H-81
many, more, most

H-82
first, last, middle

- Project the example problem from the transparency of page 232.
  Q: In this example from page 232, we are given the words *all*, *none*, and *some*. We will rank these three words as an increasing sequence. What do the three words describe?
    A: Amount

- Point to the answer line.
  Q: In an increasing sequence, we go from smallest to largest. Which word represents the smallest amount?
    A: None

- Write “none” at the beginning of the answer line.
  Q: Which word represents the largest amount?
    A: All

- Write “all” at the end of the answer line, leaving a space in the middle.
  Q: Which word represents a middle amount?
    A: Some

- Write “some” on the line between the other two answers.
  Q: If *none* is the smallest amount, *some* is a middle amount, and *all* is the largest amount represented by the words we were given, have we arranged the words from smallest to largest?
    A: Yes

- Move the covering to project exercise **H-78**.
  Q: Now let’s look at exercise **H-78**. What do the words *afternoon*, *morning*, and *night* describe?
    A: Time of day

  Q: To arrange these times into an increasing sequence, into what type of order would we put them?
    A: From earliest to latest

  Q: Which words would come first, second, and third in this sequence?
    A: Morning (earliest) would be first, afternoon (middle) second, and night (latest) third.

- Write the answers on the line.

GUIDED PRACTICE

EXERCISES: **H-78** through **H-82**

- When students have had sufficient time to complete these exercises, check answers by discussion to determine whether they have answered correctly. Students should identify the characteristic of each sequence (size, degree, rank, order, time, length, etc.).

INDEPENDENT PRACTICE

- Assign exercises **H-83** through **H-94**.

THINKING ABOUT THINKING

Q: What did you pay attention to when you decided which word came next?

1. I decided what each group of words had in common.
2. I selected the word that was smallest or lowest in the sequence.
3. I then put the next words in increasing order.
4. I checked that the words were arranged in the correct sequence.

PERSONAL APPLICATION

Q: When do you want or need to know what comes next in a given order or ranking?

A: Examples include understanding and describing test or game results; telling and understanding jokes, puns, or stories; distinguishing the size or worth of objects; playing word puzzles and games; understanding consumer product terms describing size or volume.

EXERCISES H-95 to H-112

SUPPLY A WORD THAT CONTINUES A SEQUENCE

ANSWERS H-95 through H-112 — Student book pages 235–36
Guided Practice: H-95 hour (day, week, etc.) (time, increasing) **H-96** go (increasing in terms of preparedness or decreasing in terms of time remaining before start) **H-97** May; **H-98** dinner (supper)
Independent Practice: H-99 yard (mile) **H-100** large (big, grand, super, etc.) **H-101** end; **H-102** fall; **H-103** dime (quarter, dollar, etc.) **H-104** thirty (forty if doubling) **H-105** tomorrow; **H-106** fourth; **H-107** woman (teen, lady, etc.) **H-108** ton; **H-109** gone; **H-110** black; **H-111** eat (dine, wash, etc.) **H-112** five

DETAILED SOLUTIONS

Note: The detailed solutions follow the pattern shown in the Teaching Suggestions on page 195.

H-95

1. The given words are *second* and *minute*. Both words are measures of time.
2. A minute is 60 seconds.
3. The sequence is increasing in length of time.
4. Think of a word that represents a period of time longer than a minute.
5. *Hour*, *day*, *week*, etc. are longer than a minute.

H-96

1. ready, set: preparedness (starting a race)

2. *Ready* means prepared for action; *set* means in the starting position.
3. increasing in terms of preparedness (decreasing in terms of time remaining before start)
4. Think of a word that comes after *set* when preparing for something or starting a race.
5. *Go* means the action is started.

H-97

1. March, April: months of the year
2. March is the third month of the year. April is the fourth month.
3. increasing sequence of position in time
4. Think of a word that names a month that comes after April.
5. *May* is the fifth month of the year.

H-98

1. breakfast, lunch: meals of the day
2. Breakfast is the meal eaten first (in the morning); lunch is the second meal of the day (at noon).
3. increasing sequence of time (earliest-latest)
4. Think of a word that names a meal eaten after lunch (in the evening).
5. *Dinner (supper)* is the last meal of the day (in the evening).

H-99

1. inch, foot: units of measure of length
2. A foot equals 12 inches; a foot is longer than an inch.
3. increasing sequence
4. Think of a word that names a unit of measure longer than a foot.
5. A yard or a mile is longer than a foot (a yard equals 3 feet and a mile is 5280 feet).

H-100

1. small, medium: measurement of size
2. Small is something tiny, and medium is middle-sized or average.
3. increasing sequence of size
4. Think of a word that describes something bigger than medium.
5. *Large, giant, jumbo* are all words that mean bigger than medium.

H-101

1. beginning, middle: sections or parts of something
2. Beginning indicates first in time or position. Middle is later than first, about halfway through something.
3. increasing in time or position
4. Think of a word that names a section that comes after (later than) middle.
5. *End, conclusion, last* all name times or positions after middle.

H-102

1. spring, summer: seasons of the year
2. Spring is early in the year, starting March 21. Summer follows spring and starts June 21.
3. increasing in order or time
4. Think of a word that names a season that comes later in the year after summer.
5. *Fall, autumn, winter* all name later seasons; fall or autumn starts in September, and winter begins in December

H-103

1. penny, nickel: coins
2. A penny has a value of 1¢; a nickel has a value of 5¢.
3. increasing in value
4. Think of a word that names a coin with a value greater than 5¢.
5. *Dime, quarter, half dollar* all name coins with values greater than 5¢.

H-104

1. ten, twenty: numbers ending in zero (counting by tens)
2. Ten is the lowest positive number ending in zero. Twenty is the next possible higher number ending in zero (ten is the first number when counting by tens; twenty is the second number when counting by tens).
3. increasing sequence of numbers or amounts
4. Think of a higher number than twenty that ends in zero (the third number when counting by tens).
5. *Thirty* is the next number higher than twenty that ends in zero and also the third number when counting by tens (forty, fifty, etc. might also be used if the student explains the relationship).

H-105

1. yesterday, today: generic names for days of the week
2. *Yesterday* names the day before this day; *today* names this day.
3. increasing sequence of time
4. Think of a general name for the day after this day.
5. *Tomorrow* names the day after this day.

H-106

1. second, third: ordinal numbers indicating rank or standing
2. *Second* names the rank or standing after first. *Third* names the next rank or standing after second.
3. decreasing in rank or standing
4. Think of a word that names the rank or standing after third.
5. *Fourth* is the ordinal number that names the rank or standing after third.

H-107

1. baby, girl: stages in life (female)

2. *Baby* names the youngest stage in human life. *Girl* names a female child.
3. increasing in age
4. Think of a word that names a stage in life after childhood for a female.
5. *Woman* or *mother* names a female in the next stage of life after girl. (Alternate answers might be teen or adult, but they do not specify female.)

H-108

1. ounce, pound: measure of weight
2. Ounce is the smallest common measure of weight in the United States (1/16 of a pound); pound is the next largest common measure (unit) of weight (16 ounces).
3. increasing unit of weight
4. Think of a word that names the unit of weight greater than a pound.
5. *Ton* is the next larger common unit of weight. (Note: Although products are packaged in ounces or one-pound packages, followed by 2, 5, and 10 pound packages, those are not different units of weight.)

H-109

1. go, going: state of action
2. *Go* means to start moving; *going* means the action has already started and is continuing.
3. increasing in time (earlier to later)
4. Think of a word that means an action already started and ended.
5. *Gone* indicates an action finished.

H-110

1. white, gray: colors
2. White is the lightest color given. Gray is darker than white (mixture of black and white).
3. increasing degree of darkness (decreasing degree of lightness)
4. Think of a color that is darker than gray.
5. *Black* is a color that is darker than gray.

H-111

1. cook, serve: parts of preparing a meal
2. To cook means to prepare food for a meal. To serve means the act of putting food on the plate or table and comes after cook.
3. increasing in time (earlier to later)
4. Think of a word that names the part of a meal that comes after serving.
5. *Eat, dine, wash dishes,* and *save leftovers* all name parts of a meal that come after serving the food.

H-112

1. one, three: odd numerals

2. One is the first odd number; three is the second odd number.
3. increasing by two
4. Think of a number two larger than three.
5. *Five* is the next odd number.

LESSON PREPARATION

OBJECTIVE AND MATERIALS

OBJECTIVE: Students must think of the next word that will continue an increasing sequence.

MATERIALS: Transparency of student book page 235 • washable transparency marker

CURRICULUM APPLICATIONS

Language Arts: Sequencing events by increasing importance; expressing comparative and superlative rank of adjectives or adverbs; exercises involving chronological order (i.e., writing narratives or letters, relating story plots); selecting nouns and verbs to express degree, rank, or order

Mathematics: Ordering types of measurement by size, quantity, etc.; solving word problems involving transitivity or inequality; describing geometric proportions in angle or size

Science: Identifying and sequencing stages of development, recognizing and predicting size or frequency, writing reports of science demonstrations

Social Studies: Recognizing chronological order and using it to place historical events, eras, artifacts, cultures, and people into proper time relationships; recognizing divisions and subdivisions of governmental or political structures

Enrichment Areas: Describing gradations of color or size in art; describing gradations of pitch, rhythm, or volume in music; recognizing and describing degrees of expertise in any area

TEACHING SUGGESTIONS

Encourage students to discuss and justify their answers. Discussion techniques should follow the pattern shown below:

1. Pronounce given words and state their common relationship.
2. Define the given words.
3. State characteristic of sequence, i.e., what is being ranked.
4. Think of a synonym for the answer.
5. Supply a word that fits the answer.

Students may use a dictionary or Thesaurus to search for unusual or uncommon words that would continue the sequence. Encourage students to name additional words in a given sequence.

MODEL LESSON

LESSON

Introduction

Q: You have arranged words to form an increasing sequence.

Explaining the Objective to the Students

Q: In this lesson, you will think of a word that will continue an increasing sequence.

Class Activity

- Project exercise **H-95** from the transparency of page 235.

  Q: In this sample exercise from page 235, you are given the two words *second* and *minute*. What do the two words have in common?

  A: They are measures of time.

  Q: Before we can add the next word, we need to decide whether the measures of time are increasing or decreasing in length so that we know whether we have an increasing or decreasing sequence. Which is longer, a second or a minute?

  A: A minute is longer than a second.

SUPPLY A WORD THAT CONTINUES A SEQUENCE

DIRECTIONS: The first two words in each sequence suggest a degree, rank, size, or order. Think of a word that will continue the sequence and write it on the line. You may use a dictionary to get help.

H-95 second, minute, ______

H-96 ready, set, ______

H-97 March, April, ______

H-98 breakfast, lunch, ______

H-99 inch, foot, ______

H-100 small, medium, ______

H-101 beginning, middle, ______

H-102 spring, summer, ______

H-103 penny, nickel, ______

  Q: Since *minute* comes second in the sequence and is a longer period of time, we have an increasing sequence. Can you think of a word representing an amount of time longer than a minute?

  A: Hour, day, week, month, year, etc. (*Hour* is the most probable answer.)

- Write the answers on the blank. Project exercise **H-99**.

  Q: Now look at exercise **H-99**. You are given the words *inch* and *foot*. What do these words describe?

  A: Measures of length

  Q: Are the given measures of length increasing or decreasing?

  A: Increasing

  Q: Can you think of a word representing a length longer than a foot?

  A: Yard, mile, etc.

- Write the answers on the blank.

  Q: We have shown that we must first determine what the given words represent then determine whether the sequence is an increasing or decreasing sequence before we can determine a word that will continue the sequence.

GUIDED PRACTICE

EXERCISES: **H-95** through **H-98**

- When students have had sufficient time to complete these exercises, check their answers using the discussion technique modeled in Teaching Suggestions.

INDEPENDENT PRACTICE

- Assign exercises **H-99** through **H-112**.

THINKING ABOUT THINKING

Q: What did you pay attention to when you decided which word came next?

1. I looked for a common meaning between the words.
2. I looked for a pattern of change. Was the pattern increasing or decreasing?
3. I figured out, if the change in meaning continued, what word would be next.

PERSONAL APPLICATION

Q: When do you want or need to know what comes next in a given order or ranking?

A: Examples include understanding and describing test or game results; understanding and telling jokes, puns, or stories; distinguishing the size or worth of objects; playing word puzzles and games; understanding consumer product terms describing size or volume.

EXERCISES H-113 to H-124

WARM-UP DEDUCTIVE REASONING

ANSWERS H-113 through H-124 — Student book pages 238–41
Guided Practice: H-113 The ranking (heavy–light) is Charlie, Bill, Albert; Charlie is the heaviest. **H-114** The ranking (young–old) is Betty, James, Alice; Betty is the youngest. **H-115** The ranking (most–least) is Irene, Sally, Clare; Irene has the most cats.
Independent Practice: H-116 Doug, Ivan, Lee (fast–slow); Lee, Ivan, Doug (slow–fast) **H-117** Lois, John, Sam (old–young); Sam, John, Lois (young–old) **H-118** The ranking (tall–short) is David, Fred, George, Harold; David is the tallest. **H-119** The ranking (short–tall) is June, Delores, Mary, Nancy; June is the shortest; Nancy is the tallest. **H-120** Spiro, Gina, Nick (most–few); Nick, Gina, Spiro (few–most) **H-121** Jose, Pedro, Manuel (young–old) **H-122** Larry, Emil, Carlos (high–low); Carlos, Emil, Larry (low–high) **H-123** Nina, Pedro, Sol (most–few); Sol, Pedro, Nina (few–most) **H-124** Larry, John, Marna (tall–short); Marna, John, Larry (short–tall)

LESSON PREPARATION

OBJECTIVE AND MATERIALS

OBJECTIVE: Students use words describing size, degree, rank, or order as clues to understanding the meaning of sentences.

MATERIALS: Transparency of student book page 237 • washable transparency marker

CURRICULUM APPLICATIONS

Language Arts: Comprehending subtle differences in word meaning based on context; using key words and phrases to infer a character's personality and motivation in a story; comprehending subtle distinctions in degree, size, rank, or order in reading passages; understanding and writing compare/contrast passages or papers

Mathematics: Applying key terms and phrases to solve word problems, transitivity relations, or inequality exercises

Science: Understanding directions and results of laboratory demonstrations or experiments, recognizing and analyzing variables in experiments

Social Studies: Comprehending chronological order or statistical comparisons in text materials

Enrichment Areas: Playing deductive thinking games, such as "Twenty Questions," or logic problems and games

TEACHING SUGGESTIONS

Encourage students to discuss answers and explain solutions. They should justify their ranking and point out specifically where the information came from.

MODEL LESSON

LESSON

Introduction

Q: You have identified and continued a given word sequence.

Explaining the Objective to the Students

Q: In this lesson, you will use words describing size, degree, rank, or order as clues to understanding the meaning of sentences.

Class Activity

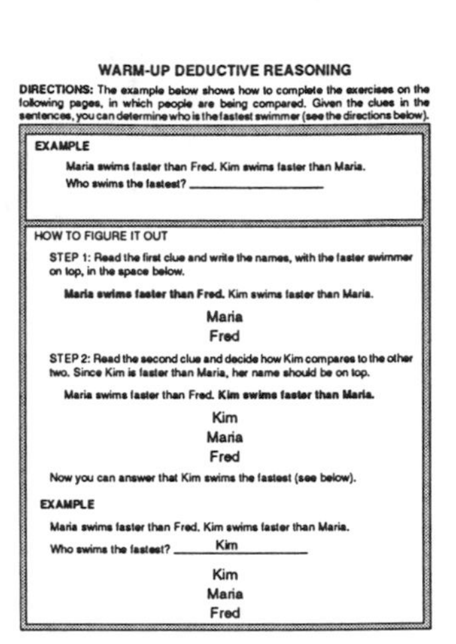

WARM-UP DEDUCTIVE REASONING

DIRECTIONS: The example below shows how to complete the exercises on the following pages, in which people are being compared. Given the clues in the sentences, you can determine who is the fastest swimmer (see the directions below).

EXAMPLE

Maria swims faster than Fred. Kim swims faster than Maria.

Who swims the fastest? ________

HOW TO FIGURE IT OUT

STEP 1: Read the first clue and write the names, with the faster swimmer on top, in the space below.

Maria swims faster than Fred. Kim swims faster than Maria.

Maria
Fred

STEP 2: Read the second clue and decide how Kim compares to the other two. Since Kim is faster than Maria, her name should be on top.

Maria swims faster than Fred. **Kim swims faster than Maria.**

Kim
Maria
Fred

Now you can answer that Kim swims the fastest (see below).

EXAMPLE

Maria swims faster than Fred. Kim swims faster than Maria.

Who swims the fastest? ___Kim___

Kim
Maria
Fred

- Project the transparency of page 237, covering the page under "Maria swims faster..." in step 1. Uncover each section as needed.

  Q: In the example on page 237, you will rank three swimmers to find out who is fastest. Step one is to read the first clue and write the names in order. The first clue says, "Maria swims faster than Fred." Write the faster swimmer's name on top.

  A: Maria above Fred

  Q: The second clue reads: "Kim swims faster than Maria." Since Kim is faster than Maria, we will put her name on top of the list above Maria's.

  A: Kim above Maria above Fred

  Q: From the list you made, can you now tell who swims the fastest?

  A: Kim

- Write "Kim" on the blank.

  Q: You should read each clue carefully and rank the information you are given in each one. Be sure the name you choose to answer the question is in the correct position in the ranking.

GUIDED PRACTICE

EXERCISES: **H-113** through **H-115**

- When the students have had sufficient time to complete these exercises, check answers by discussion to determine whether students have answered correctly. They should always be able to explain their ranking by referring to specific words in specific clues.

INDEPENDENT PRACTICE

- Assign exercises **H-116** through **H-124**.

THINKING ABOUT THINKING

Q: What did you pay attention to when you decided the order in which things rank?

1. I read the clue to decide what personal characteristic (weight, age, height, score, etc.) was being compared.

2. I compared the first two people mentioned.
3. I figured out how the third person compared with the other two.
4. I was then able to answer the question asked in the exercise.

PERSONAL APPLICATION

Q: When do you need to rank things?

A: Examples include charting test scores or sports standings; comparison shopping; organizing clues to solve puzzles; interpreting the chronology of events in newspaper articles, television shows, and stories.

EXERCISES H-125 to H-131

DEDUCTIVE REASONING

ANSWERS H-125 through H-131 — Student book pages 243–46
Guided Practice: H-125 Hernando was first; Isaac was third; Juanita was second. **H-126** Donna was first; Ernie was third; Frank was second. **Independent Practice: H-127** The one-speed is owned by Kyle; the three-speed is owned by Juan; the ten-speed is owned by Lori. **H-128** Green is in the fourth grade; Jones is in the third grade; Perez is in the second grade. **H-129** The 3 year old is a boy; the 6 year old is a girl; the 12 year old is a girl. **H-130** Amanda likes salads; Desiree likes beef; Jose likes chicken. **H-131** The basketball shoes are owned by Nick; the running shoes are owned by Tim; the sneakers are owned by Pablo.

DETAILED SOLUTIONS

	1st	2nd	3rd
H	Y	N	N
I	N	N	Y
J	N	Y	N

H-125 Clue 3 tells us that neither Isaac nor Juanita was the winner, for the winner spelled five words correctly. Isaac spelled three words correctly (clue 1), and Juanita spelled four words correctly (clue 2: Isaac's number + one). Thus, the matrix can be marked by putting a "Y" for "Yes" in the H-1 cell (Hernando—1st place), and an "N" for "No" in H-2 and H-3 (if Hernando is first, he cannot be second or third) and in I-1 and J-1 (if Hernando is first, neither Isaac nor Juanita can be first). Clue 2 tells us that Juanita correctly spelled more words than Isaac, but the matrix tells us that she did not win first place. Thus, she must have finished second. Mark a "Y" in cell J-2 and an "N" in cells I-2 and J-3. (If Juanita is second, Isaac cannot be second, and if Juanita is second, she cannot be third.) By checking the matrix, we find that Isaac is not first or second, so Isaac must be third. Mark a "Y" in cell I-3 on the matrix.

	1st	2nd	3rd
D	Y	N	N
E	N	N	Y
F	N	Y	N

H-126 Clue 1 implies that Ernie did not finish the race, although the other two did. Thus, Ernie is third. Mark a "Y" in cell E-3 and "N" in cells E-2, E-1, D-3, and F-3. (If Ernie is third, he cannot be either first or second; if he is third, then neither

Donna nor Frank is third.) Clue 2 implies that Frank is not the winner. Since the matrix tells us that he also did not finish third, he must have finished second. Mark the matrix accordingly ("Y" in cell F-2; "N" in cells F-1 and D-2). Row D now has only one unfilled cell, as does the column marked 1st. Thus, Donna must have finished in first place. (Mark a "Y" in cell D-1.)

H-127 Clue 1 implies that Juan and Lori do not own one-speed bikes. Mark an "N" in the J-1 and L-1 cells, leaving only one cell open in the "1" column. Thus, Kyle must own the one-speed. Mark a "Y" in the K-1 cell and "N" in both the K-3 and K-10 cells. (If neither Juan nor Lori own the one-speed, then Kyle does; if Kyle owns the one-speed, then he cannot own the three-speed or the ten-speed.) Clue 2 says that Lori's bike has the most gear speeds, so she must own the ten-speed. Mark a "Y" in cell L-10. If she owns the ten-speed, then she cannot own the three-speed ("N" in cell L-3), and no one else can own the ten-speed ("N" in cell J-10). The matrix is now completed except for cell J-3, so Juan must own the three-speed. (Mark a "Y" in cell J-3.)

	1	3	10
J	N	Y	N
K	Y	N	N
L	N	N	Y

H-128 Clue 1 says that Jones is in an odd-numbered grade. Since the only odd-numbered-grade choice available is third grade, Jones must be in the third grade. (Mark cell J-3 "Y.") If Jones is in the third grade, (s)he cannot be in either the second or the fourth grades. (Mark cells J-2 and J-4 "N.") If Jones is in the third grade, neither of the others can be in third grade. (Mark cells G-3 and P-3 "N.") Clue 2 says that Green is in a higher grade than Jones. If Jones is in grade three, then Green must be in grade four. (Grade two, the other choice, is a lower grade; mark cell G-4 "Y.") If Green is in grade four, (s)he cannot be in another grade, and no one else can be in grade four. (Mark cells G-2 and P-4 "N.") The only grade left for Perez is second grade. (Mark cell P-2 "Y.")

	G	J	P
2	N	N	Y
3	N	Y	N
4	Y	N	N

H-129 From clue 2, we know that the oldest grandchild is a girl who received a 10-speed bike. (The words "her sister" tell us two things: that the oldest grandchild is a girl and that two of the grandchildren are girls. "Her" refers to one girl; "sister" refers to a second girl.) Mark a "Y" in cell G-12 and an "N" in cell B-12. There are two clues left. The grandson (boy) received a tricycle (clue 1), and the second granddaughter (girl) received a bicycle with training wheels. One can infer that the youngest grandchild would logically receive the tricycle, so the 3 year old is probably a boy. (Mark a "Y" in cell B-3 and an "N" in cell G-3.) Since we know that two of the grandchildren are girls (clue 2), mark cell B-6 "N." The 6 year old must be the other girl. (Mark cell G-6 "Y.")

	B	G
3	Y	N
6	N	Y
12	N	Y

H-130 If Amanda does not eat meat, then cells A-B and A-C can be marked with an "N" and cell A-S with a "Y." If Jose does not eat beef, then J-B can be marked "N." There are two "N's" in the "B" column, so a "Y" can be written in the D-B cell. If Desiree's favorite is beef then her favorites are not chicken or salads. Mark "N's" in the D-C and D-S cells. The C column contains two "N's" and, therefore, the J-C cell should be marked "Y."

	B	C	S
A	**N**	**N**	**Y**
D	**Y**	**N**	**N**
J	**N**	**Y**	**N**

H-131 Note: You can keep track of each boy's shoe size by adding a Boy column between the two matrices. The Shoe Size column should be filled in (top to bottom) Large, Medium, Small. Clue 1 tells you that the basketball shoes are the largest. Mark a "B" (Basketball) in the large shoe-style cell. Clue 2 tells you that Tim has the smallest feet. From this you know that Tim does not own the basketball shoes (mark an "N" in the T-B cell of the large matrix and put a "T" (Tim) beside the "small" row in the Boy column. Clue 3 tells you that Pablo owns the sneakers. Mark a "Y" in the S-P cell and an "N" in cells S-N, S-T, B-P, and R-P. (If Pablo owns the sneakers, he does not own any other shoes and no one else owns the sneakers.) Clue 3 also tells you that Nick wears a larger shoe than Pablo. If you look in the Boy column, you will see that only two places are open. If Nick wears a larger shoe, then he must have the large size basketball shoes. Put an "N" (Nick) beside the Large row and a "P" (Pablo) beside the Medium row. From clue 1, we know that the basketball shoes are the largest and we now know that Nick owns the largest shoes. Therefore, Nick must own the basketball shoes. Mark a "Y" in the B-N cell and an "N" in cells R-N and T-B. (If Nick owns the basketball shoes, then he does not own any others and no one else owns the basketball shoes.) The only shoes left for Tim to own are the running shoes. Put a "Y" in the T-R cell.

	B	R	S
N	**Y**	**N**	**N**
P	**N**	**N**	**Y**
T	**N**	**Y**	**N**

Shoe size	Shoe style	Boy
L	**B**	**N**
M	**S**	**P**
S	**R**	**T**

LESSON PREPARATION

OBJECTIVE AND MATERIALS

OBJECTIVE: Students will use a set of clues to deduce identity, rank, or position.

MATERIALS: Transparency of student book page 242 • washable transparency marker

CURRICULUM APPLICATIONS

Language Arts: Reading comprehension exercises; using key words and phrases to draw conclusions about a character's personality and motivation in a literary work; comprehending subtle distinctions in degree, size, rank, or order in reading passages; understanding and writing compare/contrast passages or papers

Mathematics: Organizing and charting statistical data; using key words and phrases to solve word problems, transitivity relations, or inequality exercises
Science: Understanding directions for and results of laboratory demonstrations or experiments
Social Studies: Outlining activities using text materials, comprehending chronological order or statistical comparisons in text materials
Enrichment Areas: Deductive thinking exercises, such as "Twenty Questions," and logic puzzles or games

TEACHING SUGGESTIONS
The steps shown on page 242 in the student book should always be followed, even though brighter students may not need to complete the charts to deduce the answers. As the exercises increase in difficulty, most students will need to use the charts. Encourage students to discuss their answers.

MODEL LESSON

LESSON

Introduction

Q: You have used clues to answer questions.

Explaining the Objective to the Students

Q: In this lesson, you will use a set of clues to answer a word puzzle.

Class Activity

- Project the transparency of page 242.

  Q: In the example on page 242, you will use the clues given in the paragraph to identify the names of two pets. By making a chart and marking the clues from the paragraph on it, we can answer several questions.
- Follow the example from the transparency step by step.

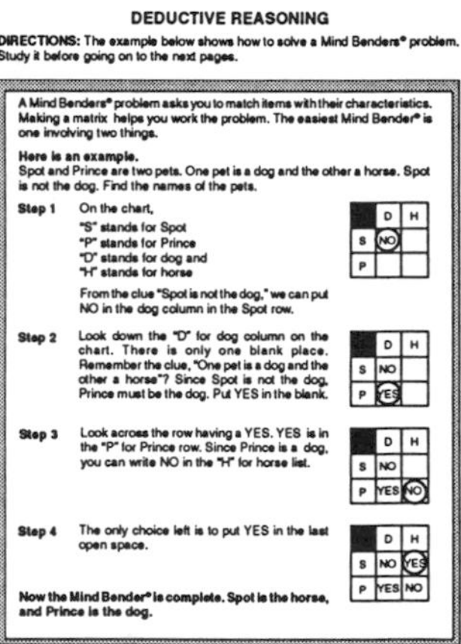

DEDUCTIVE REASONING

DIRECTIONS: The example below shows how to solve a Mind Benders® problem. Study it before going on to the next pages.

A Mind Benders® problem asks you to match items with their characteristics. Making a matrix helps you work the problem. The easiest Mind Bender® is one involving two things.

Here is an example.
Spot and Prince are two pets. One pet is a dog and the other a horse. Spot is not the dog. Find the names of the pets.

Step 1 On the chart,
"S" stands for Spot
"P" stands for Prince
"D" stands for dog and
"H" stands for horse

From the clue "Spot is not the dog," we can put NO in the dog column in the Spot row.

	D	H
S	NO	
P		

Step 2 Look down the "D" for dog column on the chart. There is only one blank place. Remember the clue, "One pet is a dog and the other a horse"? Since Spot is not the dog, Prince must be the dog. Put YES in the blank.

	D	H
S	NO	
P	YES	

Step 3 Look across the row having a YES. YES is in the "P" for Prince row. Since Prince is a dog, you can write NO in the "H" for horse list.

	D	H
S	NO	
P	YES	NO

Step 4 The only choice left is to put YES in the last open space.

	D	H
S	NO	YES
P	YES	NO

Now the Mind Bender® is complete. Spot is the horse, and Prince is the dog.

GUIDED PRACTICE

EXERCISES: **H-125, H-126**

- When students have had sufficient time to complete these exercises, check by discussion to determine whether they have answered correctly. Always have students check their solutions against the clues to make sure that all statements are true about their solutions. This process will help avoid "wild guessing" or trial-and-error solutions.

INDEPENDENT PRACTICE

- Assign exercises **H-127** through **H-131**.

THINKING ABOUT THINKING

Q: How did you use the clues to solve the problem?

1. I looked for clues that were specific to one item.
2. I looked for clues that eliminated a characteristic for an item.
3. I matched each item with its characteristics.

PERSONAL APPLICATION

Q: When do you need to figure out the correct order of something?

A: Examples include understanding test results or sports standings; comparison shopping; organizing clues to solve puzzles; interpreting chronology of events in newspaper articles, television shows, and stories.

EXERCISES H-132 to H-134

RANKING TIME/LENGTH MEASURES
RANKING IN GEOGRAPHY

ANSWERS H-132 through H-134 — Student book pages 247–49
Guided Practice: H-132 hour, day, month, year (shortest–longest)
Independent Practice: H-133 inch, foot, yard, mile (smallest–longest)
H-134 neighborhood, city, state, nation (smallest–largest)

LESSON PREPARATION

OBJECTIVE AND MATERIALS

OBJECTIVE: Students will use ranking diagrams (transitive order graphic organizers).
MATERIALS: Transparency of page 247 in student book

CURRICULUM APPLICATIONS

Language Arts: Sequencing events in story plots; sequencing steps in writing a paragraph, a report, etc.; selecting nouns and verbs to express degree, rank, or order; doing crossword puzzles
Mathematics: Sequencing numbers by relative number value, comparing units of measure, solving word problems involving a series of steps
Science: Charting the life cycles of organisms, sequencing lab instructions
Social Studies: Sequencing historical events and periods chronologically and in order of significance, ranking divisions and subdivisions of governmental or political structures
Enrichment Areas: Prioritizing decision making, charting stages in art history, describing musical progressions

TEACHING SUGGESTIONS

Direct students to find the extreme values first, i.e., the shortest length is an inch, the longest length is a mile, the smallest geographic region is a neighborhood, the largest geographic region is a nation.

MODEL LESSON

LESSON

Introduction

Q: A flowchart can be used for putting things in order of occurrence or ranking things by size.

Explaining the Objective to the Students

Q: In this lesson, you will use a diagram to rank objects in order of increasing size.

Class Activity

- Project transparency of page 247.

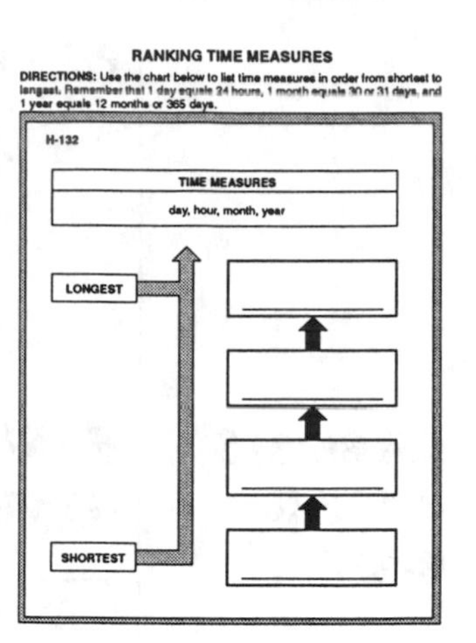

Q: This is a transparency of page 247. This diagram will be used to list four lengths of time from shortest to longest. Which of the times is the shortest?
A: An hour

- Write "hour" on the transparency in the bottom box.
  Q: Which of the times is the longest?
  A: A year
- Write "year" on the transparency in the top box.
  Q: Finish exercise **H-132**.
- Provide time for students to work.
  Q: Where did you write "day" and "month"?
- Allow time for discussion.

GUIDED PRACTICE

EXERCISE: **H-132**

INDEPENDENT PRACTICE

- Assign exercises **H-133** and **H-134**.

ADDITIONAL EXERCISES USING PICTURES

You may use the following activities to supplement exercises on ranking. Ask students to rank the items in order (size, expense, age, etc.). Students may find more than one way to rank the items. Encourage them to explain the common theme among the items and their order of choice.

- Display pictures of a bicycle, boat, bus, and ship; ask students to rank.
- Display pictures of a peanut, egg, peach, and lettuce; ask students to rank.
- Display pictures of a butterfly, duck, turkey, and ostrich; ask students to rank.
- Display pictures of a frog, fish, shark, and whale; ask students to rank.
- Display pictures of a house, garage, fire station, and school; ask students to rank.

THINKING ABOUT THINKING

Q: What did you pay attention to when you followed a transitive order graph?

1. I read very carefully and thought which word meant the "smallest" or "shortest."
2. If I didn't know a relationship, I looked it up.
3. I found the word which meant the "largest" or "longest."
4. I then ranked the other two words.

PERSONAL APPLICATION

Q: When do you need to show an order or sequence?

A: Examples include showing ages and birthdays, showing how much things cost, or ranking choices.

EXERCISES H-135 to H-137

FLOWCHART—ARITHMETIC

ANSWERS H-135 through H-137 — Student book pages 251–52
Guided Practice: H-135 2 feet 3 inches (1 foot = 12 inches; 1 ft. 9 in. = 12 + 9 = 21 in.; 21 in. + 6 in. = 27 in.; 27 ÷ 12 = 2 ft. 3 in.)
Independent Practice: H-136 3 feet 2 inches (2 ft. 10 in. = [2 x 12] + 10 = 24 + 10 = 34 in.; 34 in. + 4 in. = 38 in.; 38 in. ÷ 12 = 3 ft. 2 in. OR–Lay 2 one-foot lines plus a ten-inch and a four-inch line to make a 38-inch line. Ask a student to mark off the 38-inch line with a one-foot ruler. The student will find that it is 2 inches longer than 3 feet.) **H-137** 2 hours 15 minutes (1hr. 45 min. = 60 min. + 45 min. = 105 min.; 105 min. + 30 min. = 135 min.; 135 min. ÷ 60 = 2 hr. 15 min.) (If this exercise is too advanced for your students, assign as an optional exercise.) See below.

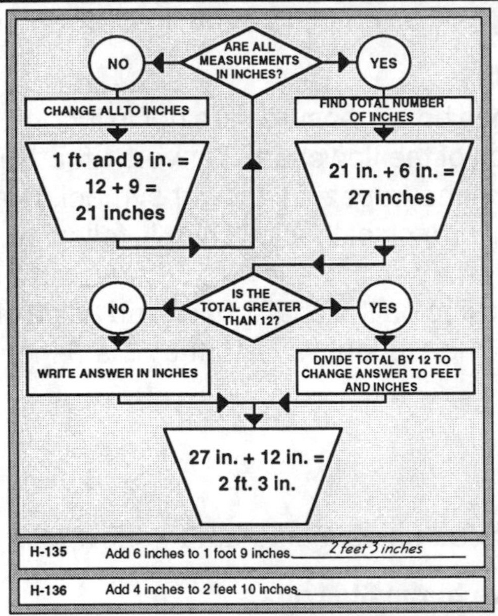

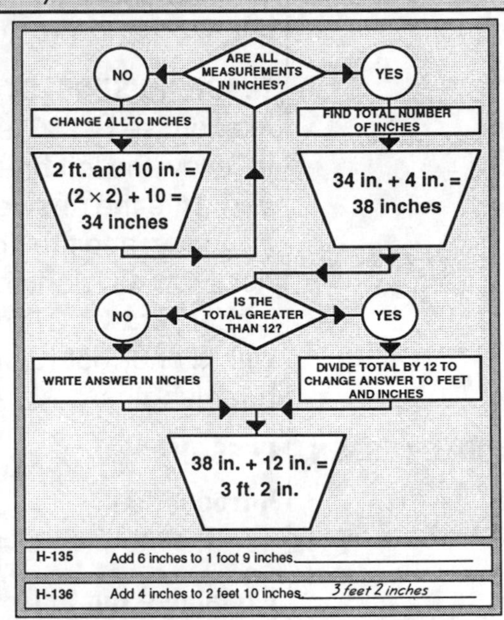

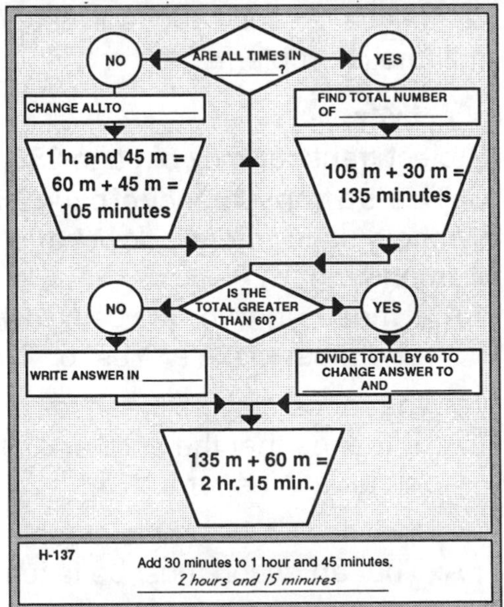

LESSON PREPARATION

OBJECTIVE AND MATERIALS

OBJECTIVE: Students will use flowcharts to organize their thinking.
MATERIALS: Transparency of page 250 in student book

CURRICULUM APPLICATIONS

Language Arts: Outlining, organizing the steps in writing a report, sequencing events in a story
Mathematics: Outlining the steps in converting measurements, fractions, etc.; solving word problems involving a series of steps
Science: Illustrating changes within the plant or animal kingdoms, sequencing the developmental stages of organisms, illustrating changes between elements in a compound or mixture
Social Studies: Analyzing historical events and periods chronologically and in order of significance; depicting steps in legislative processes, election processes, and judicial processes; recognizing divisions and subdivisions of governmental or political structures
Enrichment Areas: Prioritizing the steps in decision making, writing instructions, organizing study skills, charting a musical progression

TEACHING SUGGESTIONS

In exercise **H-135**, since both measures are not in inches, guide the students to follow the left branch of the flowchart. Convert 1 foot 9 inches to 21 inches and write the answer in the trapezoid answer symbol. Now all measures are in inches, and the right branch of the flowchart is followed. If your students do not know division, they may do the following: To convert 27 inches to feet and inches, draw a 27-inch line on the chalkboard and ask a student to mark it off with a one-foot ruler. The student will find that it is 3 inches longer than two ruler lengths.

MODEL LESSON

LESSON

Introduction

Q: You have practiced using a diagram to show rank.

Explaining the Objective to the Students

Q: In this lesson, you will use flowcharts to explain how you solve arithmetic problems.

Class Activity

- Project transparency of page 250, top part.

Q: This is a transparency of page 250 showing the symbols used to draw flowcharts. What does a diamond mean?

A: It means that you are being asked a question and must make a "Yes" or "No" decision.

Q: What does a circle mean?

A: It means that the answer to the question is either a "Yes" or a "No."

Q: What does a rectangle mean?

A: That an action must be taken

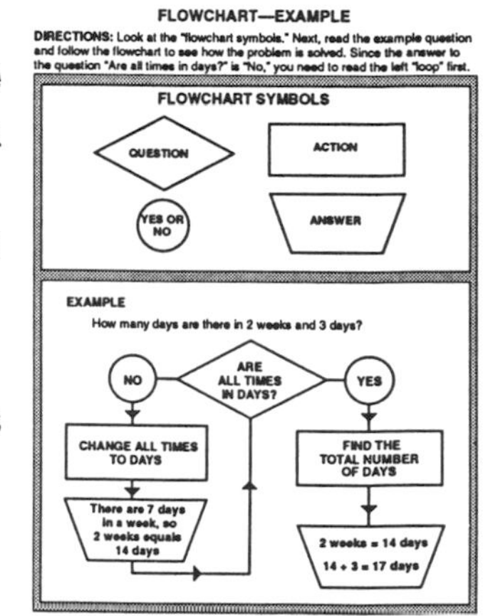

Q: What does a trapezoid mean?
A: That an answer has been found

- Without exposing the bottom part of the transparency pose the question...
  Q: Think about how you can add things that are different, like days and weeks.
- Allow time for discussion then project the bottom part of the transparency and discuss the solution.
  Q: Look at the diamond. Are all the times in days?
  A: No

  Q: Since the answer is "No," follow the left loop until you come to the instruction "Change all times to days." How many days are there in one week?
  A: Seven

  Q: How many days are there in two weeks?
  A: Seven and seven are fourteen.

  Q: Now that we have changed all the times to days, we can proceed down the right branch of the flowchart. What do you do to find a "total"?
  A: Add the two times, i.e., 2 weeks = 14 days; 14 days plus 3 days makes a total of 17 days.

GUIDED PRACTICE

EXERCISE: **H-135**

INDEPENDENT PRACTICE

- Assign exercises **H-136** and **H-137**.

THINKING ABOUT THINKING

Q: What did you pay attention to when you followed a flowchart?

1. I read very carefully and answered "Yes" or "No."
2. I followed the branch of the chart indicated by my Yes or No answer.
3. I followed the directions given in the action steps.
4. I completed the steps and wrote an answer.

PERSONAL APPLICATION

Q: When do you want or need to show the steps in doing a task?
A: Examples include writing a check, completing an assignment, scheduling your time, operating a bank machine, planning afterschool activities.

Chapter 9 — VERBAL CLASSIFICATIONS

(Student book pages 253–99)

EXERCISES I-1 to I-28

PARTS OF A WHOLE—SELECT

ANSWERS I-1 through I-28 — Student book pages 254–60
Guided Practice: I-1 WHOLE: burger PARTS: bun, patty, tomato; **I-2** WHOLE: bicycle PARTS: pedal, seat, wheel
Independent Practice: I-3 WHOLE: house PARTS: door, room, window; **I-4** WHOLE: airplane PARTS: nose, tail, wing; **I-5** WHOLE: band PARTS: drummer, trumpeter, tuba player; **I-6** WHOLE: jacket PARTS: pocket, sleeve, zipper; **I-7** WHOLE: woods PARTS: flowers, grass, tree; **I-8** WHOLE: lamp PARTS: bulb, cord, shade; **I-9** WHOLE: body PARTS: arm, head, leg; **I-10** WHOLE: tree PARTS: branch, leaves, trunk; **I-11** WHOLE: herd PARTS: bull, calf, cow; **I-12** WHOLE: car PARTS: body, engine, wheel; **I-13** WHOLE: army PARTS: trucks, soldiers, tanks; **I-14** WHOLE: market PARTS: cashier, meat, vegetables; **I-15** WHOLE: chili PARTS: beans, onions, tomatoes; **I-16** WHOLE: radio PARTS: dial, switch, volume control; **I-17** WHOLE: television PARTS: antenna, channel selector, picture tube; **I-18** WHOLE: salad PARTS: dressing, lettuce, tomato; **I-19** WHOLE: lawn mower PARTS: blade, engine, handle; **I-20** WHOLE: pen PARTS: cap, ink, point; **I-21** WHOLE: circus PARTS: acrobats, animals, clowns; **I-22** WHOLE: school PARTS: principal, students, teachers; **I-23** WHOLE: arithmetic PARTS: addition, numbers, subtraction; **I-24** WHOLE: language PARTS: adjectives, nouns, verbs; **I-25** WHOLE: government PARTS: judge, president, senator; **I-26** WHOLE: book PARTS: cover, pages, words; **I-27** WHOLE: city PARTS: houses, stores, schools; **I-28** WHOLE: map PARTS: key, mountains, rivers

DETAILED SOLUTIONS

The following answers are suggested as a model to assist in the discussion process.

I-1 The burger is the whole sandwich. The bun is the part that holds the sandwich together. The patty is the main part of the burger. The tomato is the part that adds flavor to the sandwich.

I-2 The bicycle is the whole. The pedal is the part of the bicycle that you "pump" in order to make the bike go. The seat is the part you sit on, and the wheel is the part that turns and moves the bike along.

I-3 The house is the whole. The door is the entryway into the house. A room is the part where the residents entertain, eat, sleep, or work. A window is an opening in an outside wall for people to look through.

I-4 The airplane is the whole. The nose is the front of the airplane. The tail has parts that move and make the plane go up or down or turn left or right. The wings keep the airplane in the air.

I-5 The band is the whole. The drummer is a member of the band who produces the "beat" to keep the band in step as it marches. The trumpeter and tuba player are members of the brass section. The trumpeter often plays the melody.
I-6 The jacket is the whole. The pocket is the part that is used to carry things. The sleeve is the part that keeps your arms warm. The zipper is the fastener that closes the jacket.
I-7 The woods is the whole. The flowers and grass are the parts of the woods near the ground. The tree(s) are the main part of the woods. Trees are made of wood, which suggests the name woods.
I-8 The lamp is the whole. The bulb is the part that produces light. The cord brings electricity from the wall socket to the bulb. The shade cuts down on glare and spreads out the light.
I-9 The body is the whole. The arm is the part of the body used for holding things. The head is the part of the body used for thinking. The leg is the part of the body used for walking, running, and jumping.
I-10 The tree is the whole. The branch is the part of the tree that holds the leaves. The leaves make the food for the tree. The trunk is the main body of the tree.
I-11 The herd is the whole. The bull is the male member of the herd, the cow is the female, and the calf is the offspring of the bull and the cow.
I-12 The car is the whole. The body is the outside of the car. The engine causes the wheels to turn, which moves the car.
I-13 The army is the whole. Trucks bring supplies to the soldiers so they can drive their tanks.
I-14 The market is the whole. The cashier checks out the groceries and takes in money for the meat and vegetables people buy.
I-15 The chili is the whole. Beans, onions, and tomatoes give the chili its flavor.
I-16 The radio is the whole. The dial is turned to change stations. The switch is used to turn the radio on or off. The volume control changes the sound level.
I-17 The television is the whole. The antenna is used to collect the television signal that comes in through the air. The channel selector is used to change the channel you are viewing. The picture tube produces the picture you watch.
I-18 The salad is the whole. The dressing flavors the lettuce and tomato that make up the salad.
I-19 The lawn mower is the whole. The blade is the part that cuts the grass. The engine is the part that moves the cutting blade. The handle is used to steer the lawn mower.
I-20 The pen is the whole. The cap keeps the ink from drying out. The ink is used to make marks on paper. The point is used to direct the ink onto the paper.
I-21 The circus is the whole. Acrobats entertain the people by jumping and balancing. Animals are used in many of the circus acts. Clowns, dressed in funny clothes, are circus performers who joke around with each other and the crowd.
I-22 The school is the whole. The principal is the person in charge of running the school. The students are divided into classes by age and are taught by teachers.

I-23 Arithmetic is the whole. Addition and subtraction are operations with numbers.

I-24 Language is the whole and is made up of many kinds of words. Adjectives are words that describe nouns. Nouns are names of people, places, or things. Verbs are action words.

I-25 Government is the whole. A judge is the main official of the court. The president is the chief official of the nation. A senator is one of the two officials elected from each of the 50 states of the United States of America.

I-26 Book is the whole. The outside of the book is the cover, which holds the pages together. Words are printed on the pages to make the story in the book.

I-27 City is the whole. Houses are buildings where people live. Stores are buildings where people buy things. Schools are buildings where people learn.

I-28 Map is the whole. The key is the part which gives the definitions of the symbols used on the map. Mountains and rivers are two features drawn on maps.

LESSON PREPARATION

OBJECTIVE AND MATERIALS

OBJECTIVE: Students will identify which word describes the whole object or system and which words describe parts of the whole.

MATERIALS: Transparency of student workbook page 254 • washable transparency marker

CURRICULUM APPLICATIONS

Language Arts: Identifying parts of speech, parts of a book, and parts of a letter; identifying the topic sentence and its supporting statements in a paragraph; utilizing heads and subheads in making outlines

Mathematics: Identifying components in arithmetic operations (e.g., identifying addends and sums in an addition problem), describing polygons, describing computers

Science: Identifying significant parts of living organisms; observing components of constellations, stars, the solar system, or the earth; describing equipment

Social Studies: Examining dwellings, artifacts, costumes, communities, and governments requiring identification of component parts; using keys or legends to identify component parts of maps

Enrichment Activities: Recognizing component parts of written music (e.g., concepts of measure, phrase, or stanza); utilizing concepts of positive and negative space, focal point, and basic elements of composition in art projects; describing a camera

TEACHING SUGGESTIONS

Young children may have difficulty distinguishing whether small objects are actually part of a greater whole or are totally unrelated. You can help clarify this distinction by asking students to suggest an additional part for the whole item identified in each exercise. For example, in exercise **I-9** on page 257, the whole is the body and the parts are arm, head, and leg. Students could cite other parts of the body, such as trunk, chest, or neck. It is not uncommon for young children to misidentify items of clothing or eyeglasses as parts of a body.

Part/whole analysis should also emphasize the function or purpose of each part in the value or operation of the whole. For example, when discussing the example on page 254, ask students to identify what each part adds to the whole pizza. Students will recognize that cheese provides protein and flavor. The crust serves as a base for the whole pizza and makes it easy to bake and handle. Tomatoes add flavor and color to the pizza. Detailed solutions following this format can be found at the beginning of the lesson. This lesson may require two sessions: **I-1** through **I-11** and **I-12** through **I-28**.

MODEL LESSON

LESSON

Introduction

Q: Select an object in this classroom that has several parts. Describe the object to another student.

- Give the students enough time to swap descriptions.

  Q: Sometimes descriptions are confusing because the listener isn't sure which is the whole and which are the parts.

Explaining the Objective to the Students

Q: In this lesson, you will identify the word that describes the whole and the words that describe the parts.

Class Activity

- Project the transparency of page 254.

  Q: In this example from page 254, you are to identify the whole and the parts of the whole. The pictures will help you identify the whole. We will have to find which of the given words names the one item which includes all of the others. Of the four words, which one seems to name the "whole" of which the others are parts?

  A: Pizza

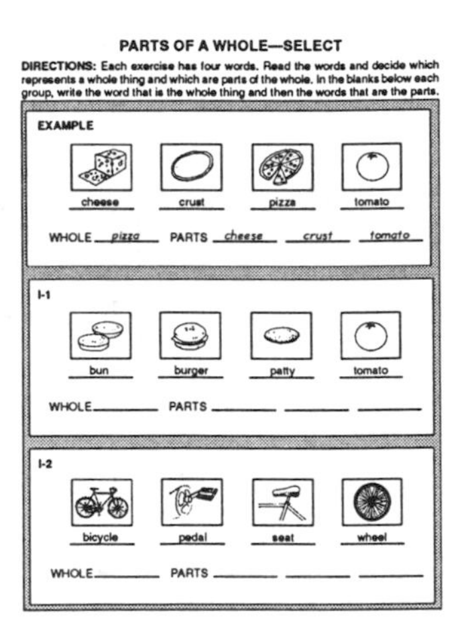

PARTS OF A WHOLE—SELECT

DIRECTIONS: Each exercise has four words. Read the words and decide which represents a whole thing and which are parts of the whole. In the blanks below each group, write the word that is the whole thing and then the words that are the parts.

EXAMPLE

cheese crust pizza tomato

WHOLE *pizza* PARTS *cheese* *crust* *tomato*

I-1

bun burger patty tomato

WHOLE ______ PARTS ______ ______ ______

I-2

bicycle pedal seat wheel

WHOLE ______ PARTS ______ ______ ______

- Write "pizza" on the line after whole.

  Q: What are the parts of the pizza shown in the example?

  A: Cheese, crust, tomato

  Q: Check that each of the words you selected as a part is really a part and that the word you selected as a whole really contains all of the other parts. Are cheese, crust, and tomato parts of a pizza?

  A: Yes

  Q: Does a pizza contain cheese, crust, and tomato?

  A: Yes

GUIDED PRACTICE

EXERCISES: **I-1**, **I-2**

- When students have had sufficient time to complete these exercises, check answers by discussion to determine whether students have answered correctly.

INDEPENDENT PRACTICE

- Assign exercises **I-3** through **I-28**. Note: You may wish to divide this assignment into two lessons, one using pictures and words (**I-3** through **I-11**, pp. 255–57) and the second using words only (**I-12** through **I-28**, pp. 258–60).

THINKING ABOUT THINKING

Q: What did you pay attention to when you decided which was the whole and which were the parts?

1. I carefully looked at each picture.
2. I checked to see which picture was the whole.
3. I checked to see whether other pictures were parts of the whole.

PERSONAL APPLICATION

Q: When do you need to describe the parts of a whole?

A: Examples include assembling or disassembling models, appliances, or construction toys; using recipes and fabric patterns; examining household objects and food preparation instruments.

EXERCISES I-29 to I-30

PARTS OF A WHOLE—GRAPHIC ORGANIZER

ANSWERS I-29 through I-30 — Student book pages 261–62
Guided Practice: I-29 WHOLE: nation LARGE: state SMALL: city SMALLEST: neighborhood; A neighborhood is part of a city which is part of a state which is part of the whole nation.
Independent Practice: I-30 WHOLE: story LARGE: paragraph SMALL: sentence SMALLEST: word; A word is part of a sentence which is part of a paragraph which is part of the whole story.

LESSON PREPARATION

OBJECTIVE AND MATERIALS

OBJECTIVE: Students will arrange items on a diagram to show how large things are made of smaller parts which in turn are made of still smaller parts.
MATERIALS: Globe or map of the world • transparency of TM 29 (p. 277) • washable transparency marker

CURRICULUM APPLICATIONS

Language Arts: Recognizing components of a play or poem; identifying parts of speech, parts of a book, and parts of a letter; identifying the topic sentence and its supporting statements from a paragraph
Mathematics: Identifying fractional parts, units of measure, geometric shapes
Science: Depicting physical anatomy of humans and animals, geologic formations such as volcanoes, cellular structures, solar systems, etc.; drawing components of equipment
Social Studies: Depicting components of a community or social group; illustrating political structures, judicial systems, etc.
Enrichment Activities: Recognizing components of a musical work such as a song or a symphony, classifying components of architectural structures, identifying elements of design in a painting

TEACHING SUGGESTIONS

First, ask the students to read the diagram from "inside to outside," i.e., a neighborhood is part of a city, a city is part of a state, and a state is part of a nation. Then, ask the students to read the diagram from "outside to inside," i.e., a nation contains states. States are made up of cities which contain neighborhoods.

MODEL LESSON

LESSON

Introduction

Q: Many things are made of parts. Some things are collections of parts which in turn have parts; for example, a brake is part of a bicycle. The brake is made of many parts.

Explaining the Objective to the Students

Q: In this lesson, you will arrange words on a diagram to show how large things are made of smaller parts which in turn are made of still smaller parts.

Class Activity

- Display a globe or map of the world.
  Q: What does this show?
  A: The world
- Project the transparency of TM 29, showing the graphic only.
  Q: Since the world is the whole, I will write world on the outer oval.

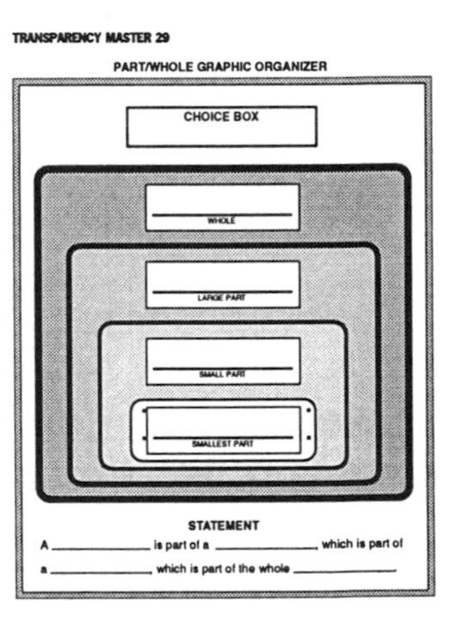

- Point to North America, circling the entire continent.
  Q: Canada, the United States of America, Mexico, and Central America are part of this big area; can you name it?
  A: North America
- Point to South America, circling the entire continent.
  Q: This large land mass contains many countries. Can you name this big area?
  A: South America

  Q: What do we call large land masses like North and South America?
  A: Continents

  Q: The land of the world is made up of seven large parts called continents.
- Write "continent" on the chart on the large part line of the chart. Point to countries within North America.
  Q: What are the parts of a continent?
  A: Countries
- Write "country" on the small part line of the chart.

Q: What are the parts of a country?
A: States, provinces, or cities

- Write the response on the smallest part line of the chart.
  Q: Read the diagram from inside to outside, or from the smallest part to the whole. Now we can fill in the statement below the diagram.
- Write the words in the blanks.

GUIDED PRACTICE
EXERCISE: **I-29**
- When students have finished the diagram, discuss their results.

INDEPENDENT PRACTICE
- Assign exercise **I-30**.

THINKING ABOUT THINKING
Q: What did you pay attention to when you decided which was the whole and which were the parts?
1. I carefully read the passage.
2. I looked for clues such as largest, smallest, is part of, contains, etc.
3. I wrote the answers.
4. I checked by reading the diagram from inside to outside saying "is part of" between each answer.

PERSONAL APPLICATION
Q: For what objects do you sometimes need to describe the parts of a whole?
A: Examples include assembling or disassembling models, appliances, or construction toys; repairing household objects; using recipes and fabric patterns.

EXERCISES I-31 to I-41

CLASS AND MEMBERS—SELECT

ANSWERS I-31 through I-41 — Student book pages 263–64
Guided Practice: I-31 CLASS: clothing MEMBERS: coat, hat, pants; **I-32** CLASS: sport MEMBERS: baseball, basketball, football; **I-33** CLASS: color MEMBERS: blue, green, red; **I-34** CLASS: tool MEMBERS: hammer, saw, screwdriver; **I-35** CLASS: school subject MEMBERS: arithmetic, reading, writing
Independent Practice: I-36 CLASS: language MEMBERS: English, French, Spanish; **I-37** CLASS: grain MEMBERS: corn, oats, wheat; **I-38** CLASS: fish MEMBERS: bass, salmon, tuna; **I-39** CLASS: plant MEMBERS: fruit, grain, vegetable; **I-40** CLASS: animal MEMBERS: bird, fish, reptile; **I-41** CLASS: tree MEMBERS: oak, palm, pine

LESSON PREPARATION

OBJECTIVE AND MATERIALS
OBJECTIVE: Students will decide which word represents a class and which words represent members of that class or group.

MATERIALS: Transparency of student book page 263 (with answers removed) • washable transparency marker

CURRICULUM APPLICATIONS

Language Arts: Writing or stating proper definitions of nouns (i.e., state the general class and the characteristics that distinguish it from others in the class), using reference books to locate information on a topic (Readers' Guide, encyclopedias, etc.)
Mathematics: Using cue words to determine functions for solving word problems, analyzing sets
Science: Identifying natural objects (name, general class, and distinguishing characteristics)
Social Studies: Writing or stating definitions of social studies terms or identifications of people, events, artifacts, or eras
Enrichment Areas: Classifying music according to type (classical, jazz, operatic, etc.), paintings according to characteristics (artist's techniques, type of medium used, etc.), or dances according to style (folk, modern, tap, jazz, ballet, ballroom, etc.)

TEACHING SUGGESTIONS

It is sometimes difficult for younger children to distinguish between parts of a whole and members of a class. In each of these relationships, a smaller item is being compared to a larger item or group. To reinforce parts of a whole and class/subclass concepts, identify these two important relationships in content lessons. Use bulletin board displays and analysis of classroom objects to show both parts/whole and class/subclass relationships.

Encourage students to define each of the terms in the assigned exercises, following this procedure to define a person, place, or thing:

1. Name the class to which the person, place, or thing belongs.
2. State the characteristics that make it different from other members of the same class.

Emphasize naming both the larger class to which the object belongs and its characteristics and special qualities whenever you or your students define terms. Identify classroom objects by general classes (e.g., tools, texts, furniture, paper products, objects which can be magnetized, and building materials) and by things that belong to each class.

Use this procedure to define objects or organisms in current lessons. Young students or children learning English may not know the names of common categories or the types of characteristics that distinguish different members of that class. For example, exercise **F-53** (on page 174 of the student book) illustrates the definition of a "frog." "A frog is an animal that has a backbone (as do all vertebrates). Frogs hatch from eggs and change appearance as they grow from tadpoles to frogs. Frogs have moist skin and spend much of their time near water." The following specific suggestions may help students define words used in the exercises in this activity:

a. coins (value, appearance, country of origin)
b. clothing (part of body it covers, material, function, ethnic group, style, shape)

c. sports (equipment needed, where played, how scored, special functions of players)
d. color (common examples, effect or reaction to)
e. tool (use of, made from, shaped like, skill needed to use)
f. school subject (what is studied, procedures used, when used)
g. language (where spoken, how different from others)
h. grain (where grown, how used, appearance)
i. fish (size, shape, color, habitat, special adaptations)
j. plants (where grown, how used, appearance)
k. animals (size, shape, color, habitat, special adaptations)
l. tree (where grown, how used, appearance)

MODEL LESSON

LESSON

Introduction

Q: You have identified the whole object or system and its parts. Now, you will identify a class of things and its members. We call ourselves a "class" of students. What do all the students in this class have in common? In this class, everyone is about the same age, meets in the same place, studies the same things, and has me for a teacher. *Class* means more than just a school room; it also means a group which has a common characteristic. When we describe the group by naming that common characteristic, we are classifying.

Explaining the Objective to the Students

Q: In this lesson, you will decide which word represents a class and which words represent members of that class or group.

Class Activity

- Project the transparency of page 263.

Q: In the example on page 263, *class* means a group of things which have something important in common. When we discussed a class in the figural exercises, we always identified the common characteristic of the class. Each member of a class always has that common characteristic, but it also has special qualities which distinguish it from other members in that class. When we discuss members, we identify both the common characteristic and the special qualities that make each member different. Given these four words (*coin*, *dime*, *nickel*, *penny*), identify the word which names the class and the words which name members of the class. Which of these words seems to be the most general?

CLASS AND MEMBERS—SELECT

DIRECTIONS: In each exercise are four words. Read the words and decide which represents a class to which the others belong. In the blanks below each group, write the word that represents the class and then the words that are its members.

EXAMPLE
coin dime nickel penny
CLASS *coin* MEMBERS *dime* *nickel* *penny*

I-31
clothing coat hat pants
CLASS ______ MEMBERS ______ ______ ______

I-32
baseball basketball football sport
CLASS ______ MEMBERS ______ ______ ______

I-33
blue color green red
CLASS ______ MEMBERS ______ ______ ______

I-34
hammer saw screwdriver tool
CLASS ______ MEMBERS ______ ______ ______

I-35
arithmetic reading school subject writing
CLASS ______ MEMBERS ______ ______ ______

A: Coin

Q: What is the definition of a coin?

A: A coin is metal money.

Q: If we say that the class is coin, then we must see if each of the other words describes a member of this class. Is a dime a coin?

A: Yes

Q: Is a nickel a coin? Is it a form of metal money?
A: Yes

Q: Is a penny a coin? Is it a form of metal money?
A: Yes

Q: We know that the class "coin" includes all metal money and that dimes, nickels, and pennies are coins because they are all types of metal money. What is the special quality that makes dimes, nickels, and pennies different from each other?
A: The values of the coins are different. A dime is worth ten cents, a nickel is worth five cents, and a penny is worth one cent.

Q: We can write the answers in the correct blanks now because the class "coin" and the members dime, nickel, and penny fit our definitions. Coin is the general name for dimes, nickels, and pennies, and the members of the class are all similar in that they are all forms of metal money and different in that they all have different values.

- Write the words on the proper blanks. (Option: Either project exercise **I-31** and do a second example or assign **I-31** as guided practice.)
Q: In exercise **I-31**, you are given the words *clothing*, *coat*, *hat*, and *pants*. Which word seems to be the most general?
A: Clothing

- Write "clothing" on the Class line.
Q: What is clothing?
A: Clothing is something that is worn to cover the body.

Q: Is a coat an item of clothing?
A: Yes

Q: How is a coat different from other items of clothing?
A: It is worn over other items of clothing such as a shirt.

- Write "coat" on the first Members line.
Q: Is a hat an item of clothing?
A: Yes

Q: How is a hat different from other items of clothing?
A: It is worn on the head.

- Write "hat" on the second Members line.
Q: Are pants an item of clothing?
A: Yes

Q: How are pants different from other items of clothing?
A: Pants are worn over the legs.

- Write "pants" on the last Member line.

GUIDED PRACTICE

EXERCISES: **I-31** through **I-35**

- When students have had sufficient time to complete these exercises, check answers by discussion to determine whether they have answered correctly. Follow the process modeled in the Teaching Suggestions to help students differentiate part/whole relationships clearly. Reinforce this process in stating definitions and in preparing for the verbal analogies in the next chapter.

INDEPENDENT PRACTICE

- Assign exercises **I-36** through **I-41**.

THINKING ABOUT THINKING

Q: What did you pay attention to when you figured out the class to which the members belonged?

1. I thought about the characteristics of each thing.
2. I asked myself two questions about each word: (1) Was this a group of many different kinds of things? and (2) Was this something that was described by another word among the choices?
3. I checked to see that the rest of the words were members of the class I chose.

PERSONAL APPLICATION

Q: When do you need to identify the class to which something belongs?

A: Examples include finding items in the supermarket, hardware store, mall directory, telephone book yellow pages, or classified ads in newspapers; answering identification-type questions on essay tests; locating information in a library.

EXERCISES I-42 to I-52

WHAT IS TRUE OF BOTH WORDS?—SELECT

ANSWERS I-42 through I-52 — Student book pages 265–67
Guided Practice: I-42 b, c (You learn both at school and at home.) **I-43** a, d
Independent Practice: I-44 a, b; **I-45** b, c; **I-46** a, b, d; **I-47** c, d; **I-48** a, b, d; **I-49** b, d; **I-50** c; **I-51** b, c; **I-52** b, c

LESSON PREPARATION

OBJECTIVE AND MATERIALS

OBJECTIVE: Students will select the characteristics that two words have in common.

MATERIALS: Transparency of student book page 265 • washable transparency marker

CURRICULUM APPLICATIONS

Language Arts: Choosing proper reference books when looking up facts for reports, recognizing parts of speech or types of literature

Mathematics: Distinguishing between types of arithmetic problems, recognizing numerical properties, grouping numbers according to place and face values

Science: Naming and recognizing attributes of different phyla of plants or

Enrichment Areas: Naming the attributes of types of dance, art, or music;

naming the functions and attributes of different tools in art, shop, or home economics

TEACHING SUGGESTIONS

This exercise develops precise language and specificity in describing attributes. Urge students to describe attributes of terms they use from everyday life and academic subjects. "What is true of both words?" should become a signal to students to use the most precise words they know in order to compare two terms. Encourage students to follow the principles below in order to define a person, place, or thing:

1. Name the class to which it belongs.
2. State the characteristics that make it different from other members of the same class.

Apply these definition principles to all curriculum areas.

MODEL LESSON

LESSON

Introduction

Q: You have selected and defined a class then described the members of that class. To classify items, you need to decide how they are alike.

Explaining the Objective to the Students

Q: In this lesson, you will select the characteristics that two items have in common.

Class Activity

- Project the example from the transparency of page 265.

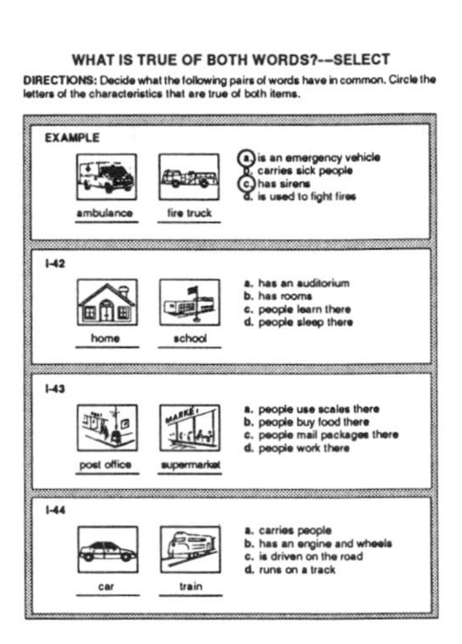
WHAT IS TRUE OF BOTH WORDS?—SELECT

DIRECTIONS: Decide what the following pairs of words have in common. Circle the letters of the characteristics that are true of both items.

EXAMPLE
ambulance, fire truck
(a.) is an emergency vehicle
b. carries sick people
(c.) has sirens
d. is used to fight fires

I-42
home, school
a. has an auditorium
b. has rooms
c. people learn there
d. people sleep there

I-43
post office, supermarket
a. people use scales there
b. people buy food there
c. people mail packages there
d. people work there

I-44
car, train
a. carries people
b. has an engine and wheels
c. is driven on the road
d. runs on a track

Q: In this example from page 265, you are given two words, *ambulance* and *fire truck*. You are asked to select characteristics that are true for both words. Description *a* asks if both an ambulance and a fire truck are emergency vehicles.

A: Yes, circle answer *a*.

Q: Description *b* asks if both an ambulance and a fire truck carry sick people.

A: No, do not circle answer *b*.

Q: Do both an ambulance and a fire truck have sirens?

A: Yes, circle answer *c*.

Q: Are both an ambulance and a fire truck used to fight fires?

A: No, do not circle answer *d*.

GUIDED PRACTICE

EXERCISES: **I-42**, **I-43**

- When students have had sufficient time to complete these exercises, check answers by discussion to determine whether they have answered correctly.

INDEPENDENT PRACTICE

- Assign exercises **I-44** through **I-52**.

THINKING ABOUT THINKING

Q: What did you pay attention to when you decided whether or not a statement was true of more than one thing?

1. I carefully read each statement.
2. I asked myself whether the statement applied to the first word.
3. I asked myself whether the statement applied to the second word.
4. If I got two "yes" answers, I circled the statement.

PERSONAL APPLICATION

Q: When do you need to decide whether or not a characteristic is true of more than one thing?

A: Examples include comparison shopping; reading food labels; shopping from a catalog; comparing movie, record, or book reviews; deciding which person someone is describing; reading directions for assembling parts of a model.

EXERCISES I-53 to I-74

HOW ARE THESE WORDS ALIKE?—SELECT

ANSWERS I-53 through I-74 — Student book pages 268–72
Guided Practice: I-53 c; **I-54** a; **I-55** b
Independent Practice: I-56 c; **I-57** c; **I-58** b; **I-59** c; **I-60** c; **I-61** a; **I-62** c; **I-63** b; **I-64** b; **I-65** b; **I-66** a; **I-67** b; **I-68** c; **I-69** c; **I-70** b; **I-71** a; **I-72** b; **I-73** b; **I-74** c

LESSON PREPARATION

OBJECTIVE AND MATERIALS

OBJECTIVE: Students will again find similar characteristics among items then pick the word that best describes the class to which all three objects belong.

MATERIALS: Transparency of student book page 268 • washable transparency marker

CURRICULUM APPLICATIONS

Language Arts: Classifying parts of speech or types of literature, choosing proper reference books when looking up facts for reports, using an index to look up facts in a book

Mathematics: Distinguishing between types of polygons or types of angles, recognizing numerical properties, grouping numbers according to place and face values

Science: Identifying different phyla of plants or animals, naming and recognizing various elements and compounds

Social Studies: Classifying types of architectural structures, governmental functions, or community institutions according to their functions or other attributes

Enrichment Areas: Naming types of dance, art, or music; classifying tools in art, shop, or home economics

TEACHING SUGGESTIONS

Discussion should follow the model provided in the lesson. In addition to justifying their choice, encourage students to be specific as to why the wrong answers are *not* the best descriptors. This exercise develops specificity in describing attributes. Urge students to describe attributes of terms they use in everyday life and in academic subjects. "How are these things alike?" should become a signal to students to carefully compare two terms.

Encourage students to define each of the terms in the assigned exercises, following this procedure to define a person, place, or thing:

1. Name the class to which the person, place, or thing belongs.
2. State the characteristics that make it different from other members of the same class.

Apply these definition principles to all curriculum areas.

MODEL LESSON

LESSON

Introduction

Q: You have identified similar characteristics that are true of a pair of words.

Explaining the Objective to the Students

Q: In this lesson, you will again find similar characteristics among items then pick the word that best describes the class to which all three objects belong.

Class Activity

- Project exercise **I-53** from the transparency of page 268.

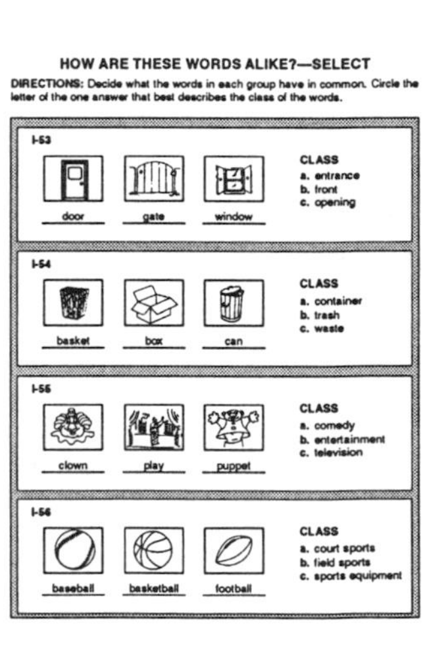

HOW ARE THESE WORDS ALIKE?—SELECT

DIRECTIONS: Decide what the words in each group have in common. Circle the letter of the one answer that best describes the class of the words.

I-53
door gate window
CLASS
a. entrance
b. front
c. opening

I-54
basket box can
CLASS
a. container
b. trash
c. waste

I-55
clown play puppet
CLASS
a. comedy
b. entertainment
c. television

I-56
baseball basketball football
CLASS
a. court sports
b. field sports
c. sports equipment

Q: In this example from page 268, you are given three words and a choice of answers. You are asked to select the one answer that best describes the class of given words. In exercise **I-53**, the three words are *door*, *gate*, and *window*. Answer *a* is entrance. Are a door, a gate, and a window examples of entrances?

A: No, ordinarily people do not enter through a window.

Q: Since a window is not an entrance, are a door, a gate, and a window examples of fronts?

A: No, all could be at a side or the rear of a building.

Q: Are a door, a gate, and a window examples of openings?

A: Yes, a door enters into a building, a gate enters a yard, and a window is an opening in a wall. All open, shut, or let in light.

Q: The best answer is *c*, opening, because it is the class to which all of the words belong.

- Circle answer *c*.

GUIDED PRACTICE

EXERCISES: **I-53** through **I-55**

- When students have had sufficient time to complete these exercises, check answers by discussion to determine whether they have answered correctly.

INDEPENDENT PRACTICE

- Assign exercises **I-56** through **I-74**. Note: You may want to introduce the example on page 271 of the student book before assigning exercises **I-65** through **I-74**. These exercises use words only.

THINKING ABOUT THINKING

Q: What did you pay attention to when you chose the class for several things?

1. I thought about the characteristics of each thing.
2. I asked myself, "What characteristics do these three words have in common?"
3. I named the group that has those characteristics.

PERSONAL APPLICATION

Q: When do you need to know the name of the group to which several things belong?

A: Examples include finding items in the supermarket, hardware store, mall directory, telephone book yellow pages, or classified ads in newspapers.

EXERCISES I-75 to I-96

HOW ARE THESE WORDS ALIKE?—EXPLAIN

ANSWERS I-75 through I-96 — Student book pages 273–77

Guided Practice: I-75 All are materials used for tying. **I-76** All are items worn to protect and warm people's feet.

Independent Practice: I-77 All are tools. **I-78** They are all emergency vehicles. **I-79** All are buildings, commonly found in cities, where people work or do business. **I-80** They are all vehicles that transport people. **I-81** They are all places where people live. **I-82** All are small, heat-producing appliances. (Note that not all stoves are electric.) **I-83** They are all types of animals' feet. **I-84** All are parts of plants. **I-85** All are land features. (geography terms, parts of the earth) **I-86** All are removable tops used to close jars or bottles. **I-87** All are unintentional releases of liquid from a container. **I-88** All are used to hold things together (adhesives). **I-89** All are things (movements) done in or on water. **I-90** All are words naming sounds. (All are also words that reflect the sounds they name.) **I-91** All produce heat. **I-92** All are citrus fruits that can be either eaten or used to make juice. **I-93** All are sensory parts of the head. **I-94** All are parts that a plant uses to exchange moisture and gases with the atmosphere (photosynthesis). **I-95** All are forms of water (clouds are water vapor, rain is liquid water, and ice is solid water). **I-96** All are large cats.

LESSON PREPARATION

OBJECTIVE AND MATERIALS

OBJECTIVE: Students will describe in their own words how the three words in the class are alike.

MATERIALS: Transparency of student workbook page 273 • washable transparency marker

CURRICULUM APPLICATIONS

Language Arts: Classifying and describing parts of speech or types of literature

Mathematics: Distinguishing between types of arithmetic problems, recognizing numerical properties, describing polygons or angles

Science: Describing different phyla of plants or animals; categorizing types of rocks, weather formations, cells, etc.

Social Studies: Linking common trends in various cultures; describing types of architectural structures, governmental functions, or community institutions according to their common attributes

Enrichment Areas: Classifying types of dance, art, or music

TEACHING SUGGESTIONS

This exercise develops precise language and specificity in describing attributes. Urge students to find as many similarities as they can, although the specificity of characteristics will vary according to grade and developmental level. Encourage students to define each of the terms in the assigned exercises, following this procedure to define a person, place or thing:

1. Name the class to which the person, place, or thing belongs.
2. State the characteristics that make it different from other members of the same class.

Apply these definition principles to all curriculum areas.

MODEL LESSON

LESSON

Introduction

Q: In the last exercise, you picked a word or phrase which best described the class to which three objects belonged.

Explaining the Objective to the Students

Q: In this lesson, you will describe how three words in a class are alike.

Class Activity

- Project the example from the transparency of page 273.

  Q: In this example from page 273, you are given three words, *bat*, *hammer*, and *racket*, and are asked to describe how these things are alike. To tell how they are alike, we must make sure that we can describe each item. What is a bat and how is it used?

  A: A bat is a long, slender club made of aluminum or wood and is used to hit a ball.

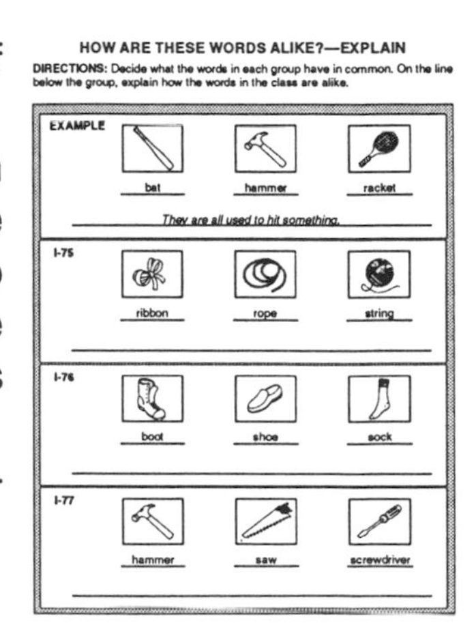
HOW ARE THESE WORDS ALIKE?—EXPLAIN

DIRECTIONS: Decide what the words in each group have in common. On the line below the group, explain how the words in the class are alike.

EXAMPLE: bat, hammer, racket

They are all used to hit something.

I-75: ribbon, rope, string

I-76: boot, shoe, sock

I-77: hammer, saw, screwdriver

Q: What is a hammer and how is it used?

A: A hammer is a T-shaped tool, normally made of metal and wood, used to pound nails, break stones, or shape metal.

Q: What is a racket and how is it used?

A: A racket is an oval-shaped frame strung with nylon, normally made of aluminum or wood, and is used to hit a ball in several games (tennis, racquetball, squash, etc.).

Q: What do all three things have in common?

A: They are all used to hit something. (Alternate or additional answers might be that all are made of some type of metal, wood, or combination of both.)

GUIDED PRACTICE

EXERCISES: **I-75**, **I-76**

- When students have had sufficient time to complete these exercises, check by discussion to determine whether they have answered correctly.

INDEPENDENT PRACTICE

- Assign exercises **I-77** through **I-96**. Note: You may want to review the example on page 276 of the student book before assigning exercises **I-86** to **I-96**.

THINKING ABOUT THINKING

Q: What did you pay attention to when you named the class for a group of items?

1. I defined each thing.
2. I looked for a common class that each definition fit.
3. I named the common class.

PERSONAL APPLICATION

Q: When do you identify how several items are the same?

A: Examples include finding items in the supermarket, hardware store, mall directory, telephone book yellow pages, or classified ads in newspapers.

EXERCISES I-97 to I-113

EXPLAIN THE EXCEPTION

ANSWERS I-97 through I-113 — Student book pages 278–82

Guided Practice: I-97 Nails: the other objects are all kind of tools. **I-98** Guitar: the other objects are blown to make noise.

Independent Practice: I-99 Orange: the other three are dairy products. **I-100** Eggs: the other three are vegetables. **I-101** Fish: the other three are birds. **I-102** Rocket: the other three are natural objects found in the sky (astronomical objects or heavenly bodies). **I-103** Umbrella: the other three are forms of moisture or water. (Alternate answer: Ice is the exception; the other three are associated with rainy weather.) **I-104** Television: the other three are printed forms of communication.

I-105 Sleep: the other three are types of exercise or speeds of movement by foot. **I-106** Ruler: the other three are used for writing or shading. **I-107** Scissors: the other three are eating utensils. **I-108** Recess: the other three are school subjects. **I-109** Write: the other three are art processes. **I-110** None: the other three represent multiple members. **I-111** Banana: the other three are citrus fruits. **I-112** Bicycle: the other three are motorized (powered) vehicles. **I-113** Cook: the other three are methods of food consumption.

LESSON PREPARATION

OBJECTIVE AND MATERIALS

OBJECTIVE: Students will explain how three words are alike and how the fourth word is different.

MATERIALS: Transparency of student workbook page 278 • washable transparency marker

CURRICULUM APPLICATIONS

Language Arts: Recognizing sentences that do not support a topic, making an outline as a prewriting exercise (choosing only those subheads or points which support the main idea)

Mathematics: Distinguishing between types of polygons, grouping numbers according to place and face values

Science: Classifying foods into basic food groups or determining nutritional values, sorting plants and animals into the correct phyla

Social Studies: Sorting artifacts according to their historical periods

Enrichment Areas: Sorting musical instruments according to categories

TEACHING SUGGESTIONS

In each exercise, students should name the characteristics which the three similar words have in common. Encourage students to discuss their answers using specific terms to support their decisions. Remember the words that students use to describe their choices and use these same words to remind students of the key characteristics of items in these lessons and subsequent lessons. Often a child's words will communicate with other children more meaningfully than the words a teacher may choose. Encourage students to define each of the terms in the assigned exercises, following this procedure to define a person, place or thing:

1. Name the class to which the person, place, or thing belongs.
2. State the characteristics that make it different from other members of the same class.

Apply these definition principles to all curriculum areas.

MODEL LESSON

LESSON

Introduction

Q: You have identified how a group of things are alike.

Explaining the Objective to the Students

Q: In this lesson, you will explain how three words are alike and how the fourth word is different.

Class Activity

- Project the transparency of page 278.

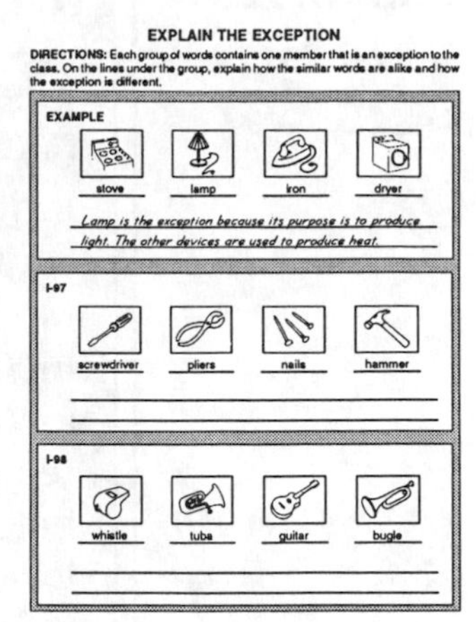

Q: In this example from page 278, you are given four words, *dryer*, *iron*, *lamp*, and *stove*. You are then asked to describe how three of these words are alike and how the fourth word is different. First, tell me the meaning of each word. What is a dryer?

A: An appliance that makes heat to dry clothes

Q: What is an iron?

A: A hand-held electric appliance that makes heat to press clothes

Q: What is a lamp?

A: An electric device used to produce light

Q: What is a stove?

A: An appliance used to cook food

Q: Which word is least like the others in meaning? Which is the exception to the class?

A: Lamp

Q: Why is lamp the exception?

A: A lamp is used to produce light; the other devices are used to produce heat.

GUIDED PRACTICE

EXERCISES: **I-97**, **I-98**

- When students have had sufficient time to complete these exercises, check answers by discussion to determine whether they have answered correctly.

INDEPENDENT PRACTICE

- Assign exercises **I-99** through **I-113**.

ADDITIONAL EXERCISES USING PICTURES

You can provide additional practice for students by having them classify and sort actual objects or pictures of things.

THINKING ABOUT THINKING

Q: What did you pay attention to when you identified something that did not fit into the same category or class as other things?

1. I thought about the meaning of each word.
2. I looked for three things that had something in common.
3. I named the class of three things.
4. I checked that the fourth word did not fit into this class.

PERSONAL APPLICATION

Q: When must you name or identify something that does not fit into the same group as other things?

A: Examples include putting away items at home or in school; replacing items in a supermarket, warehouse, workshop, or library.

EXERCISES I-114 to I-119

PICTURE DICTIONARY—SORTING INTO CLASSES

ANSWERS I-114 through I-119 — Student book pages 283–91
Guided Practice: I-114 See below.
Independent Practice: I-115 The animals may be classified in many ways. Some sample classes are Animals That Float/Swim in Water, Move on Land, Are Birds, Are Fish, Are Mammals, Are Reptiles, Are Insects, etc. See next page. **I-116** HAPPY: bright, cheerful, glad, jolly, joyous, merry SAD: crying, gloomy, grumpy, upset; **I-117** WHEN: already, always, never, often, seldom WHERE: above, across, behind, below, between; **I-118** See next page. **I-119** OPEN: clear, entrance, free, lower (a barrier), on, outdoors, raise (a window), start, wide CLOSED: end, fence, finish, indoors, locked, lower (a window), off, raise (a barrier), shut, stop, wall See below.

I-114

PLACES WHERE PEOPLE		
BUY THINGS	LIVE	GET SERVICES
gas station post office (stamps) restaurant supermarket	apartment farm house mobile home	barber shop fire station gas station (repairs) hospital police station post office supermarket (some have film developing)

I-115

ANIMALS THAT	
FLOAT/SWIM IN WATER	
frog jellyfish snake (some)	shark whale

ANIMALS THAT		
MOVE ON LAND		
deer duck frog kangaroo lion	lizard monkey mouse ostrich	snail snake spider squirrel

I-118

THINGS THAT			
MAKE MUSIC	ARE USED FOR INFORMATION	ARE USED FOR ENTERTAINMENT	WE RIDE ON
drum	letter	baseball	airplane
horn	newspaper	(bike)	bike
piano	radio	(boat)	boat
	telephone	circus	bus
	television	(horse)	helicopter
		(jeep)	horse
		(motorcycle)	jeep
		(piano)	motorcycle
		play	ship
		puppet	sled
		radio	swing
		(sled)	train
		swing	
		television	

LESSON PREPARATION

OBJECTIVE AND MATERIALS

OBJECTIVE: Students will sort words into three classes or groups.

MATERIALS: Transparency of student book page 283 (Cut the pictures apart and enlarge the bottom half.) • washable transparency marker

CURRICULUM APPLICATIONS

Language Arts: Classifying types of books; classifying examples into various figures of speech, parts of speech, or paragraph (essay) types

Mathematics: Differentiating polygons, graphs, or measurement systems

Science: Differentiating between phyla of plants and animals, classifying elements or compounds according to properties

Social Studies: Classifying types of dwellings, weapons, household articles, or tools belonging to various eras or cultures

Enrichment Areas: Recognizing music, art, architecture, or dance by era, culture, or type

TEACHING SUGGESTIONS

Encourage students to discuss their answers. Remember the words that students use to describe their choices and use these same words to remind students of the key characteristics of items in these lessons.

MODEL LESSON

LESSON

Introduction

Q: Suppose you want to give the clothes that you have outgrown to your younger sister or brother. You divide your clothes into two groups: clothes that fit and clothes that are too small.

Explaining the Objective to the Students

Q: In this lesson, you will sort groups of words into classes.

Class Activity

- Place the pictures from the transparency of page 283 in two rows at the top of the overhead projector. Put the blank transparency with the column headings under the pictures.

  Q: These pictures are from exercise **I-114** on page 283. You are given twelve place names which you are to sort into three groups: Places Where People Buy Things, Places Where People Live, and Places Where People Get Services. Look at the first word, *apartment*. Do people buy things in apartments, live in apartments, or get services in apartments?

  A: People live in apartments.

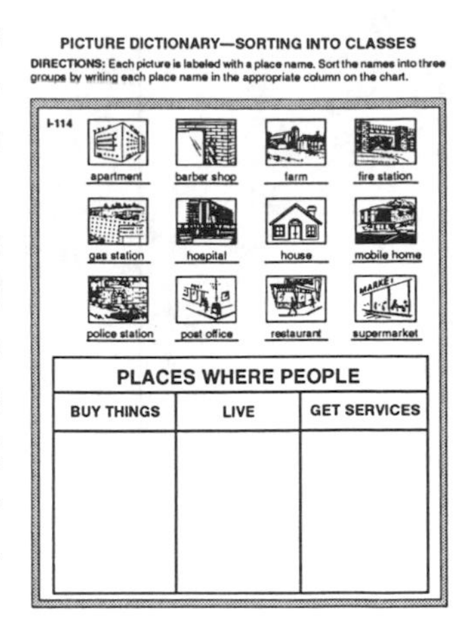
PICTURE DICTIONARY—SORTING INTO CLASSES

DIRECTIONS: Each picture is labeled with a place name. Sort the names into three groups by writing each place name in the appropriate column on the chart.

I-114

apartment, barber shop, farm, fire station

gas station, hospital, house, mobile home

police station, post office, restaurant, supermarket

PLACES WHERE PEOPLE		
BUY THINGS	LIVE	GET SERVICES

- Move the apartment piece to the column headed Places Where People Live. Note: Some students might say that people can also buy things in large apartment buildings. Should this argument arise, write the word "apartment" in the other column also.

  Q: Look at the second word, *barber shop*. Do people buy things in barber shops, live in barber shops, or get services in barber shops?

  A: People get services (haircuts, shampoos, shaves, etc.) in barber shops.

- Move the barber shop to the column headed Places Where People Get Services. Note: Some students might say that people can also buy things such as shampoo and other hair products in barber shops. Should this argument arise, write the word "barber shop" in the other column also.

  Q: Look at the third word, *farm*. Do people buy things, live on, or get services on farms?

  A: People live on farms.

- Move the farm piece to the column headed Places Where People Live. Note: Some students might say that people can also buy things on farms (i.e., eggs, strawberries or other produce, milk or other dairy products). Should this argument arise, write the word "farm" in the other column also.

  Q: The fourth word is *fire station*. Do people buy things, live in, or get services in fire stations?

  A: People obtain fire protection service from the fire station.

- Move the fire station piece to the Places Where People Get Services column.

  Q: Complete this exercise by writing the name of each place in the correct column on your paper. We will talk about your answers when you finish.

GUIDED PRACTICE

EXERCISE: **I-114**

- When students have had sufficient time to complete this exercise, check their answers by discussion to determine whether they have answered correctly.

INDEPENDENT PRACTICE

- Assign exercises **I-115** through **I-119**.

SUPPLEMENTAL ACTIVITY

Assemble 12–14 food product labels or coupons for each group of three students. Ask students to sort the labels/coupons by product type to determine where they can be found in the supermarket. List the types of products (classification) before students begin.

THINKING ABOUT THINKING

Q: What did you pay attention to when you chose the class for each picture?

1. I thought about the characteristics of each picture.
2. I asked myself, "Which class matches these characteristics?"
3. I selected the class that had a similar characteristic.

PERSONAL APPLICATION

Q: When might you need to sort a group of things into specific classes?

A: Examples include sorting tools, toys, clothes, coupons, records, or books for storage and easy retrieval; making a shopping list; making lists for special projects or occasions (birthday gifts, planning a party or school project, etc.); filing records or letters in an office.

EXERCISES I-120 to I-122

WORD CLASSES—SELECT/SORTING BY CLASS AND SIZE

ANSWERS I-120 through I-122 — Student book pages 292–96
Guided Practice: I-120 Row 1: trunk, tail, head, tail; Row 2: head, trunk, tail, trunk; Row 3: tail, trunk
Independent Practice: I-121 Row 1: track, blades, painting or paint, basket; Row 2: painting, wave, basket, track; Row 3: blades, basket, paint, wave; Row 4: track, wave, blades; **I-122** Small Building: cabin, hut, shed; Small Container: cup, glass, pint; Small fish: guppy, minnow; Medium Building: fast food restaurant, house, laundromat; Medium Container: gallon, quart; Medium Fish: perch, trout; Large Building: apartment building, school, shopping center, supermarket; Large Container: barrel; Large Fish: tuna, shark (Answers may vary.)

LESSON PREPARATION

OBJECTIVE AND MATERIALS

OBJECTIVE: Students will sort words according to meaning.

MATERIALS: Transparency of student book page 292 • washable transparency marker

CURRICULUM APPLICATIONS

Language Arts: Doing multiple-meaning exercises in reading or spelling; choosing exact words to express meaning in both descriptive and persuasive writing and speaking; using connotation and denotation exercises; understanding, using, and interpreting figures of speech

Mathematics: Recognizing and using cue words to determine processes for solving word problems

Science: Following directions in conducting experiments, writing laboratory reports

Social Studies: Classifying historical documents by type (speeches, letter, records, etc.)

TEACHING SUGGESTIONS

Collect jokes, comic strips, or puns with double-meaning "punch lines" to share with students. Students will find that the same word can have more than one meaning and be used in more than one way, often with humorous results.

MODEL LESSON

LESSON

Introduction

Q: Sometimes the same word can mean many different things. When I say "classes," you may think of the other classes of students in this school. However, we have been using the word "class" to mean a group of words which have something in common.

Explaining the Objective to the Students

Q: In this lesson, you will sort words according to their meanings.

Class Activity

- Project the transparency of page 292.

  Q: In this exercise from page 292, you are given three words inside a choice box. Each of the pictures on the page can be named by writing one of these three words in the blank below the picture. Look at the first picture. Do you recognize the nose of an elephant? Which of the words in the choice box is another name for an elephant's nose?

  A: Trunk

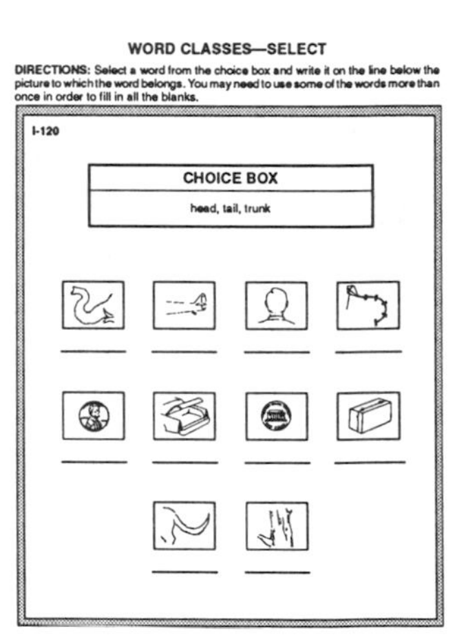

WORD CLASSES—SELECT

DIRECTIONS: Select a word from the choice box and write it on the line below the picture to which the word belongs. You may need to use some of the words more than once in order to fill in all the blanks.

I-120

CHOICE BOX

head, tail, trunk

- Write "trunk" on the line provided.

  Q: Look at the second picture. This is the back end of an airplane. Which word from the choice box names the rear section of an airplane?

  A: Tail

- Write "tail" on the blank.

  Q: Look at the third picture. Which word belongs with this picture?

  A: Head

- Write "head" on the blank.

Q: The fourth picture is a picture of a kite. Which word belongs with kite?
A: Tail

- Write "tail" on the blank.

GUIDED PRACTICE
EXERCISE: **I-120** (remaining rows)

INDEPENDENT PRACTICE
- Assign exercises **I-121**, **I-122**. Note: In exercise **I-122**, students will be sorting words by two categories, class and size. You may want to review page 137, Overlapping Classes–Matrix, in the student book.

THINKING ABOUT THINKING
Q: What did you pay attention to when you matched words with pictures?
1. I looked at a picture.
2. I read the words in the choice box.
3. I selected the word that matched the picture.

PERSONAL APPLICATION
Q: When might you need to know different meanings or connotations for a word?
A: Examples include telling or understanding jokes and puns, playing board games or word games where clues or hints are given, understanding advertising techniques.

EXERCISES I-123 to I-127

RECOGNIZING CLASSES—GRAPHIC ORGANIZER

ANSWERS I-123 through I-127 — Student book pages 297–99
Guided Practice: I-123 Largest–Smallest class: vehicle, airplane, jet
Independent Practice: I-124 Largest–Smallest class: truck, van, ambulance; **I-125** Largest–Smallest class: animal, bird, chicken; **I-126** Largest–Smallest class: bird, duck, mallard; **I-127** General–Specialized class: parallelograms, rectangles, squares

LESSON PREPARATION

OBJECTIVE AND MATERIALS
OBJECTIVE: Students will classify words by identifying a class and its members then sorting the members into smaller classifications and recording their answers on a diagram.
MATERIALS: Transparency of page 297 exercise **I-123**

CURRICULUM APPLICATIONS
Language Arts: Writing or stating proper definitions of nouns (i.e., state the general class and the characteristics that distinguish it from others in the class), using reference books to locate information on a topic (Readers' Guide, encyclopedias, etc.)
Mathematics: Sorting by sets, sorting geometric shapes, using cue words to determine functions for solving word problems

Science: Identifying natural objects (name, general class, and distinguishing characteristics)
Social Studies: Writing or stating definitions of social studies terms or identifications of people, events, artifacts, or eras
Enrichment Areas: Classifying music according to type (classical, jazz, operatic, etc.), paintings according to characteristics (artist's techniques, type of medium used, etc.), or dance according to style (folk, modern, tap, jazz, ballet, ballroom, etc.)

TEACHING SUGGESTIONS

Ask students to explain the diagram. Possible responses: A jet is a kind of airplane. An airplane is a kind of vehicle. OR A jet belongs to the class airplane which belongs to the class vehicle. OR All jets are airplanes and all airplanes are vehicles. Apply the same process to classes and subclasses that students study in social studies and science.

MODEL LESSON

LESSON

Introduction

Q: In the last lesson, you classified pictures by sorting them.

Explaining the Objective to the Students

Q: In this lesson, you will identify smaller classes within larger ones.

Class Activity

Q: Which of the words *airplane*, *jet*, or *vehicle* represents the largest group of things? Are there more kinds of airplanes than jets? Are there more kinds of vehicles than airplanes? Let's start by writing definitions of each word.

- Allow time for discussion. Write answers on the chalkboard.
  A: An airplane is a vehicle used to fly people and packages from place to place. A jet is a kind of airplane that is propelled by hot gases coming out a nozzle. A vehicle is a device that usually has wheels and is used to carry people and things from place to place.

  Q: Which of the words represents the largest class?
  A: Vehicles. It was necessary to use the word *vehicle* to define airplane; both airplanes and jets are examples of vehicles.

- Project the transparency of page 297 exercise **I-123**.
  Q: Since *vehicle* is the largest class, write the word on the line in the largest oval.

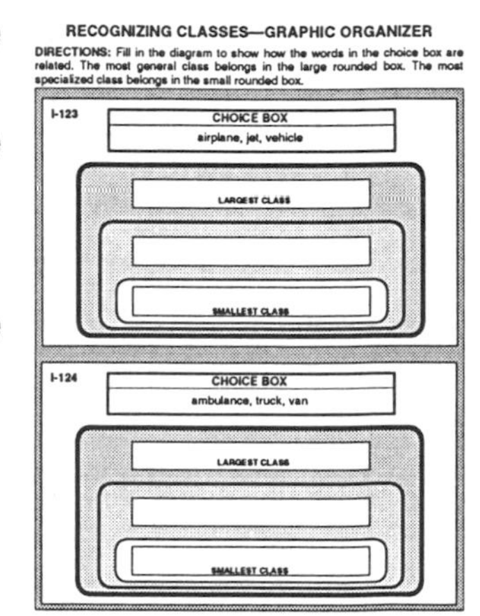

- Allow time for writing.
  Q: Which word in the choice box belongs in the smallest class?
  A: Jet

- Write the word on the line in the smallest oval.

GUIDED PRACTICE

EXERCISE: **I-123**

- Finish the diagram.

INDEPENDENT PRACTICE

- Assign **I-124** through **I-126** (**I-127** optional).

THINKING ABOUT THINKING

Q: What did you pay attention to when you figured out the class to which the members belonged?

1. I thought about the characteristics of each thing.
2. I asked myself two questions about each word: (1) Which is the largest group of many kinds of things? and (2) Which things belong in a larger category?
3. I checked to see that the rest of the words were members of the class I chose.

PERSONAL APPLICATION

Q: When is it helpful to find or name things by identifying the class to which they belong?

A: Examples include finding items in the supermarket, hardware store, mall directory, telephone book yellow pages, or classified ads in newspapers; answering identification-type questions on essay tests; locating information in a library.

Chapter 10

VERBAL ANALOGIES

(Student book pages 301–327)

EXERCISES J-1 to J-8

PICTURE ANALOGIES—SELECT

ANSWERS J-1 through J-8 — Student book pages 303–5
Guided Practice: J-1 woman (opposite) **J-2** boy (age)
Independent Practice: J-3 doghouse (kind of home) **J-4** barn (kind of home) **J-5** nest (kind of home) **J-6** hand (part of) **J-7** elbow (function/how it is used) **J-8** arm (extension of)

LESSON PREPARATION

OBJECTIVE AND MATERIALS

OBJECTIVE: Students will practice recognizing how two pairs of words are related in an analogy.
MATERIALS: Transparencies of student book pages 302 and 303 • washable transparency marker

CURRICULUM APPLICATIONS

Language Arts: Recognizing and completing word analogies, using context clues to infer meaning for unfamiliar words, recognizing metaphors and similes in poetry
Mathematics: Recognizing and using the relationship between graphic information and numerical information, using cue words to decide process in word problems (e.g., the words *in all, and, total* indicate addition; *have left, less, difference* indicate subtraction; etc.)
Science: Observing and describing analogous structures in animals or plants, using analogies to explain results in experiments
Social Studies: Recognizing and describing parallel or similar events in history, using analogies to explain concepts, recognizing historic or cultural patterns
Enrichment Areas: Comparing or contrasting eras, mediums, artists, types, or styles in music, drama, dance, art, etc.

TEACHING SUGGESTIONS

Encourage students to discuss their answers using the skills they learned in the similarities and differences, sequences, and classification strands. They should be able to discuss how the first and third words and the second and fourth words are similar and different. Any sequential or classification relationships should be discussed (e.g., girl to woman and boy to man). Remember the words that students use to describe their choices and use these same words to remind students of the key characteristics of items in these lessons and subsequent lessons.

MODEL LESSON

LESSON

Introduction

Q: You have practiced identifying relationships between words. You have examined how words are alike and how they are different, how to put them in order, and how to separate them into classes.

Explaining the Objective to the Students

Q: In this lesson, you will learn to recognize how two pairs of words are related.

Class Activity

- Project the top of the transparency of page 302. Point to the first two pictures.
  Q: Look at the first two figures. How are they alike?
  A: They are alike in that both of them are men.
  Q: How are they different?
  A: The picture of the father has a child in it.

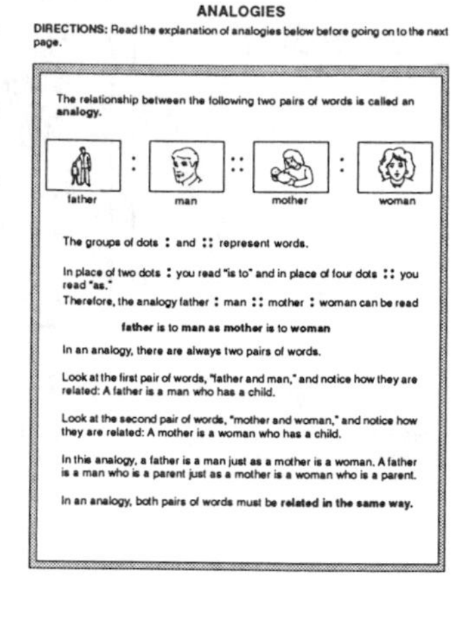

ANALOGIES

DIRECTIONS: Read the explanation of analogies below before going on to the next page.

The relationship between the following two pairs of words is called an **analogy.**

father : man :: mother : woman

The groups of dots : and :: represent words.

In place of two dots : you read "is to" and in place of four dots :: you read "as."

Therefore, the analogy father : man :: mother : woman can be read

father is to **man as mother** is to **woman**

In an analogy, there are always two pairs of words.

Look at the first pair of words, "father and man," and notice how they are related: A father is a man who has a child.

Look at the second pair of words, "mother and woman," and notice how they are related: A mother is a woman who has a child.

In this analogy, a father is a man just as a mother is a woman. A father is a man who is a parent just as a mother is a woman who is a parent.

In an analogy, both pairs of words must be **related in the same way.**

- Point to the second two figures.
  Q: How are these last two figures alike?
  A: They are both women.
  Q: How are these pictures different?
  A: The picture of a mother has a child in it.
- Point to the first word pair, then to the second.
  Q: These two pairs of words are called an analogy. The groups of dots represent words.

- Uncover the bottom half of the transparency.
  Q: The two dots (:) are commonly read *is to*, and the four dots (::) are commonly read *as*. We can now read the analogy.
- Point as you read.
  Q: Father is to man as mother is to woman. In these exercises, you will be looking at the relationship between two pairs of words. Like figural analogies, however, both pairs of words must be related in the same way. In this sample analogy, the relationship shown is "kind of." A father is a kind of man just as a mother is a kind of woman.
- Project the first three pictures from the example on page 303.
  Q: Let's look at the example on page 303. What is the relationship between the first two words?
  A: They are opposites. The man and woman are adults of the opposite sex.
  Q: The first word in the second pair of words is *boy*. So far our analogy reads, a man is the opposite sex from a woman as a boy is the opposite sex from.....

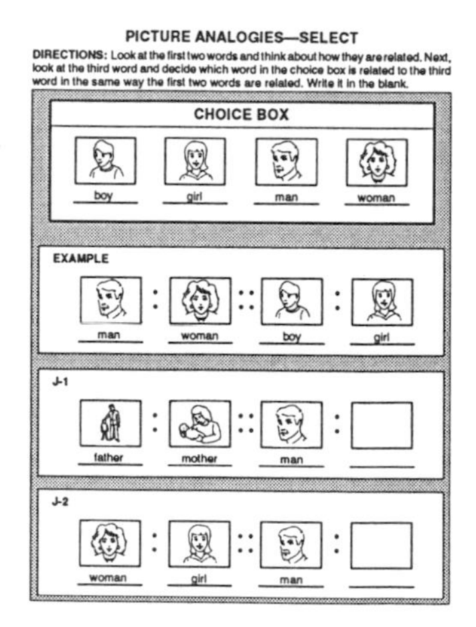

PICTURE ANALOGIES—SELECT

DIRECTIONS: Look at the first two words and think about how they are related. Next, look at the third word and decide which word in the choice box is related to the third word in the same way the first two words are related. Write it in the blank.

CHOICE BOX

boy girl man woman

EXAMPLE

man : woman :: boy : girl

J-1

father : mother :: man : ____

J-2

woman : girl :: man : ____

- Point to the choice box at the top of the transparency.
  Q: Which word in the choice box has the same relationship to boy as man does to woman?
  A: Girl

- Uncover the fourth picture in the example to confirm the choice.

  Q: What is the relationship shown by the last pair of words?

  A: Opposites. The boy and girl are children of the opposite sex.

GUIDED PRACTICE

EXERCISES: **J-1**, **J-2**

- When students have had sufficient time to complete these exercises, check answers by discussion to determine whether they have answered correctly.

INDEPENDENT PRACTICE

- Assign exercises **J-3** through **J-8**.

THINKING ABOUT THINKING

Q: What did you pay attention to when you found analogies between pairs of words?

1. I thought carefully about the meaning of each word.
2. I decided how the words in the first pair were different in meaning.
3. I checked whether the words in the second pair were different in the same way as the first pair of words.
4. I explained the relationship as an analogy: "A" is to "B" as "C" is to "D."

PERSONAL APPLICATION

Q: When do you have to choose the correct word to complete or illustrate a particular relationship?

A: Examples include giving directions for finding or constructing something, using words that show relationships, describing uses for tools or utensils.

EXERCISES J-9 to J-16

PICTURE ANALOGIES—NAME THE RELATIONSHIP

ANSWERS J-9 through J-16 — Student book pages 306–8

Guided Practice: J-9 A paw is part of (the foot of) a lion, and a hoof is part of (the foot of) a horse. **J-10** A tree is part of the woods (a member of a group of trees) just as a boy is part of the children (a member of a group of children).

Independent Practice: J-11 A bill is part of a duck's head just as a beak is part of a chicken's head. (Both are the mouth of a bird.) **J-12** Pine is a kind of tree, and daisy is a kind of flower. **J-13** An arm is a part of a body, and a branch is a part of a tree. (Alternate or additional answers might include that both are referred to as limbs and both stem from a trunk.) **J-14** A cap is used to close a bottle, and a lid is used to close a jar. (Both may also be viewed as part of or as located at the top of their respective containers.) **J-15** A glove is used to cover, warm, or protect a hand just as a sock is used to cover, warm, or protect a foot. **J-16** Up is the opposite of down just as left is the opposite of right.

LESSON PREPARATION

OBJECTIVE AND MATERIALS

OBJECTIVE: Students will explain how the word pairs in an analogy are related.

MATERIALS: Transparency of student book page 306 • washable transparency marker

CURRICULUM APPLICATIONS

Language Arts: Using paraphrasing skills, interpreting imagery or figures of speech

Mathematics: Changing numerical information to graphic information (and vice versa); recognizing and using "part-to-whole" analogies in measurements of time, weight, size, or volume; identifying the relationship between fractional parts and fractions

Science: Distinguishing among phyla of plants or animals

Social Studies: Explaining historic or cultural parallels

Enrichment Areas: Recognizing and using comparative note values in music, choosing complementary colors or styles in decorating or art projects

TEACHING SUGGESTIONS

Encourage students to discuss their answers. Remember the words that students use to describe their choices and use these same words to remind students of the key characteristics of items in this lesson.

MODEL LESSON

LESSON

Introduction

Q: You have practiced selecting a word that completes an analogy.

Explaining the Objective to the Students

Q: In this lesson, you will explain how word pairs in an analogy are related.

Class Activity

- Project the example at the top of the transparency of page 306, keeping the answer covered.

  Q: In this set of exercises, you will decide how the words in each pair of an analogy are related. How is a bat related to a baseball?

  A: A bat is used to hit a baseball.

  Q: How is a racket related to a tennis ball?

  A: A racket is used to hit a tennis ball.

  Q: So, we can say that bat and baseball are related in the same way as racket and tennis ball. A bat is used to hit a baseball, and a racket is used to hit a tennis ball.

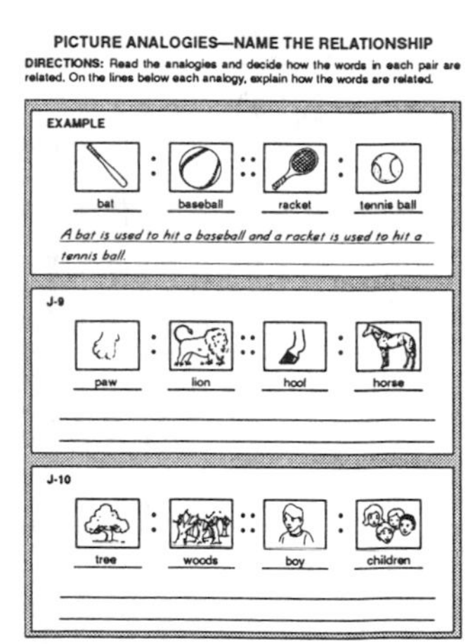

PICTURE ANALOGIES—NAME THE RELATIONSHIP

DIRECTIONS: Read the analogies and decide how the words in each pair are related. On the lines below each analogy, explain how the words are related.

EXAMPLE

bat : baseball :: racket : tennis ball

A bat is used to hit a baseball and a racket is used to hit a tennis ball.

J-9

paw : lion :: hoof : horse

J-10

tree : woods :: boy : children

- Show the example answer on the transparency.

GUIDED PRACTICE

EXERCISES: **J-9**, **J-10**

- When students have had sufficient time to complete these exercises, check by discussion to determine whether they have answered correctly.

INDEPENDENT PRACTICE

- Assign exercises **J-11** through **J-16**.

THINKING ABOUT THINKING

Q: What did you pay attention to when you found analogies between pairs of words?

1. I thought carefully about the meaning of each word.
2. I decided how the words in the first pair were different in meaning.
3. I checked whether the second pair was different in the same way as the first pair.
4. I explained the relationship as an analogy: "A" is to "B" as "C" is to "D."

PERSONAL APPLICATION

Q: When do you need to recognize or explain the relationship between two items?

A: Examples include explaining family relationships, teaching someone to use a tool or construct a model, explaining or recognizing relationships between items in a test question, recognizing different items which can be used for similar purposes or functions.

EXERCISES J-17 to J-73

ANALOGIES—SELECT

ANSWERS J-17 through J-73 — Student book pages 309–19

Guided Practice: J-17 under (opposites) **J-18** rear (synonyms) **J-19** under (opposites) **J-20** far (opposites) **J-21** bottom (synonyms) **J-22** follow (action/situation)

Independent Practice: J-23 empty (opposites) **J-24** whole (part/whole) **J-25** full (opposites) **J-26** small (synonyms) **J-27** tall (opposites) **J-28** low (opposites) **J-29** past (synonyms) **J-30** present (opposites) **J-31** today (opposites) **J-32** past (opposites) **J-33** often (opposites) **J-34** after (time relationship) **J-35** horse (offspring) **J-36** lion or horse (part of) **J-37** lion (sound of) **J-38** bird (action to object relationship) **J-39** fish (part of) **J-40** deer (kind of) **J-41** minute (whole/part) **J-42** month or year (part/whole) **J-43** year (part/whole) **J-44** week or day (whole/part) **J-45** day or week (whole/part) **J-46** minute (part/whole) **J-47** sound (association) **J-48** heat (association) **J-49** sound (product) **J-50** heat (product) **J-51** odor (association) **J-52** light (cause and effect) **J-53** gallon (equivalence) **J-54** quart (one-half) **J-55** pint (four times as much) **J-56** pint (two times as much) **J-57** cup (four times as much) **J-58** quart (equivalence) **J-59** radio (action/object) **J-60** pound (something used to) **J-61** water (something used for) **J-62** those (opposites) **J-63** fruit (kind of) **J-64** song (action/object) **J-65** air (action/situation) **J-66** give (opposites) **J-67** that (opposites) **J-68** leave (opposites) **J-69** barn (association) **J-70** taste (something used to) **J-71** wood (something used to cut) **J-72** river (something used to cross) **J-73** supper (time relationship)

LESSON PREPARATION

OBJECTIVE AND MATERIALS

OBJECTIVE: Students will recognize how two pairs of words are related in a verbal analogy.

MATERIALS: Transparency of student book page 309 • washable transparency marker

CURRICULUM APPLICATIONS

Language Arts: Recognizing and using word analogies, using paraphrasing skills

Mathematics: Changing numerical information to graphic information (and vice versa); recognizing and using "part-to-whole" analogies in measurements of time, weight, size, or volume; identifying the relationship between fractional parts and fractions

Science: Distinguishing between phyla of plants or animals, conducting laboratory experiments and writing reports on them

Social Studies: Explaining historic or cultural parallels

Enrichment Areas: Recognizing and using comparative note values in music, choosing complementary colors or styles in decorating or art projects

TEACHING SUGGESTIONS

This lesson may require three sessions: (1) **J-17** through **J-40**, (2) **J-41** through **J-58**, and (3) **J-59** through **J-73**. Pages 313 and 315 may require some preteaching of time and volume measurement concepts. Encourage students to discuss their answers. Remember the words that students use to describe their choices and use these same words to remind students of the key characteristics of the relationships in this lesson. Stating why or how incorrect answers were eliminated is as valuable as supporting the choice of a correct answer. Students should be encouraged to do both.

MODEL LESSON

LESSON

Introduction

Q: You have worked with picture analogies, selecting the picture that completes the relationship. You have also practiced naming and explaining the relationship between two pairs of words.

Explaining the Objective to the Students

Q: In this lesson, you will supply a word from memory to complete an analogy.

Class Activity

- Project the transparency of page 309.

  Q: In this exercise from page 309, you will find the relationship between two pairs of words. Each pair of words must be related in the same way. In exercise **J-17**, you are given three words. The first two words make a pair with a particular relationship. Your job will be to find a word in the choice box that will make the same relationship between a second pair of words. The relationship

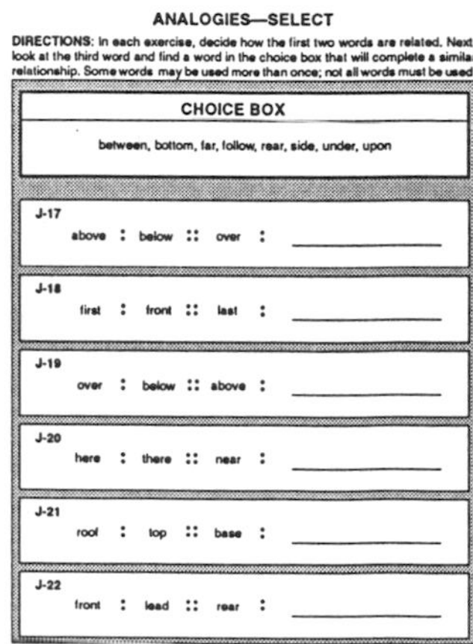

ANALOGIES—SELECT

DIRECTIONS: In each exercise, decide how the first two words are related. Next, look at the third word and find a word in the choice box that will complete a similar relationship. Some words may be used more than once; not all words must be used.

CHOICE BOX

between, bottom, far, follow, rear, side, under, upon

J-17 above : below :: over : ________

J-18 first : front :: last : ________

J-19 over : below :: above : ________

J-20 here : there :: near : ________

J-21 root : top :: base : ________

J-22 front : lead :: rear : ________

shown in the first pair, *above* and *below*, is that of opposites. Above means the opposite of below. The third word is *over*. Can you find a word in the choice box that means the opposite of over?

A: Under

- Write "under" in the blank provided for **J-17**.

Q: Look at exercise **J-18**. How are *first* and *front* related?

A: They are synonyms (have similar meanings).

Q: Can you find a word in the choice box that is similar to last?

A: Rear

- Write "rear" in the answer blank.

GUIDED PRACTICE

EXERCISES: **J-17** through **J-22**

- When students have had sufficient time to complete these exercises, check answers by discussion to determine whether they have answered correctly.

INDEPENDENT PRACTICE

- Assign exercises **J-23** through **J-73**.

THINKING ABOUT THINKING

Q: What did you pay attention to when you found analogies between pairs of words?

1. I thought carefully about the meaning of each word.
2. I decided how words in the first pair were different in meaning.
3. I checked whether the second pair was different in the same way as the first pair.
4. I explained the relationship as an analogy: "A" is to "B" as "C" is to "D."

PERSONAL APPLICATION

Q: When must you recognize, explain, or duplicate the relationship between two items?

A: Examples include explaining family relationships to someone outside the family, teaching someone to use a tool or construct a model, explaining or recognizing relationships between items in a test question, recognizing different items that can be used for similar purposes or functions.

EXERCISES J-74 to J-109

ANALOGIES—SUPPLY

ANSWERS J-74 through J-109 — Student book pages 320–25
Guided Practice: J-74 shirt, blouse, sweater (worn under) **J-75** shoe, boot, sneaker (used to fasten) **J-76** finger (worn on) **J-77** hand (worn on) **J-78** head (worn on) **J-79** arms (used to cover)
Independent Practice: J-80 floor (opposites) **J-81** any interior room of the house (opposites) **J-82** glass (made of) **J-83** walls (location or

direction) **J-84** floor (cover) **J-85** wall (location or part of) **J-86** hot (degree of) **J-87** light (characteristic) **J-88** bill, currency (type of) **J-89** slow (characteristic) **J-90** window (part of) **J-91** navy (member of) **J-92** red (form of) **J-93** car, truck (part of) **J-94** below (synonyms) **J-95** last (synonyms) **J-96** water (property of) **J-97** cut (used to) **J-98** go (symbols) **J-99** walk (opposites) **J-100** floor (location) **J-101** go (synonyms) **J-102** hard (characteristic) **J-103** lid (top part of) **J-104** first (synonyms) **J-105** most (opposites) **J-106** winter (characteristic) **J-107** write (used to) **J-108** stove (used to) **J-109** skillet, pan (used to)

LESSON PREPARATION

OBJECTIVE AND MATERIALS

OBJECTIVE: Students will supply a word from memory to complete an analogy.

MATERIALS: Transparency of student book page 320 • washable transparency marker

CURRICULUM APPLICATIONS

Language Arts: Recognizing and using word analogies, using paraphrasing skills

Mathematics: Changing numerical information to graphic information (and vice versa); recognizing and using "part-to-whole" analogies in measurements of time, weight, size, or volume; identifying the relationship between fractional parts and fractions

Science: Distinguishing between phyla of plants or animals, writing science reports

Social Studies: Explaining historic or cultural parallels

Enrichment Areas: Recognizing and using comparative note values in music, choosing complementary colors or styles in decorating or art projects

TEACHING SUGGESTIONS

You may wish to divide the lesson into two parts: (1) **J-80** through **J-91** and (2) **J-92** through **J-109**. Encourage students to discuss their answers and develop other possible answers. This type of exercise usually has a variety of different, but correct, answers. Remember the words that students use to describe their choices and use these same words to remind students of the key characteristics of the relationships in this lesson.

MODEL LESSON

LESSON

Introduction

Q: You have selected the correct word to complete an analogy.

Explaining the Objective to the Students

Q: In this lesson, you will select a pair of words to complete an analogy.

Class Activity

- Project exercise **J-74** from the transparency of page 320.

  Q: This is an exercise from page 320. As before,

ANALOGIES—SUPPLY

DIRECTIONS: In each exercise, decide how the first two words are related. Next, look at the third word and pick a word from your memory that belongs in the blank. In this exercise, all the words have to do with things you wear.

J-74 shoe : sock :: jacket : ________

J-75 button : shirt :: lace or string : ________

J-76 bracelet : arm :: ring : ________

J-77 shoe : foot :: glove : ________

J-78 scarf : neck :: hat : ________

J-79 pants : legs :: sweater : ________

you will be looking at the relationship between two pairs of words. Each pair of words must be related in the same way. In this analogy, shoe: sock :: jacket: ________, the relationship in the first pair is that a sock is worn underneath or inside a shoe. Can you think of something that is worn underneath or inside a jacket?

A: Sweater, shirt, or blouse

- Write the answer(s) in the blank of exercise **J-74**.

Q: Look at exercise **J-75**. How are button and shirt related?

A: A button is used to fasten or close a shirt.

Q: Can you think of something that a lace or string is used to fasten?

A: Shoe, boot, or sneaker

- Write the answer(s) in the blank.

GUIDED PRACTICE

EXERCISES: **J-74** through **J-79**

- When students have had sufficient time to complete these exercises, check answers by discussion to determine whether they have answered correctly. Note that there may be alternate answers.

INDEPENDENT PRACTICE

- Assign exercises **J-80** through **J-109**.

THINKING ABOUT THINKING

Q: What did you pay attention to when you found analogies between pairs of words?

1. I thought carefully about the meaning of each word.
2. I decided how the words in the first pair were different in meaning.
3. I checked whether the second pair was different in the same way as the first pair.
4. I explained the relationship as an analogy: "A" is to "B" as "C" is to "D."

PERSONAL APPLICATION

Q: When do you need to know a word that would show a particular relationship to another word?

A: Examples include explaining family relationships, teaching someone to use a tool or construct a model, explaining or recognizing relationships between items in a test question, recognizing different items that can be used for similar purposes or functions.

EXERCISES J-110 to J-117

ANALOGIES—SELECT THE RIGHT PAIR

ANSWERS J-110 through J-117 — Student book pages 326–27
Guided Practice: J-110 above : over (synonyms) **J-111** cat : kitten (age) **J-112** present : past (opposites)
Independent Practice: J-113 right : wrong (opposites) **J-114** over : under (opposites) **J-115** plus : minus (opposites in amount) **J-116** parent : baby (adult to young) **J-117** seldom : often (opposites)

LESSON PREPARATION

OBJECTIVE AND MATERIALS

OBJECTIVE: Students will select a pair of words to complete an analogy.
MATERIALS: Transparency of student book page 326 • washable transparency marker

CURRICULUM APPLICATIONS

Language Arts: Recognizing and using analogies; paraphrasing; stating relationships shown by words, sentences, or passages
Mathematics: Changing numerical information to graphic information (and vice versa); recognizing and using "part-to-whole" analogies in measurements of time, weight, size, or volume; identifying the relationship between fractional parts and fractions
Science: Distinguishing between phyla of plants or animals, writing science reports
Social Studies: Explaining historic or cultural parallels
Enrichment Areas: Recognizing and using comparative note values in music, choosing complementary colors or styles in decorating or art projects, taking tests

TEACHING SUGGESTIONS

Encourage students to discuss their answers. Remember the words that students use to describe their choices and use these same words to remind students of the key characteristics of the relationships in this lesson.

MODEL LESSON

LESSON

Introduction

Q: You have selected or supplied a word to complete an analogy.

Explaining the Objective to the Students

Q: In this lesson, you will select a pair of words to complete an analogy.

Class Activity

- Project exercise **J-110** from the transparency of page 326.

ANALOGIES—SELECT THE RIGHT PAIR

DIRECTIONS: In each exercise, decide how the first two words are related. Next, look at the pairs of words in the list at right; find the pair with a similar relationship. Fill in the blanks to complete the analogy.

J-110
below : under ::
_____ : _____
above : behind
above : over
between : over
beneath : over

J-111
dog : puppy ::
_____ : _____
calf : cow
cat : kitten
child : adult
colt : horse

J-112
now : then ::
_____ : _____
give : take
here : there
present : past
this : that

J-113
true : false ::
_____ : _____
lie : wrong
love : promise
right : answer
right : wrong

Q: In exercise **J-110** on page 326, you will find the relationship between two pairs of words. Each pair of words must be related in the same way. In this analogy, below : under :: ____ : ____, the relationship in the first pair is that *below* and *under* are synonyms. They name similar locations. Can you find a pair of words on the right that name similar locations? Let's look at all the choices to see the relation in each pair. How are *above* and *behind* related?

A: They are neither similar nor opposite; each word describes a different position or location.

Q: How are *above* and *over* related?

A: They are synonyms; they both describe a similar position.

Q: How are *between* and *over* related?

A: They are neither similar nor opposite; each word describes a different position or location.

Q: How are *beneath* and *over* related?

A: They are opposites; they describe opposite positions or locations.

Q: Which pair of words seems to have the same relationship as *below* and *under?*

A: Above : over

- Write "above" and "over" in the blanks of exercise **J-110**.

Q: Look at exercise **J-111**. How are *dog* and *puppy* related?

A: A dog is an adult animal and a puppy is a baby dog. The two words are related by age.

Q: We need to decide which pair of words on the right has the same kind of relationship. How are *calf* and *cow* related?

A: They are related by age; a calf is a baby cow.

Q: It is important to note the order of the words in an analogy. The words in each pair of an analogy should be related in the same way and in the same order. In the dog and puppy word pair, the older animal is listed first. Are a calf and cow related in the same order?

A: No, a calf is younger than a cow.

Q: How are *cat* and *kitten* related?

A: They are related by age; a cat is an adult animal and a kitten is a baby cat.

Q: How are *child* and *adult* related?

A: They are related by age; a child is a young adult. They are also both people and not animals.

Q: How are *colt* and *horse* related?

A: They are related by age; a colt is a baby horse.

Q: Which of the four pairs of opposites is the best answer?

A: Cat : kitten. Just as in the dog and puppy word pair, the older animal is listed first.

- Write "cat" and "kitten" in the blanks of exercise **J-111**.

GUIDED PRACTICE

EXERCISES: **J-110** through **J-112**

- When students have had sufficient time to complete these exercises, check answers by discussion to determine whether they have answered correctly. Discuss also why the incorrect answers were eliminated.

INDEPENDENT PRACTICE

- Assign exercises **J-113** through **J-117**.

THINKING ABOUT THINKING

Q: What did you pay attention to when you found analogies between pairs of words?

1. I thought carefully about the meaning of each word.

2. I decided how the words in the first pair were different in meaning.
3. I checked whether the second pair was different in the same way as the first pair.
4. I explained the relationship as an analogy: "A" is to "B" as "C" is to "D."

PERSONAL APPLICATION

Q: When do you need to find a pair of words that expresses a relationship similar to another pair of words?

A: Examples include explaining family relationships, teaching someone to use a tool or construct a model, explaining or recognizing relationships between items in a test question, recognizing different items that can be used for similar purposes or functions.

TRANSPARENCY MASTER 1

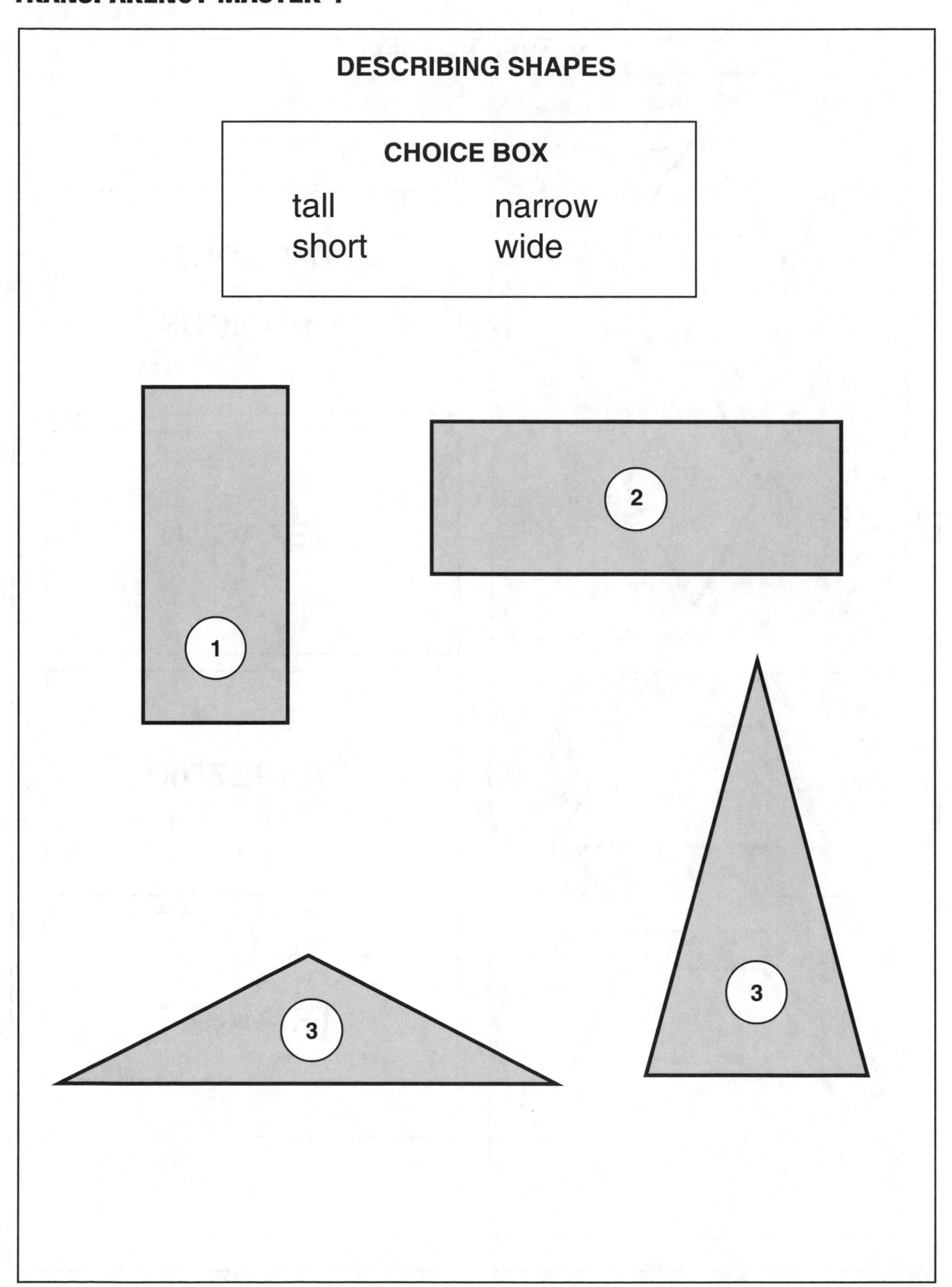

TRANSPARENCY MASTER 2

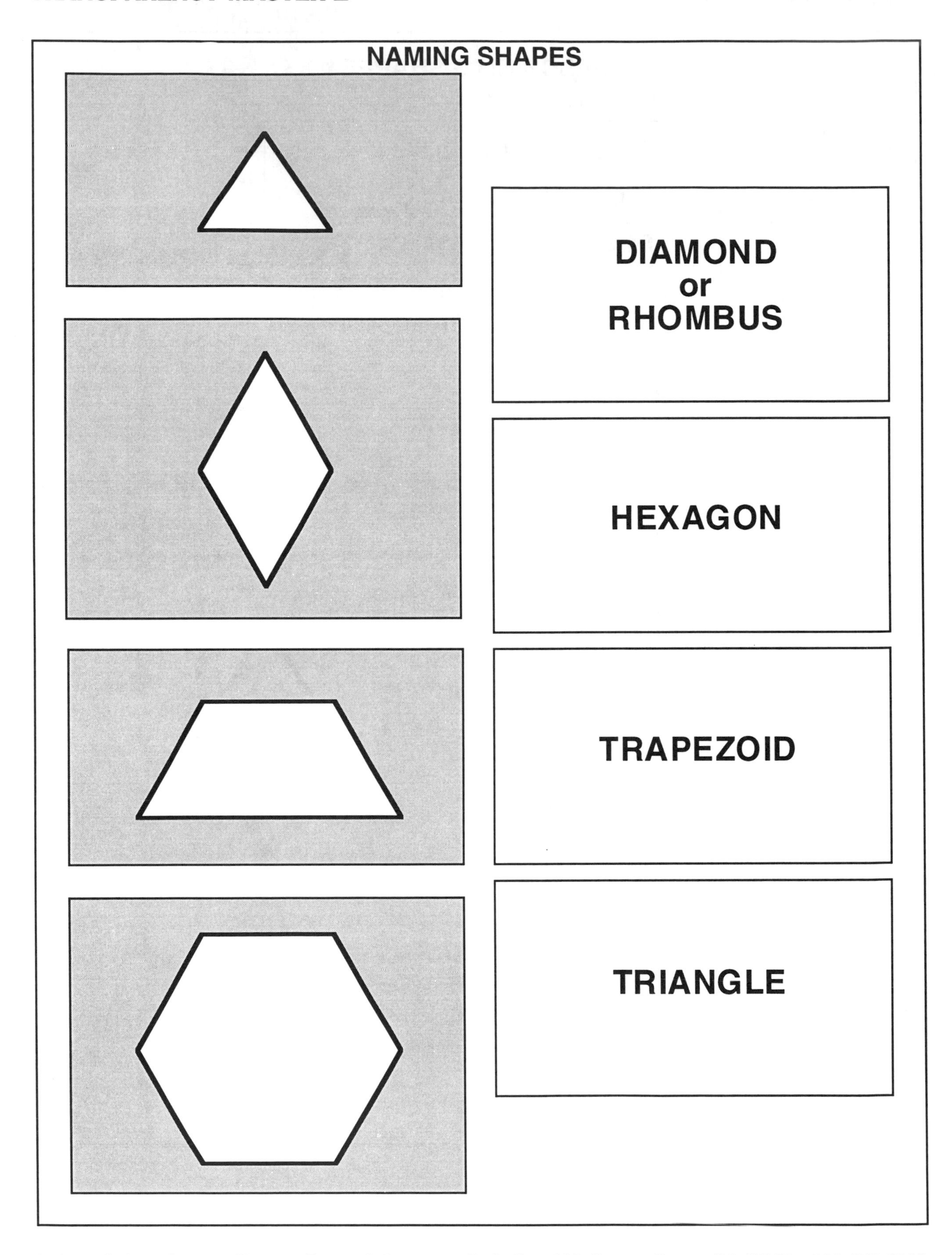

TRANSPARENCY MASTER 3

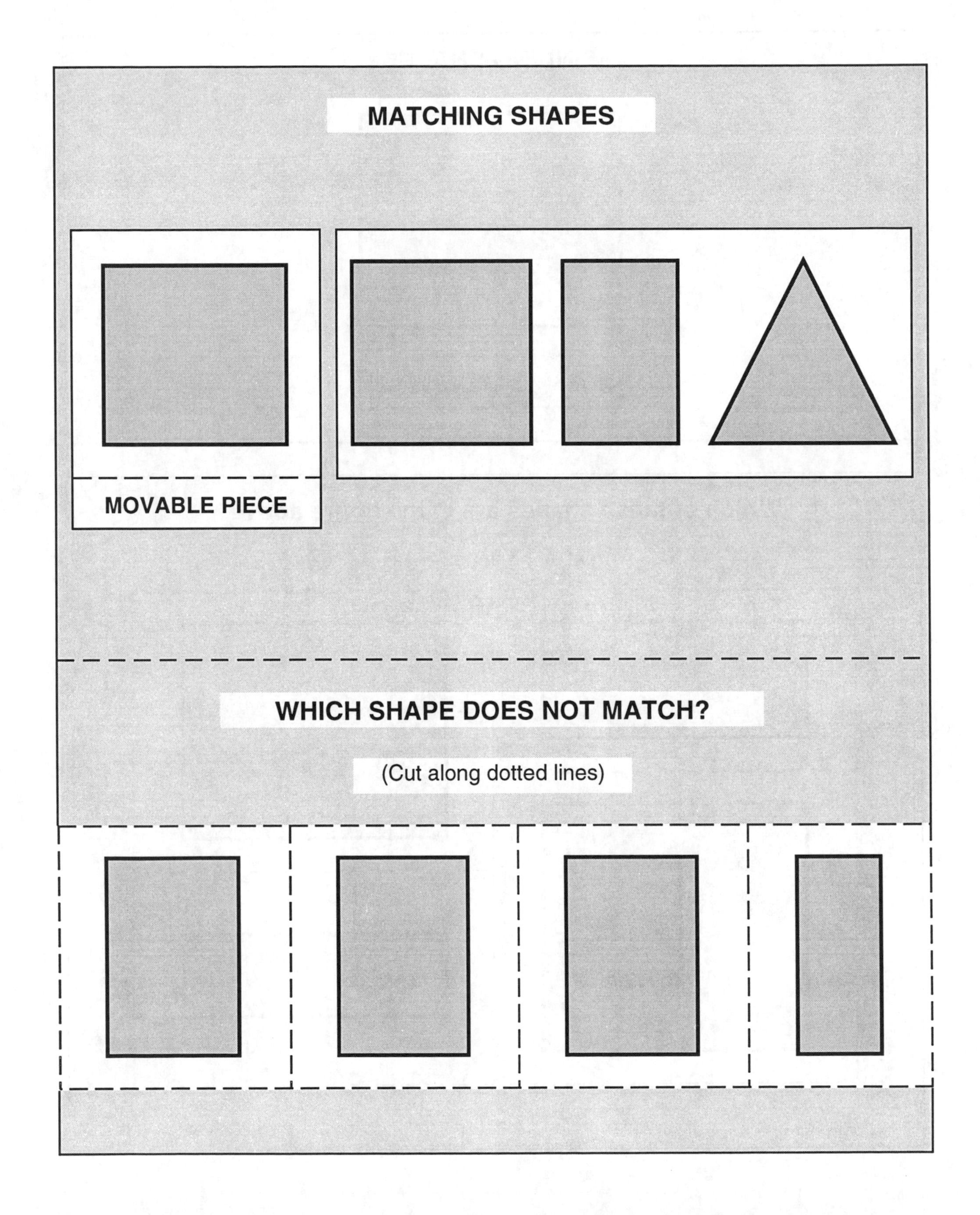

TRANSPARENCY MASTER 4

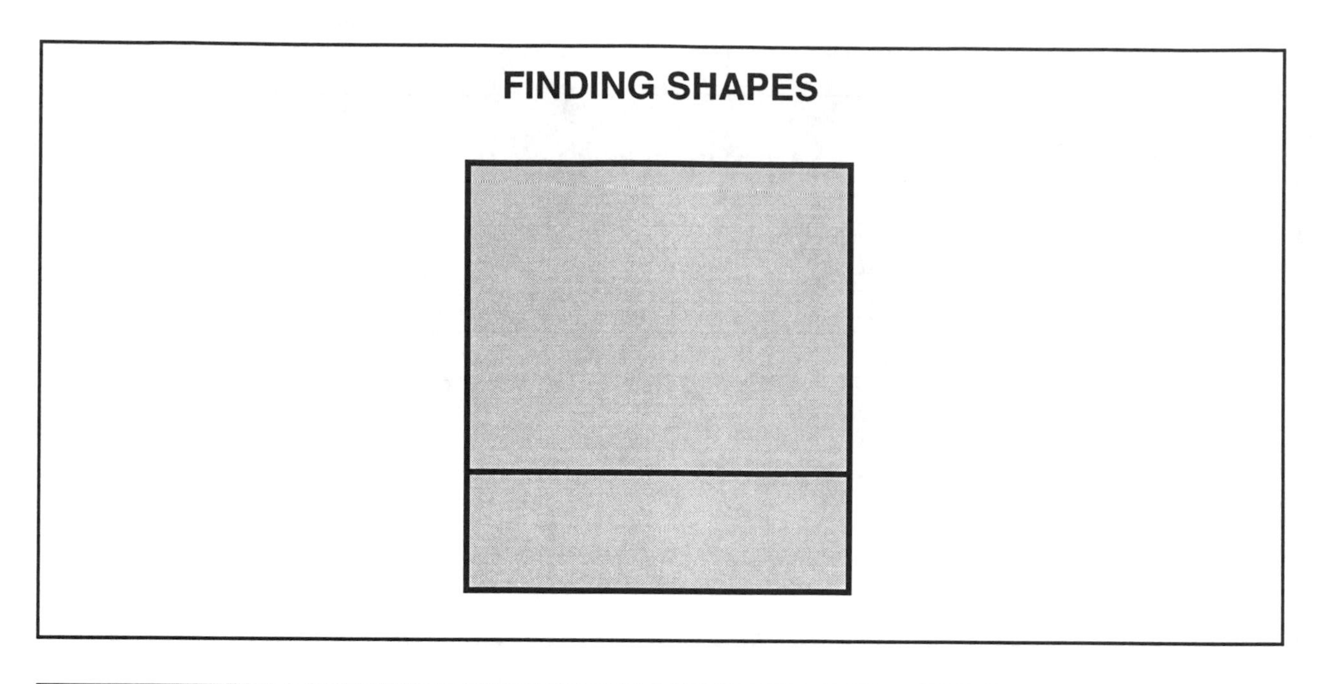

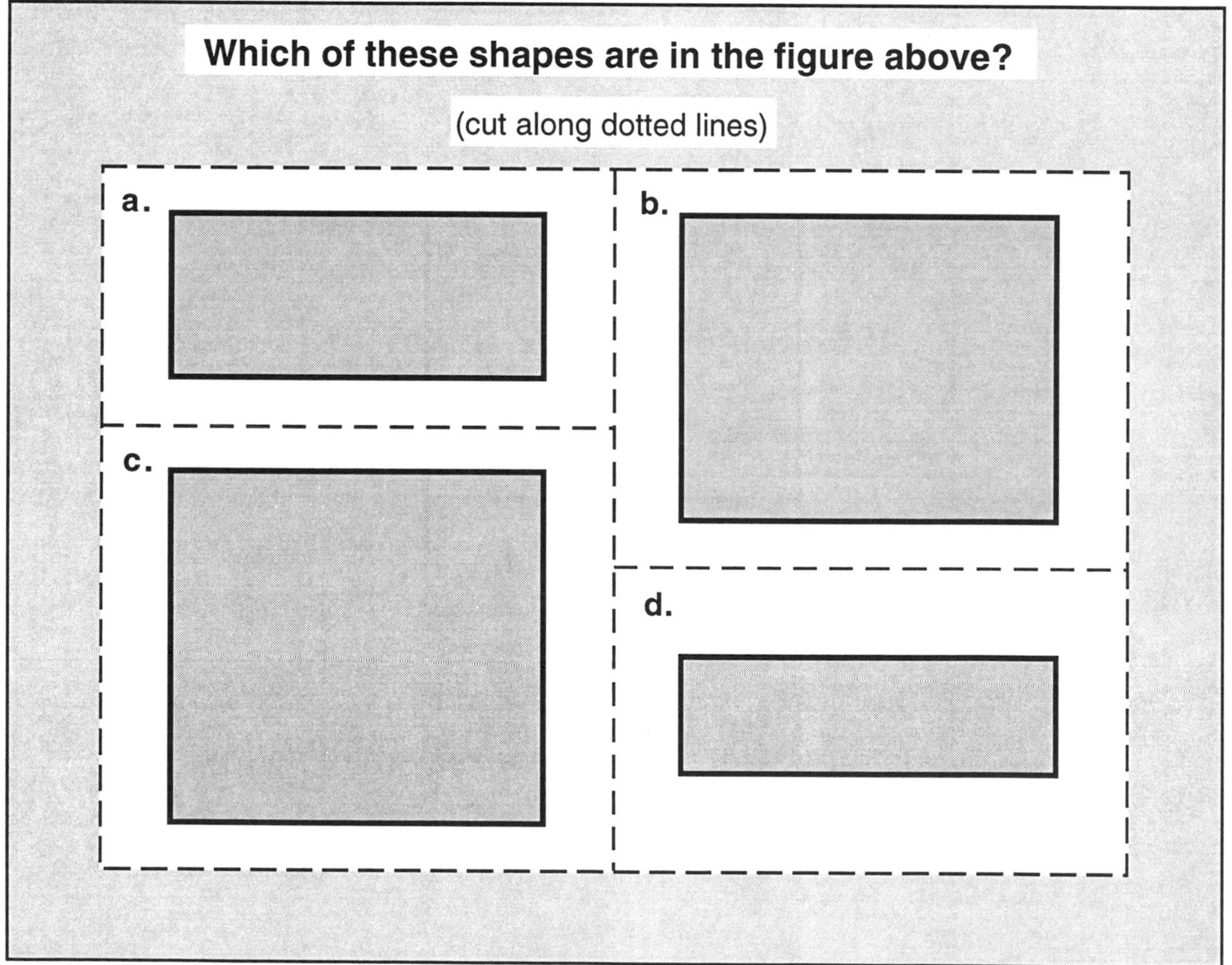

TRANSPARENCY MASTER 5

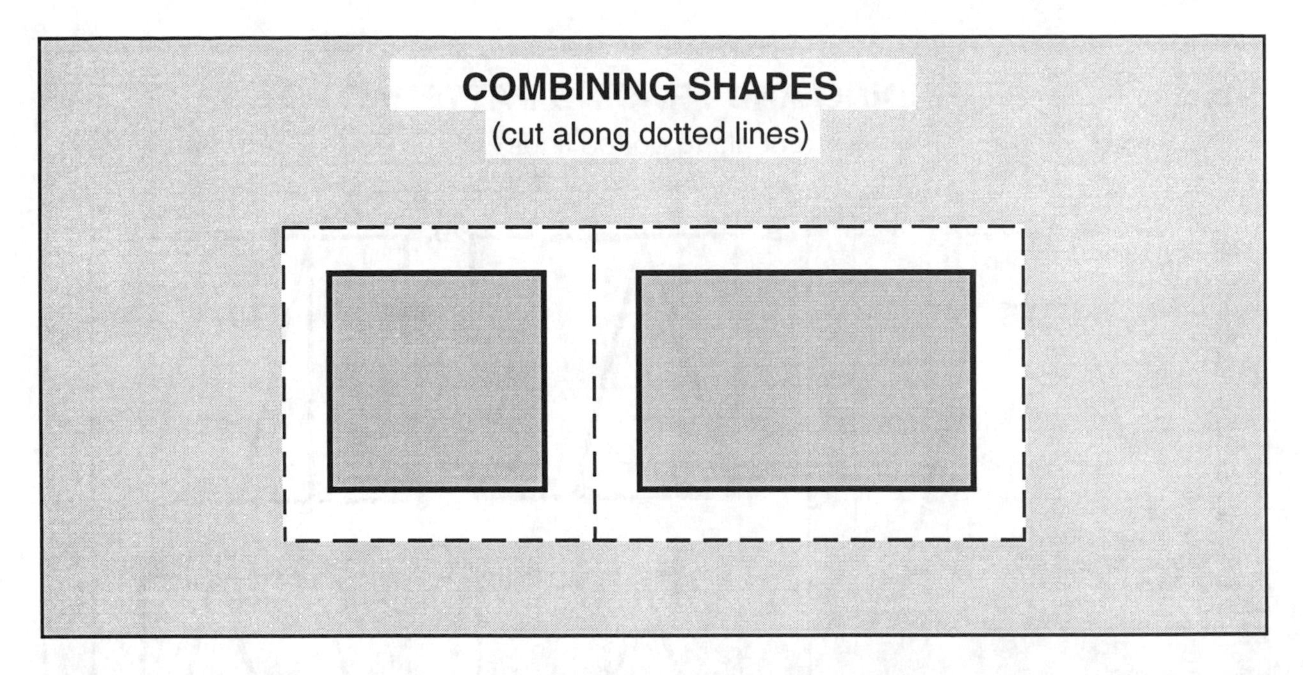

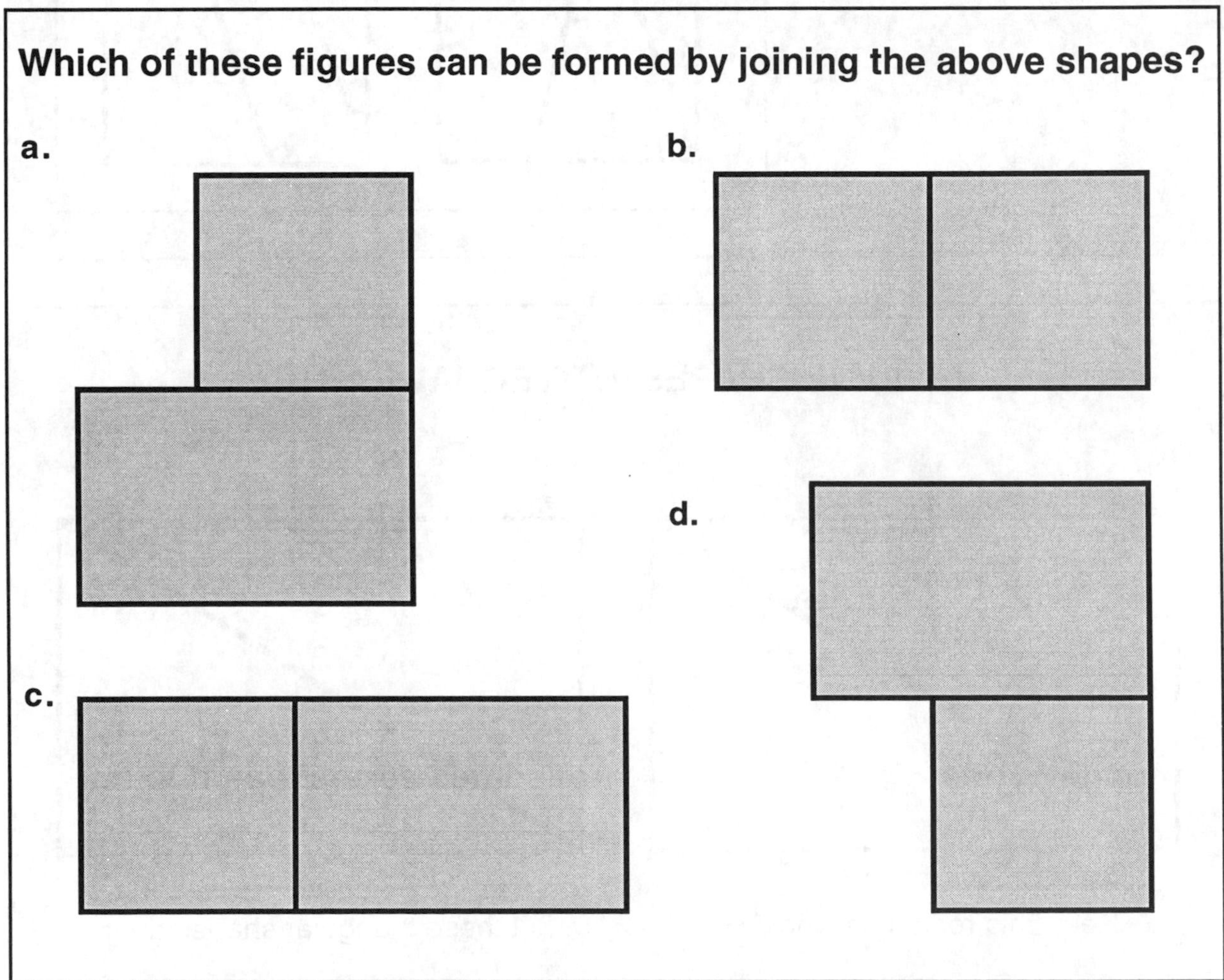

TRANSPARENCY MASTER 6

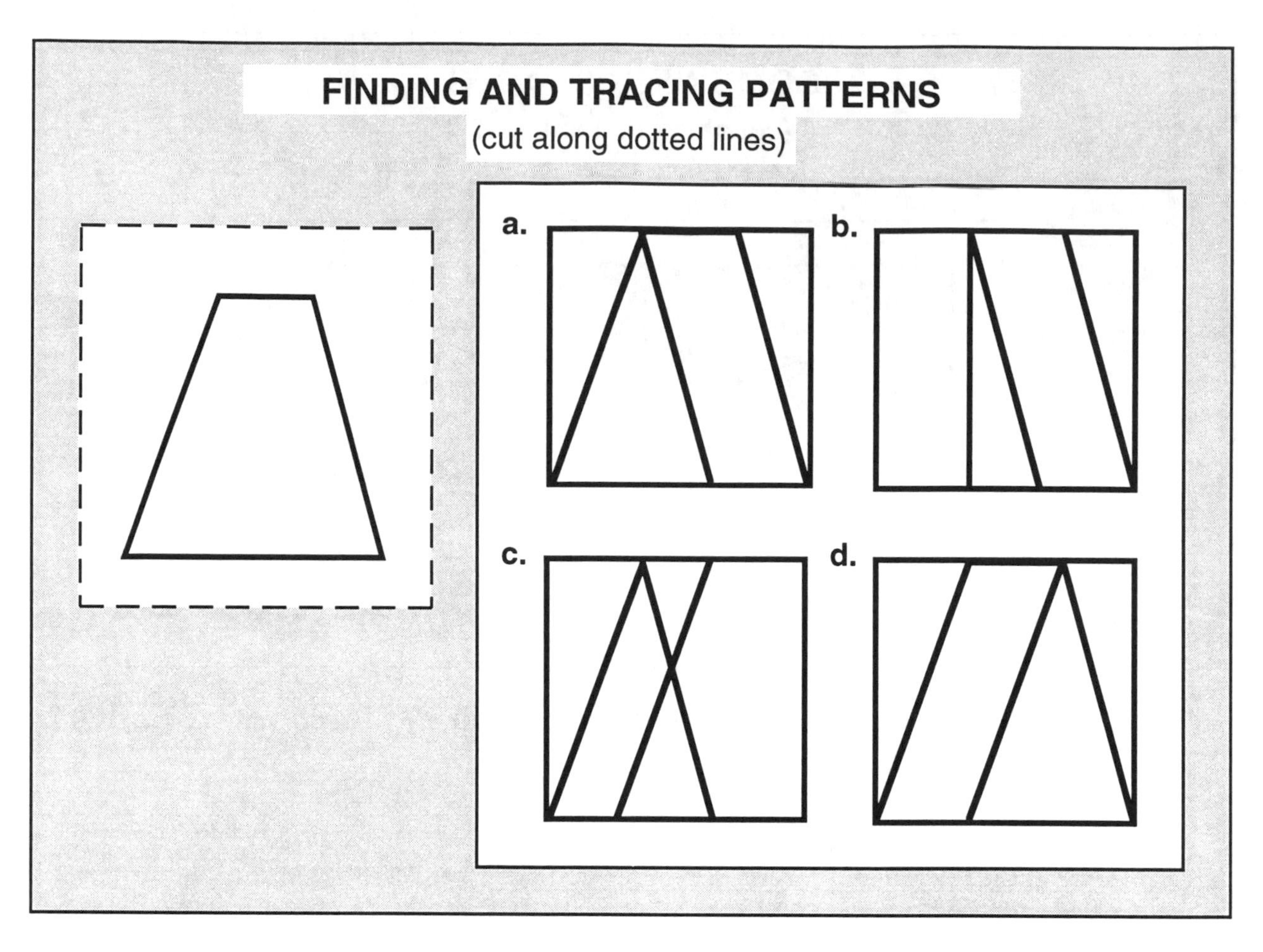

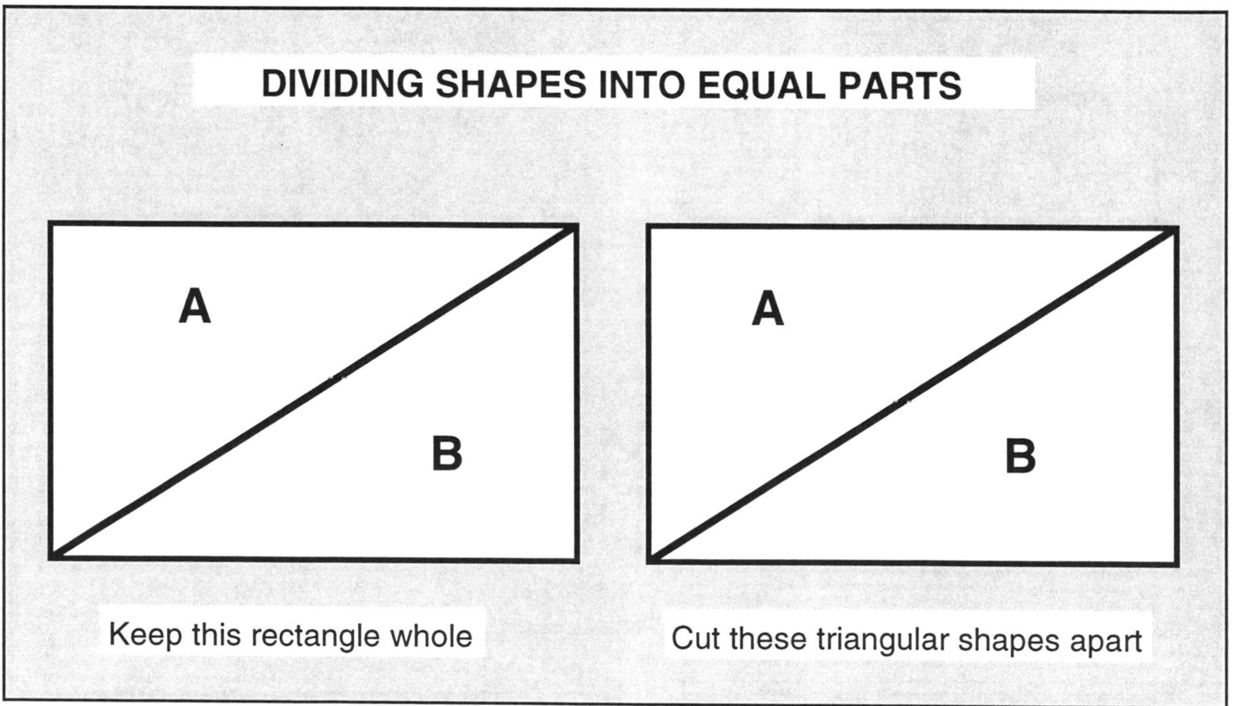

TRANSPARENCY MASTER 7

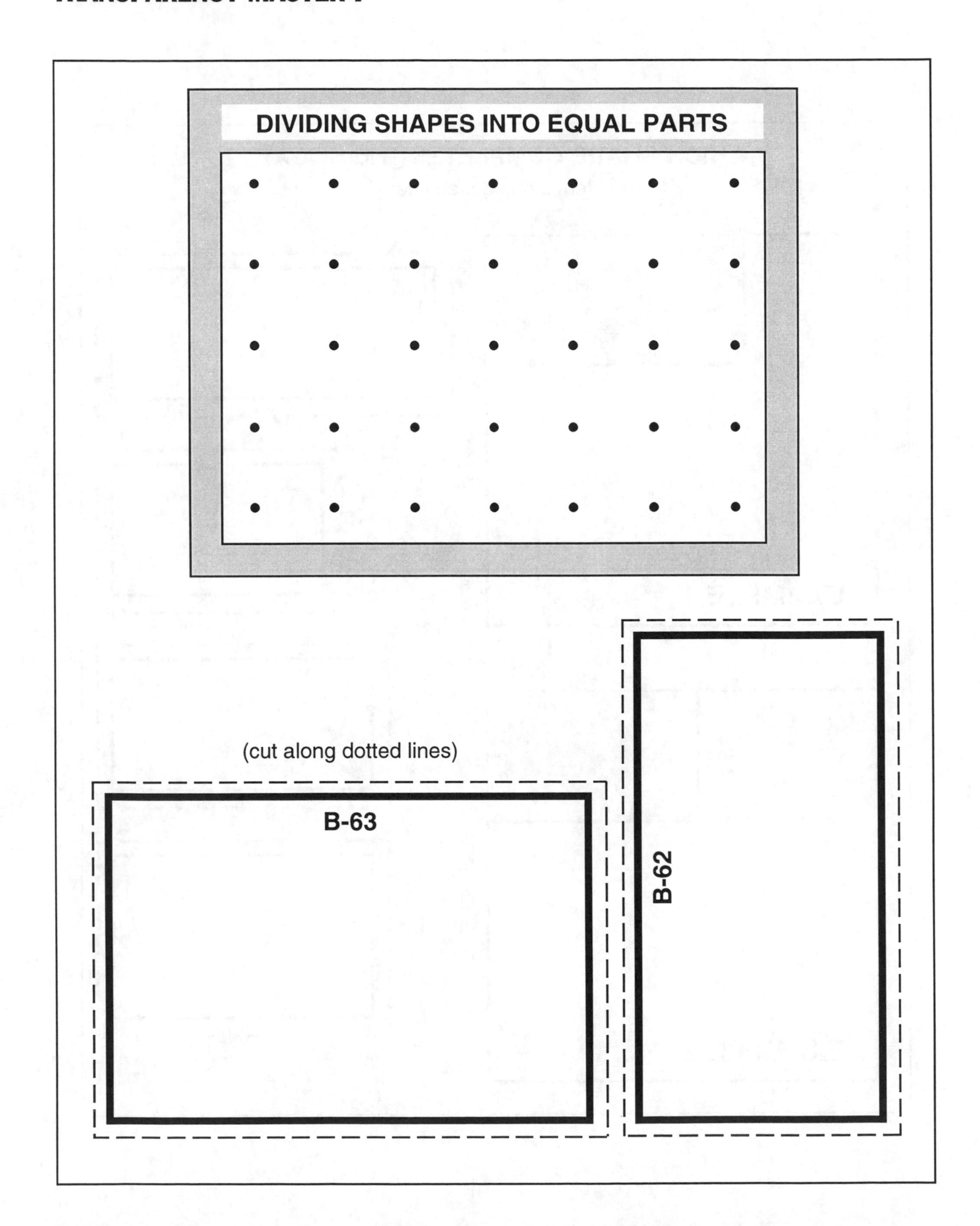

TRANSPARENCY MASTER 8

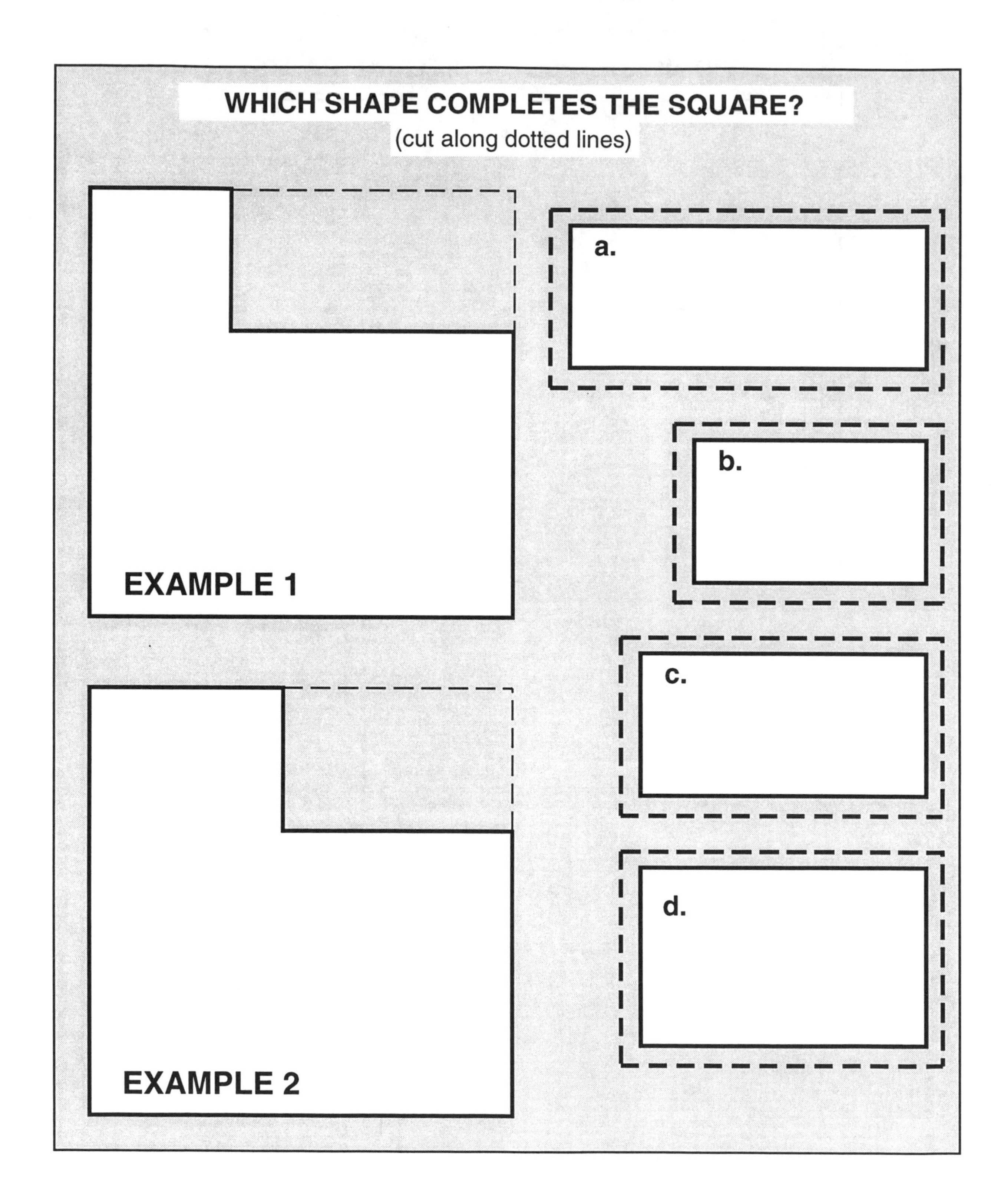

TRANSPARENCY MASTER 9

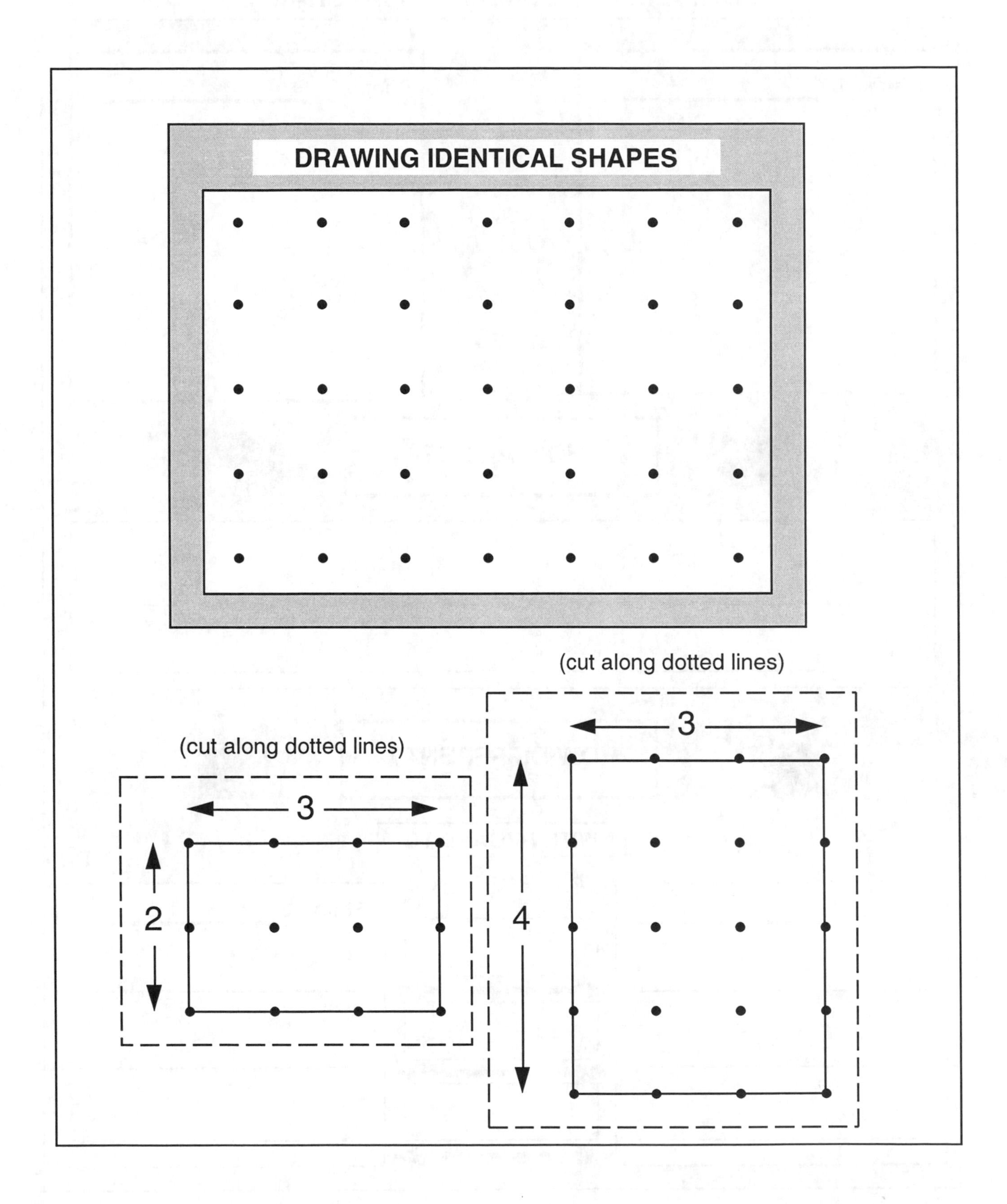

TRANSPARENCY MASTER 10

COMPARE AND CONTRAST TWO SHAPES

Shape 1 ______________

Shape 2 ______________

HOW ALIKE?

____

____

HOW DIFFERENT?

WITH REGARD TO

Shape 1 ______________

____

____

____

Shape 2 ______________

____

____

____

TRANSPARENCY MASTER 11

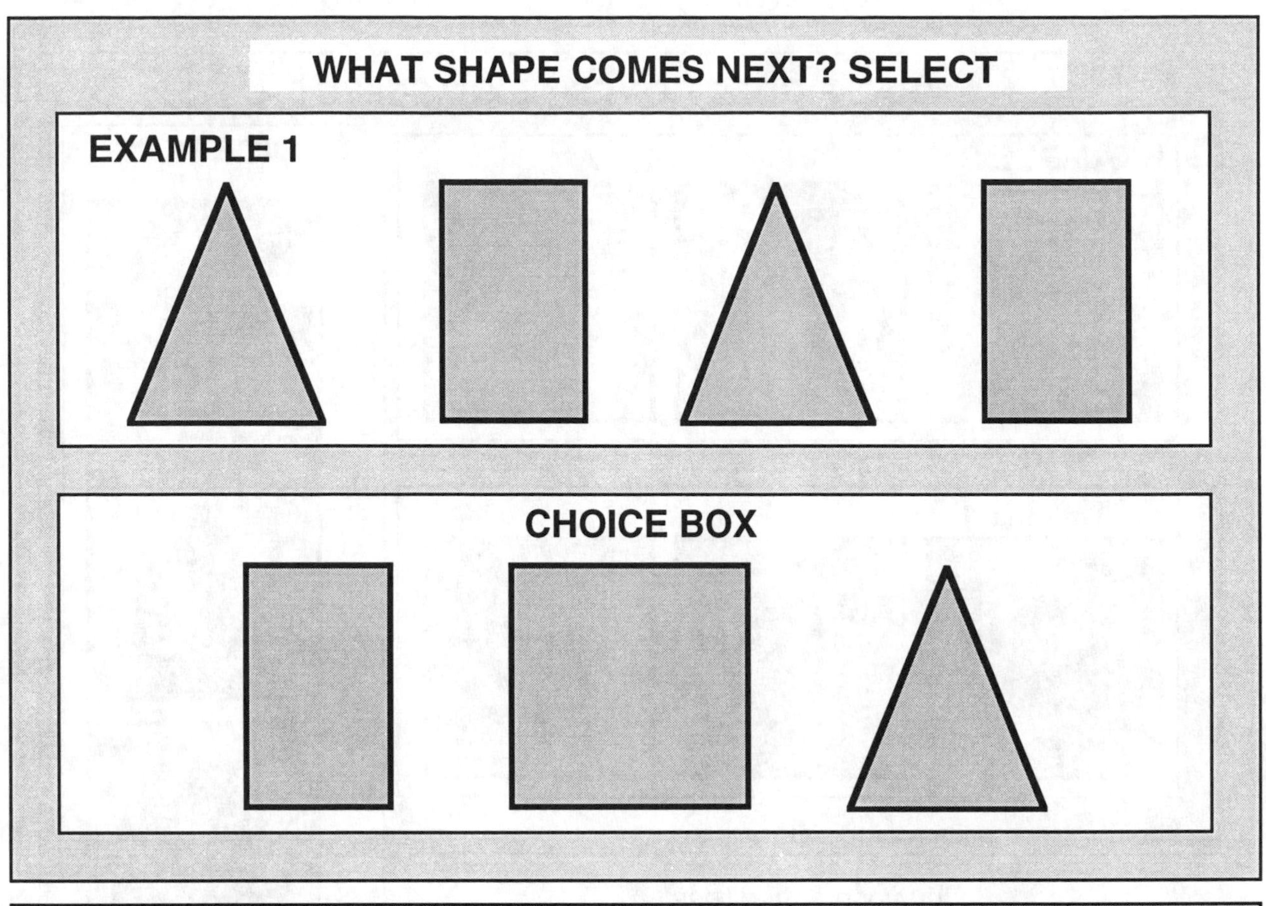

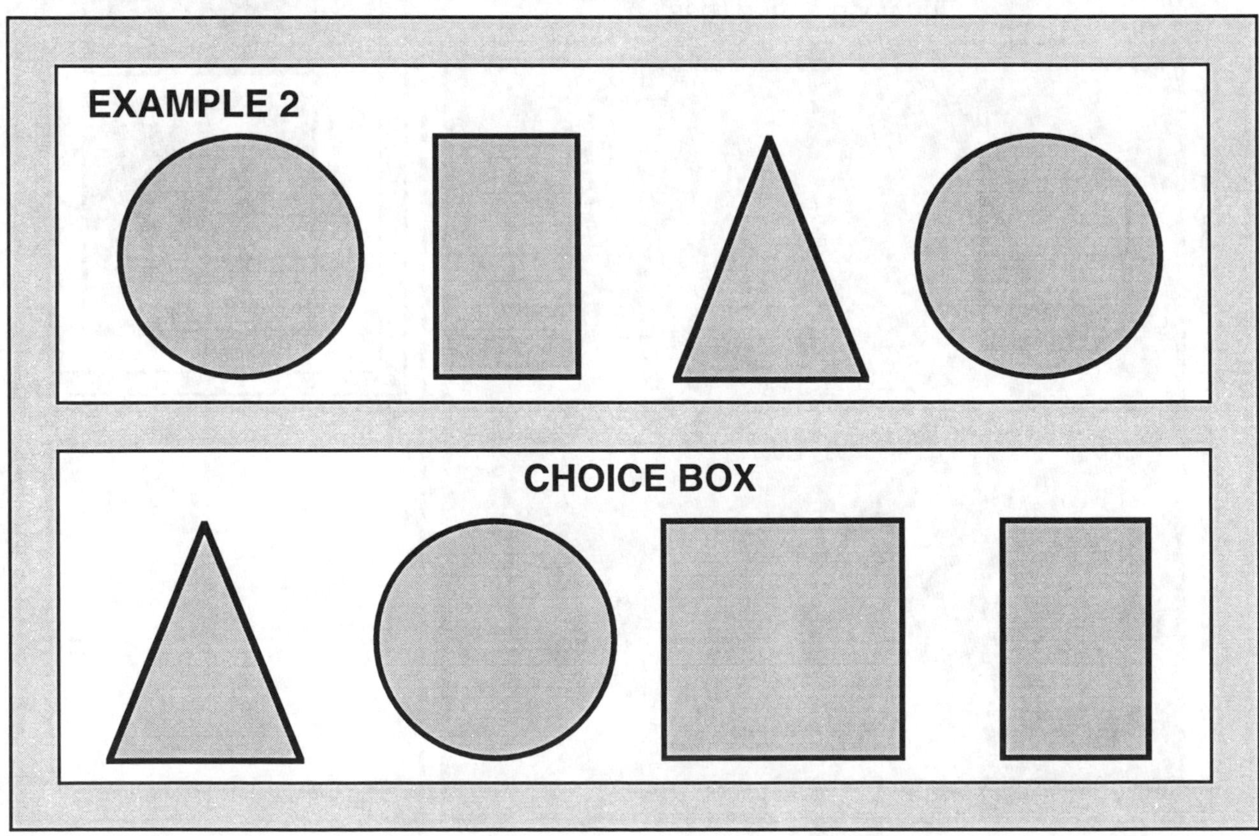

TRANSPARENCY MASTER 12

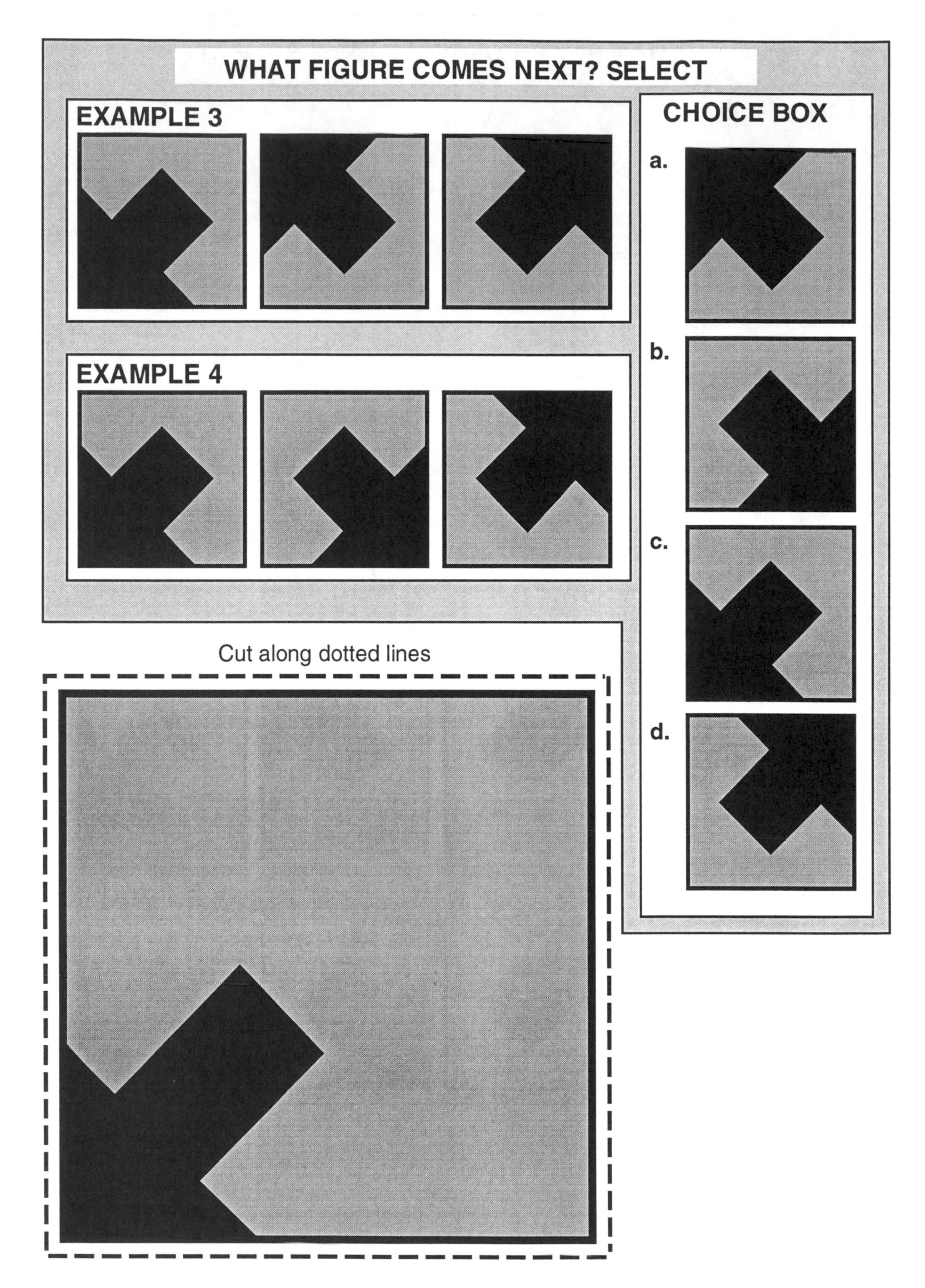

TRANSPARENCY MASTER 13

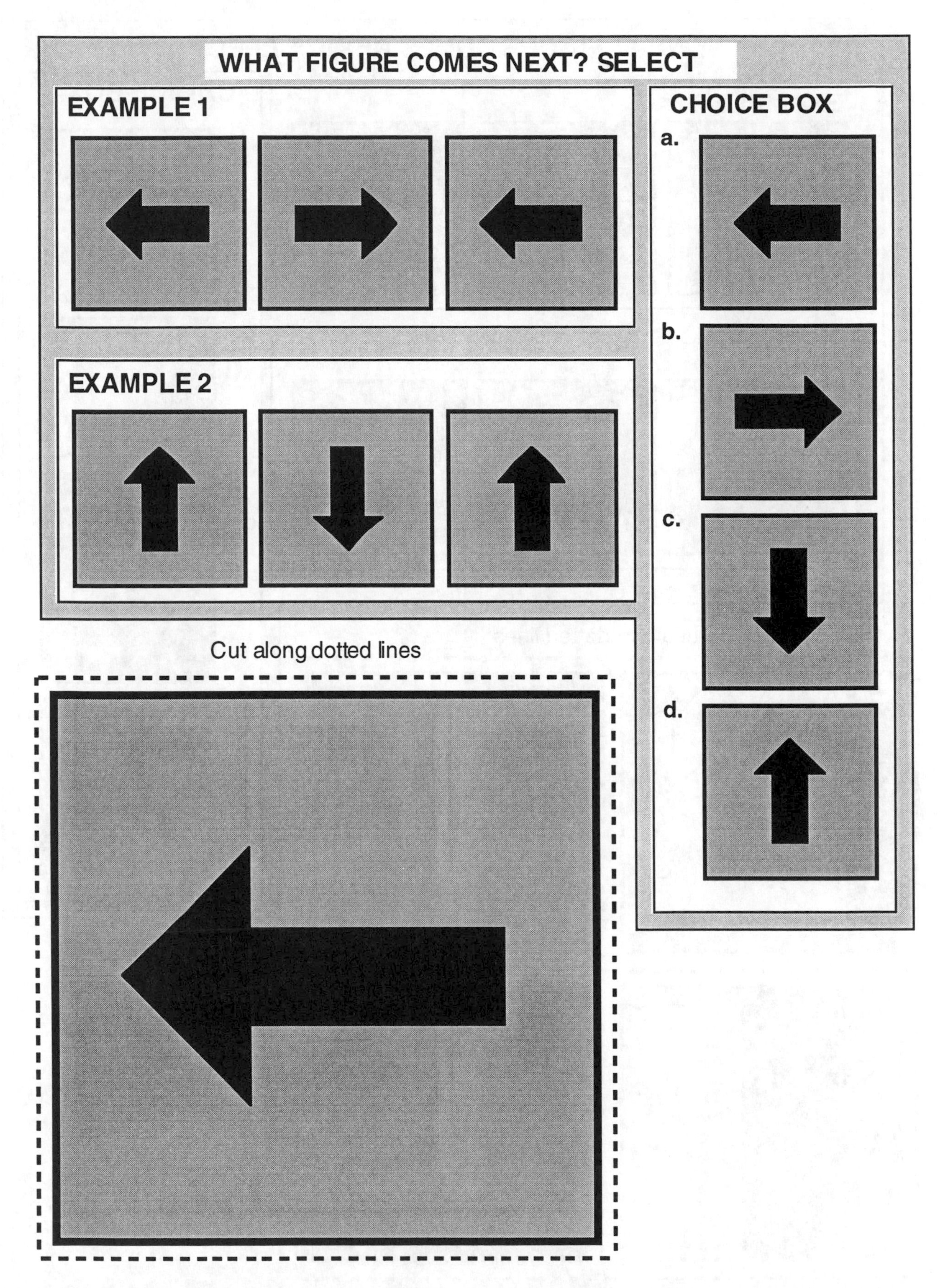

TRANSPARENCY MASTER 14

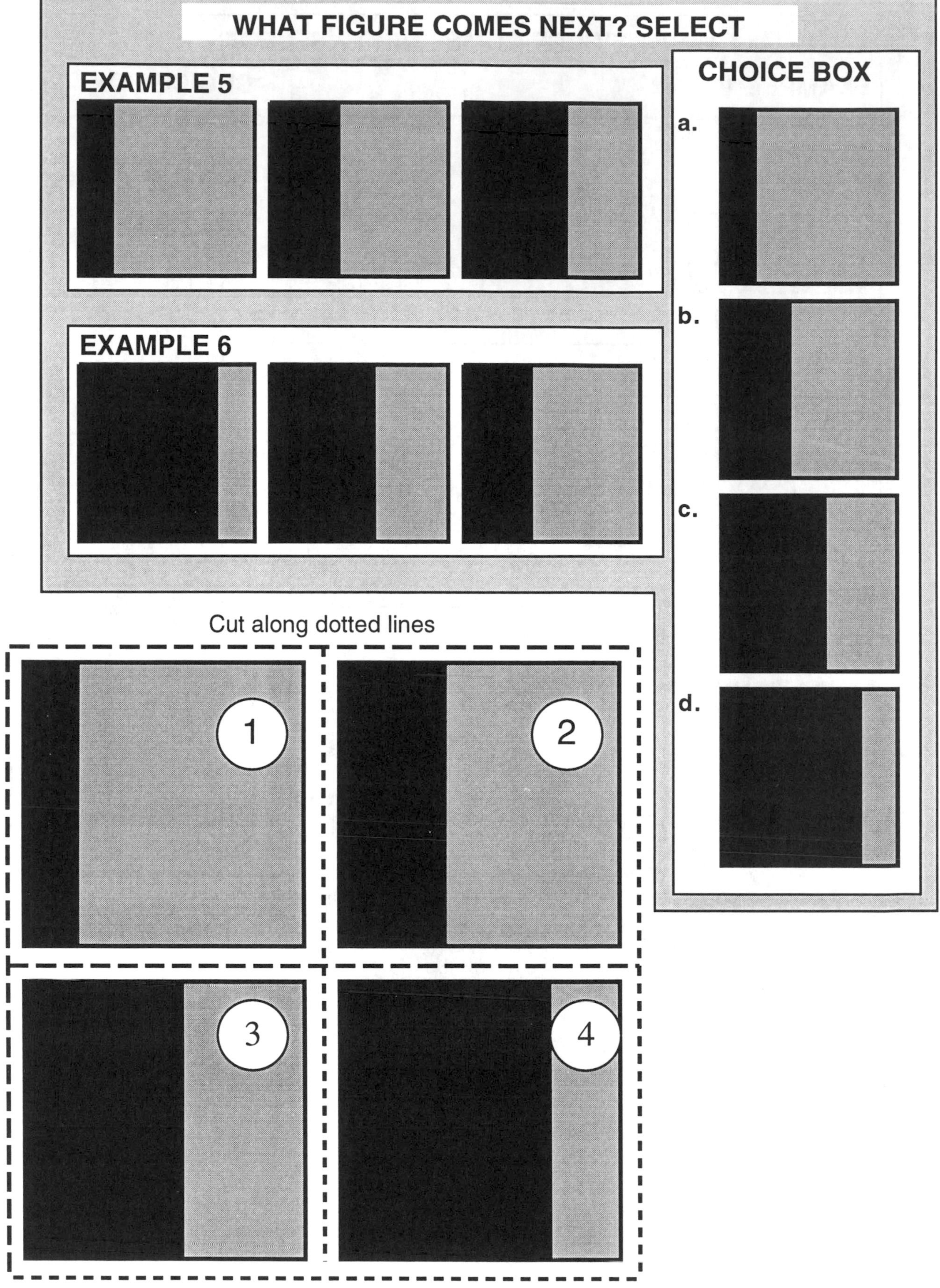

TRANSPARENCY MASTER 15

WHAT FIGURE COMES NEXT? DRAW IT

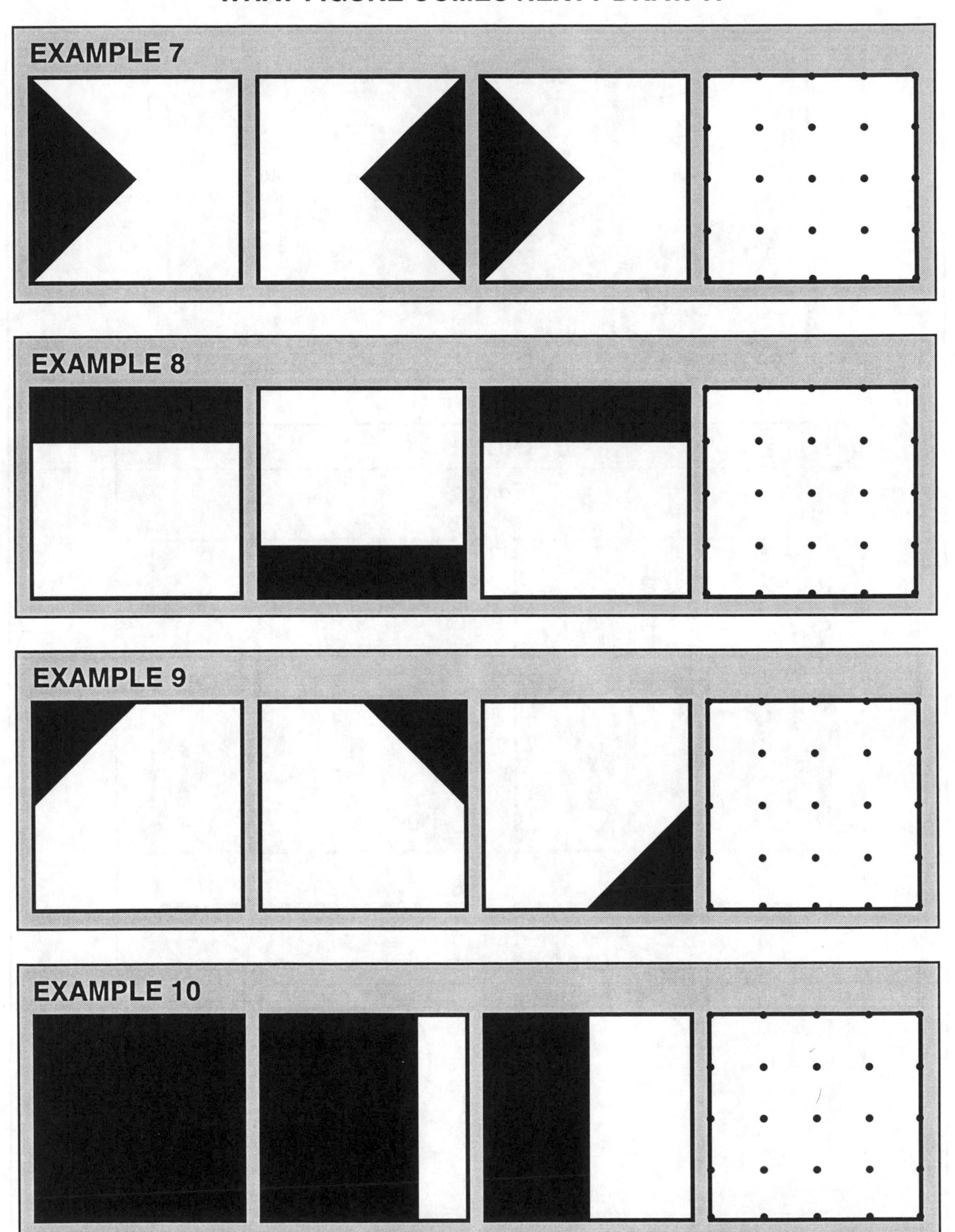

TRANSPARENCY MASTER 16

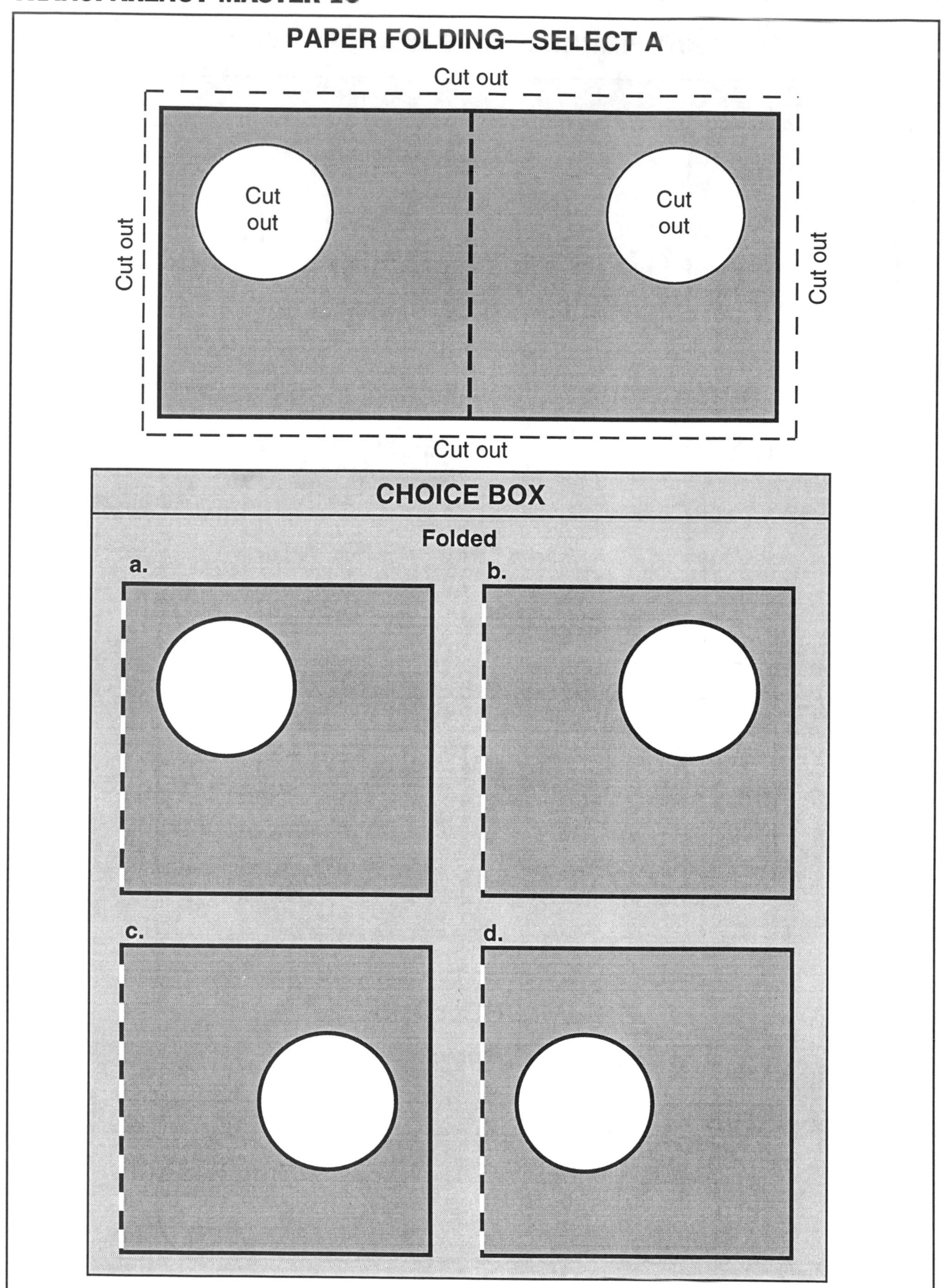

TRANSPARENCY MASTER 17

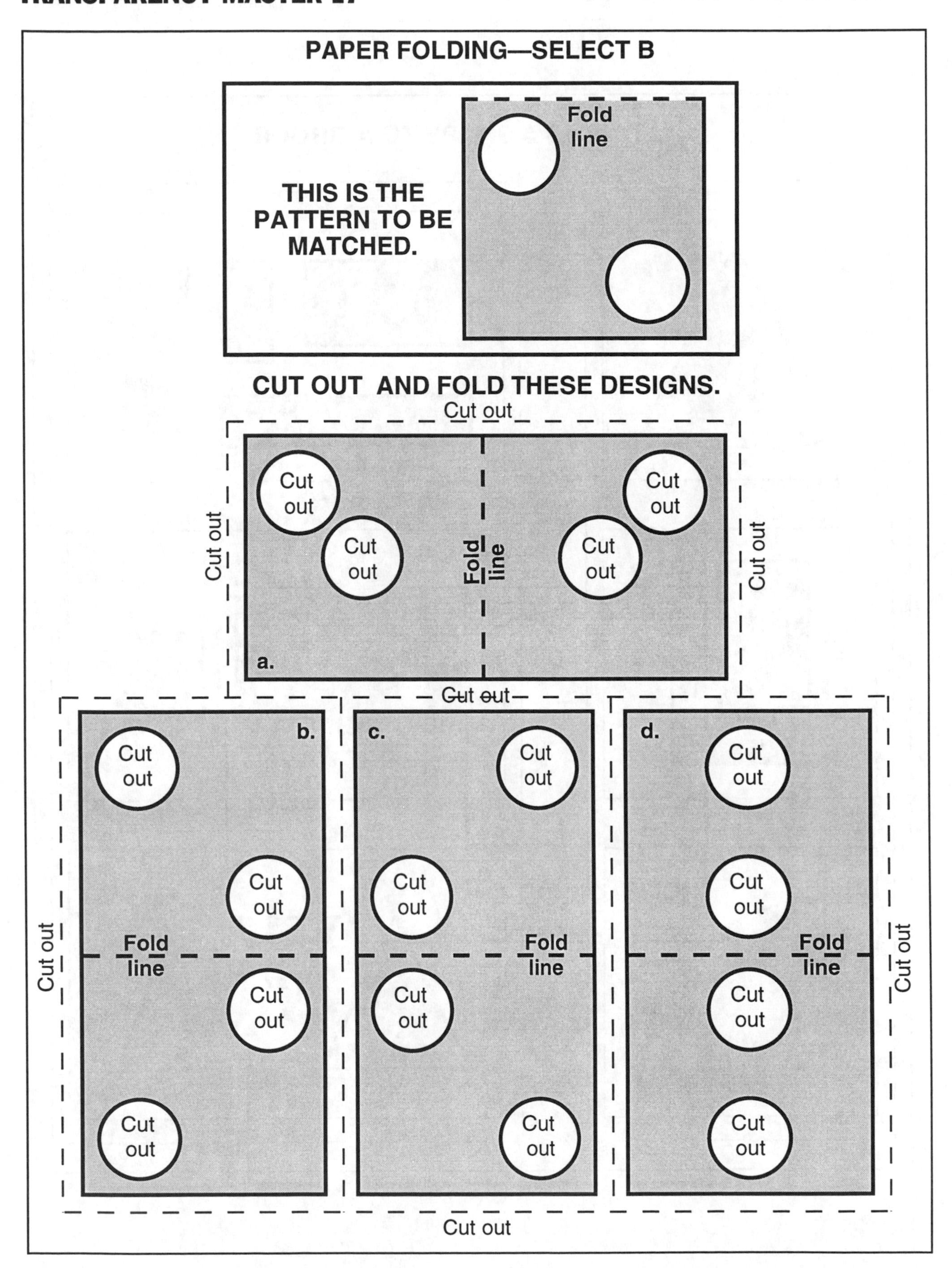

TRANSPARENCY MASTER 18

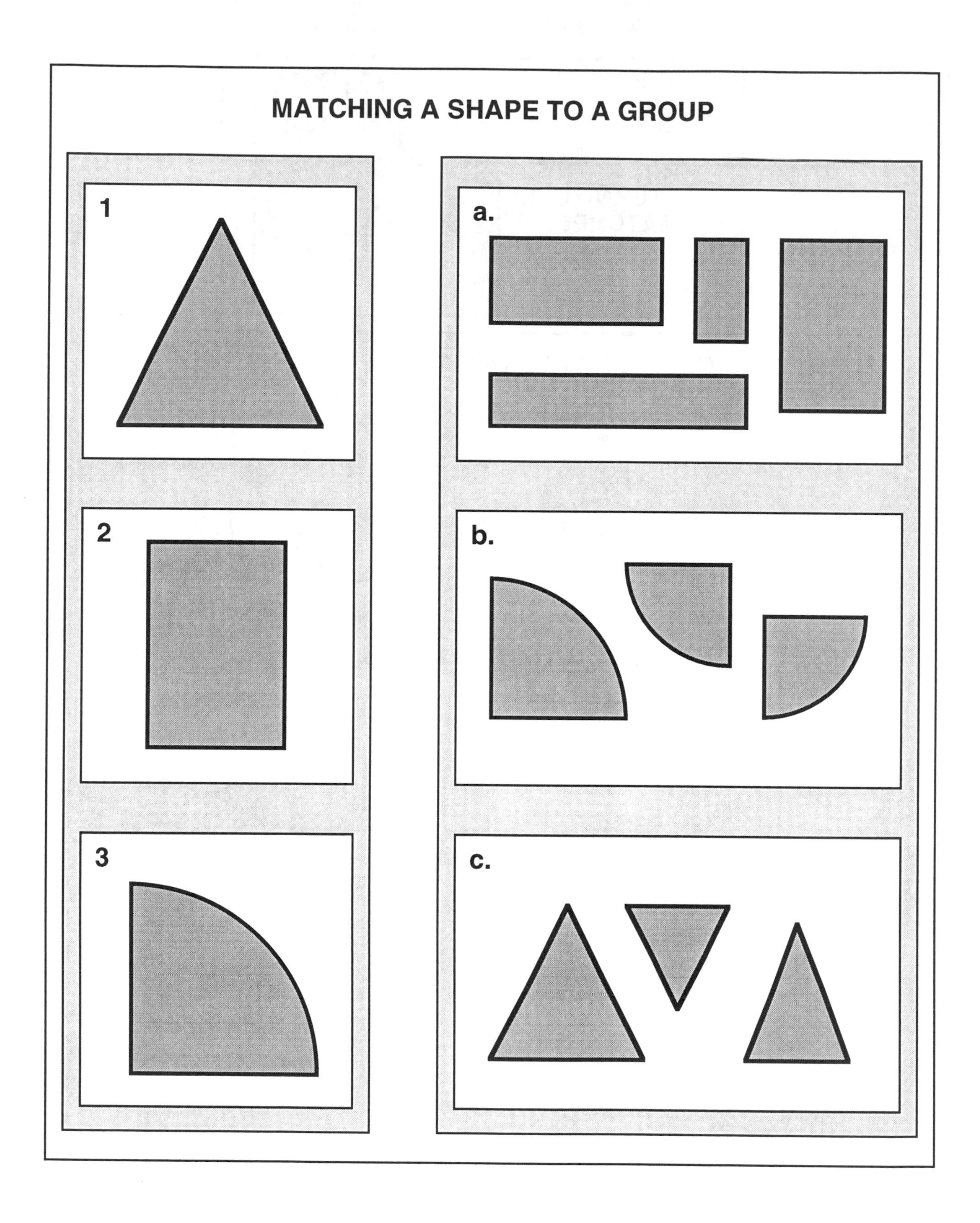

TRANSPARENCY MASTER 19

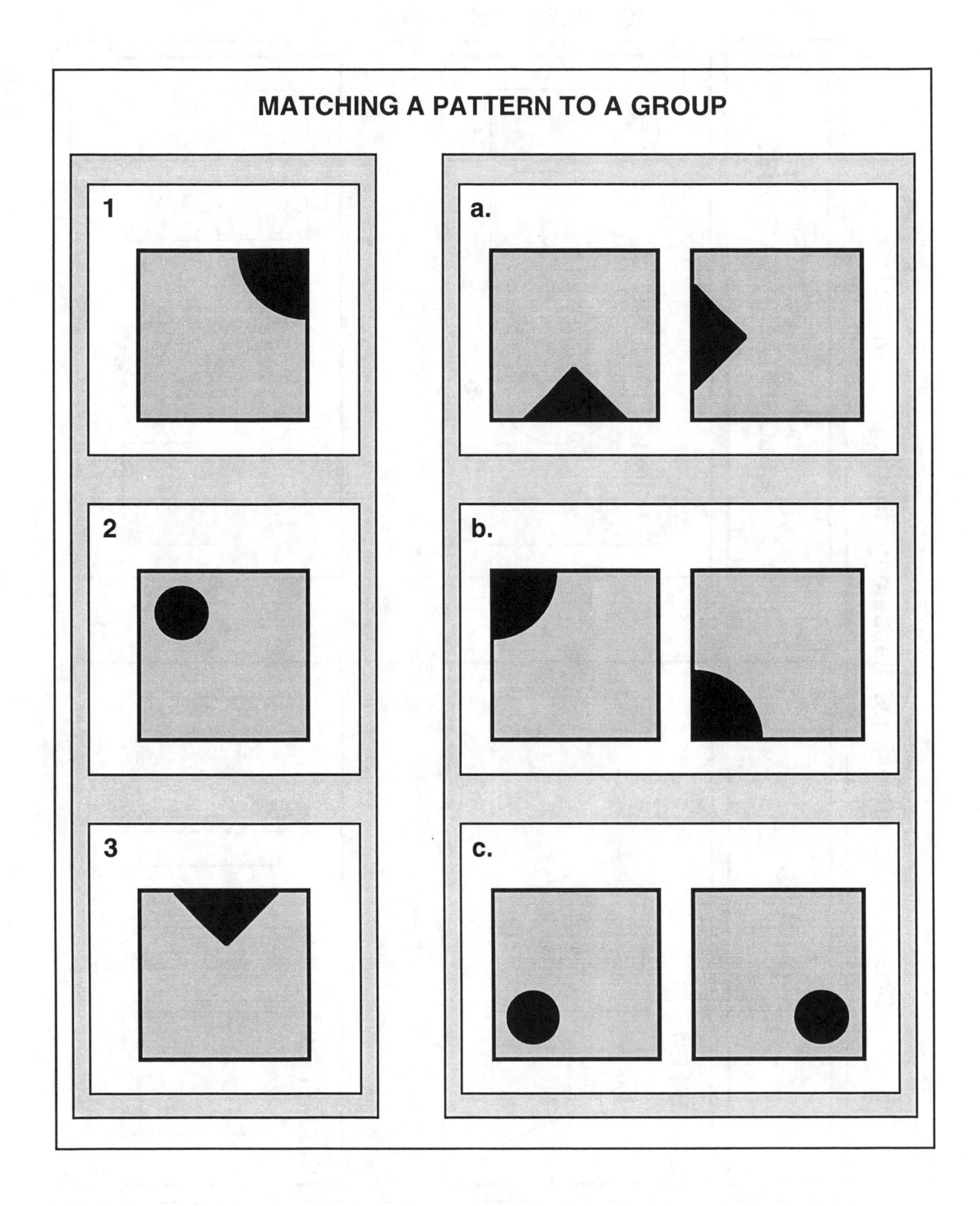

TRANSPARENCY MASTER 20

COMPLETE THE CLASS A

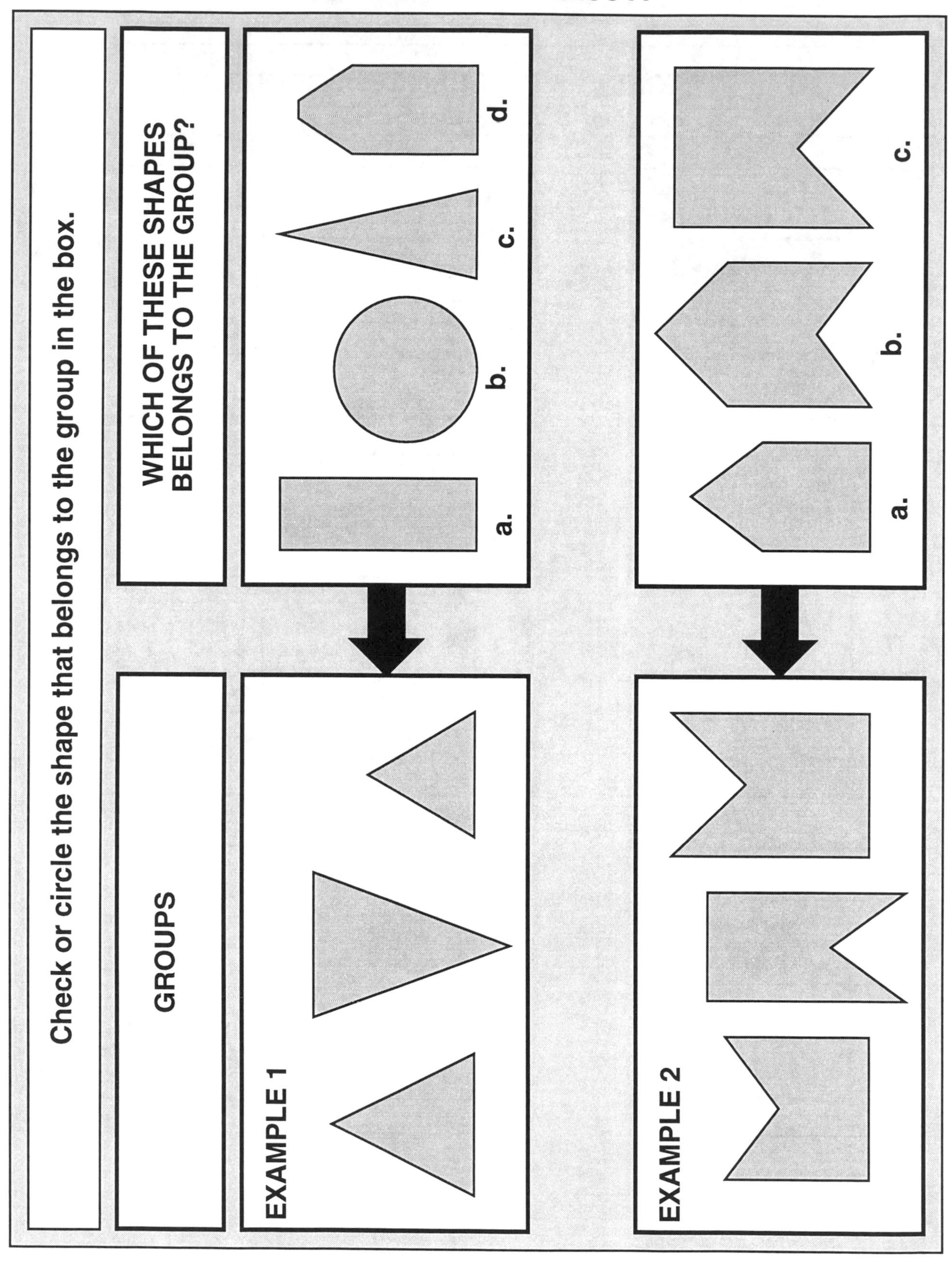

TRANSPARENCY MASTER 21

COMPLETE THE CLASS B

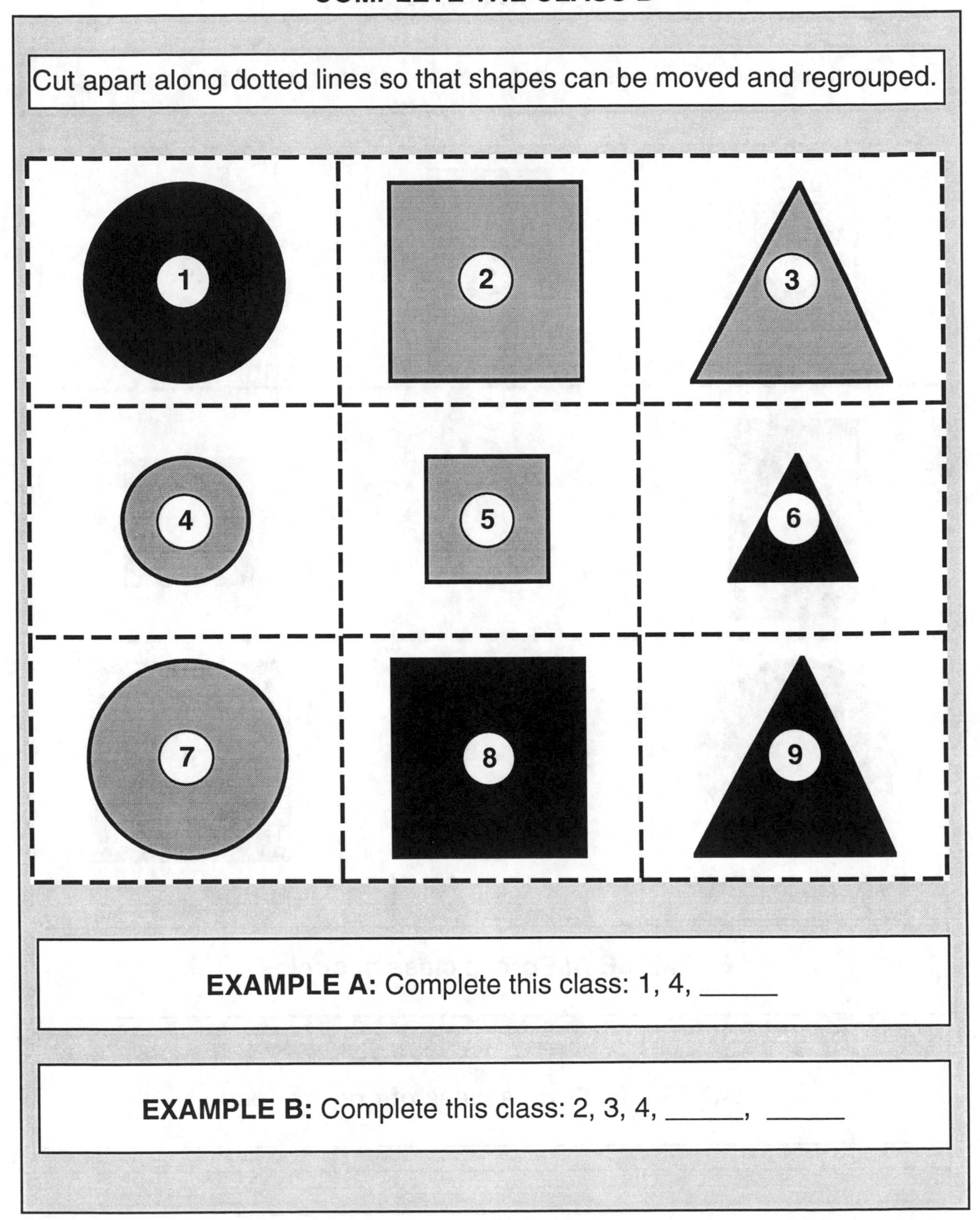

TRANSPARENCY MASTER 22

FORM A CLASS

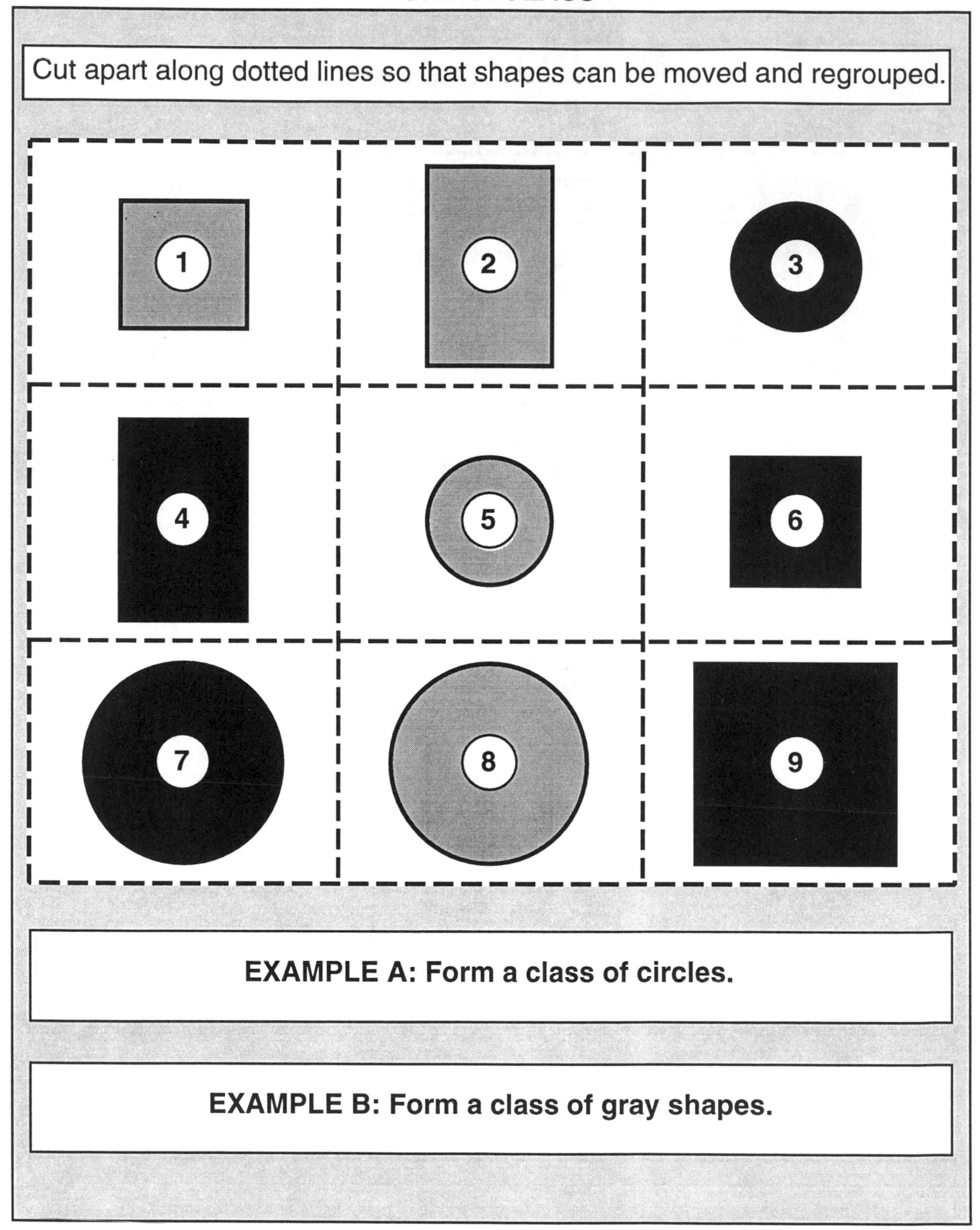

TRANSPARENCY MASTER 23

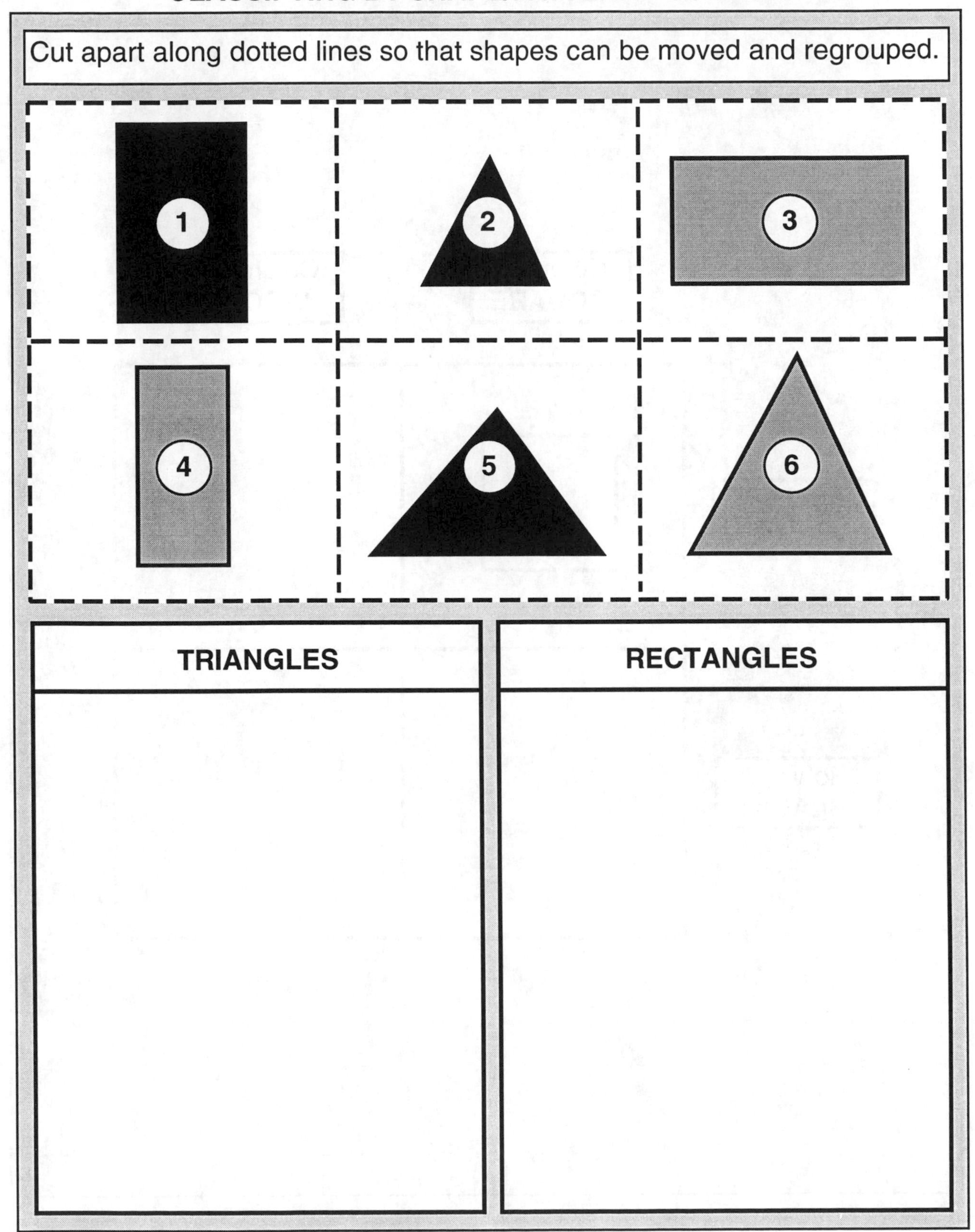

TRANSPARENCY MASTER 24

OVERLAPPING CLASSES—MATRIX

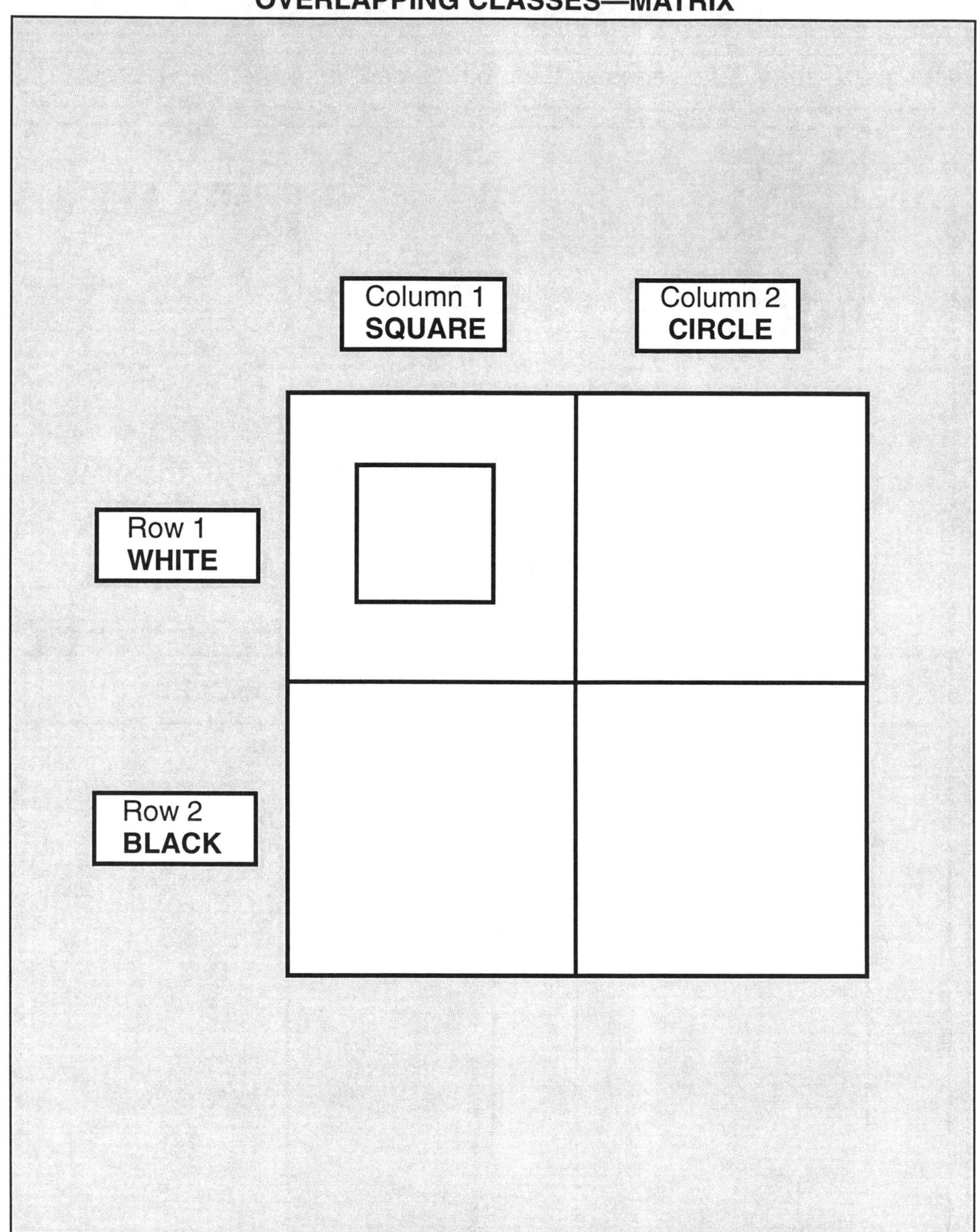

TRANSPARENCY MASTER 25

FIGURAL ANALOGIES—COMPLETE

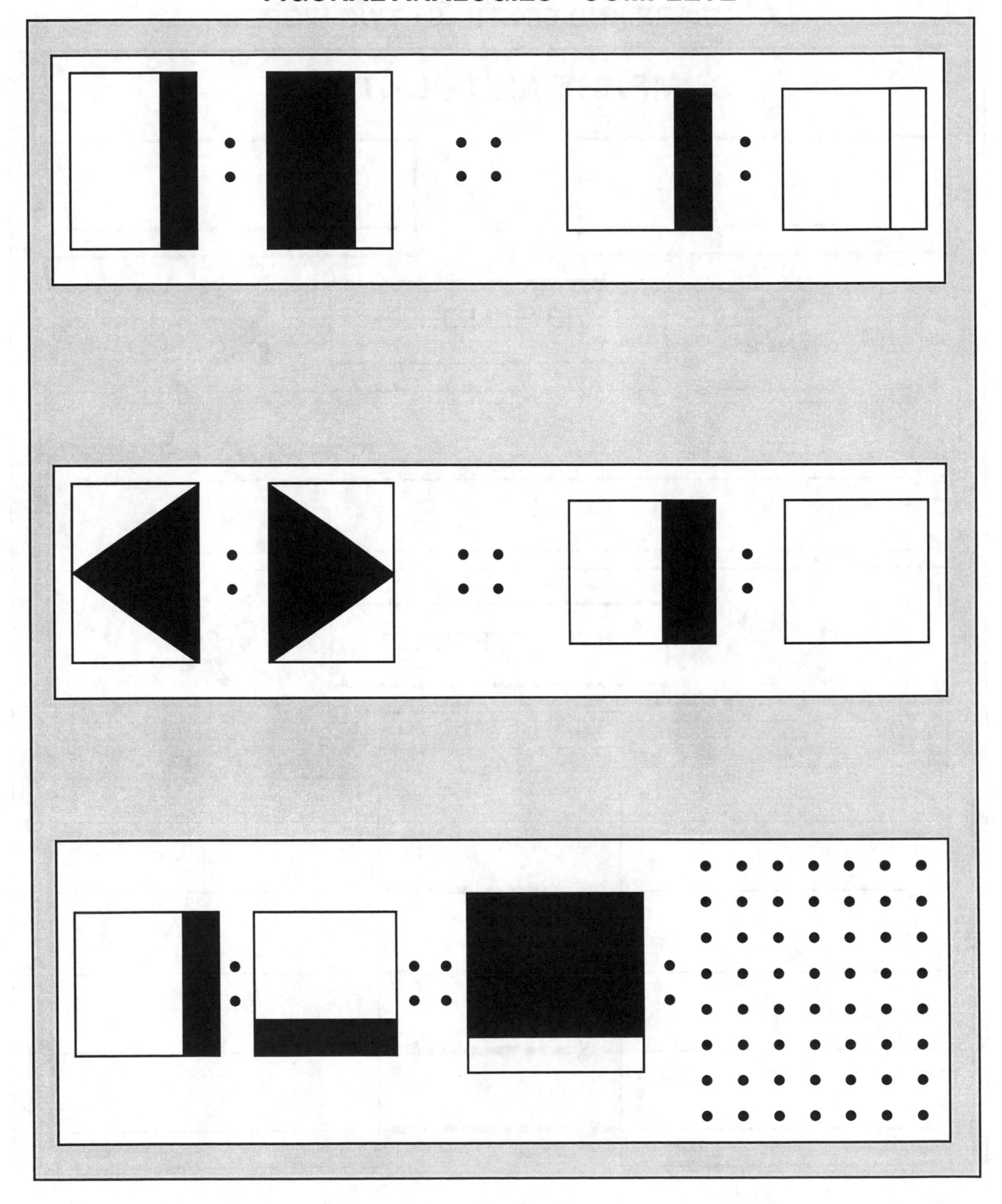

TRANSPARENCY MASTER 26

COMPARE AND CONTRAST TWO WORDS

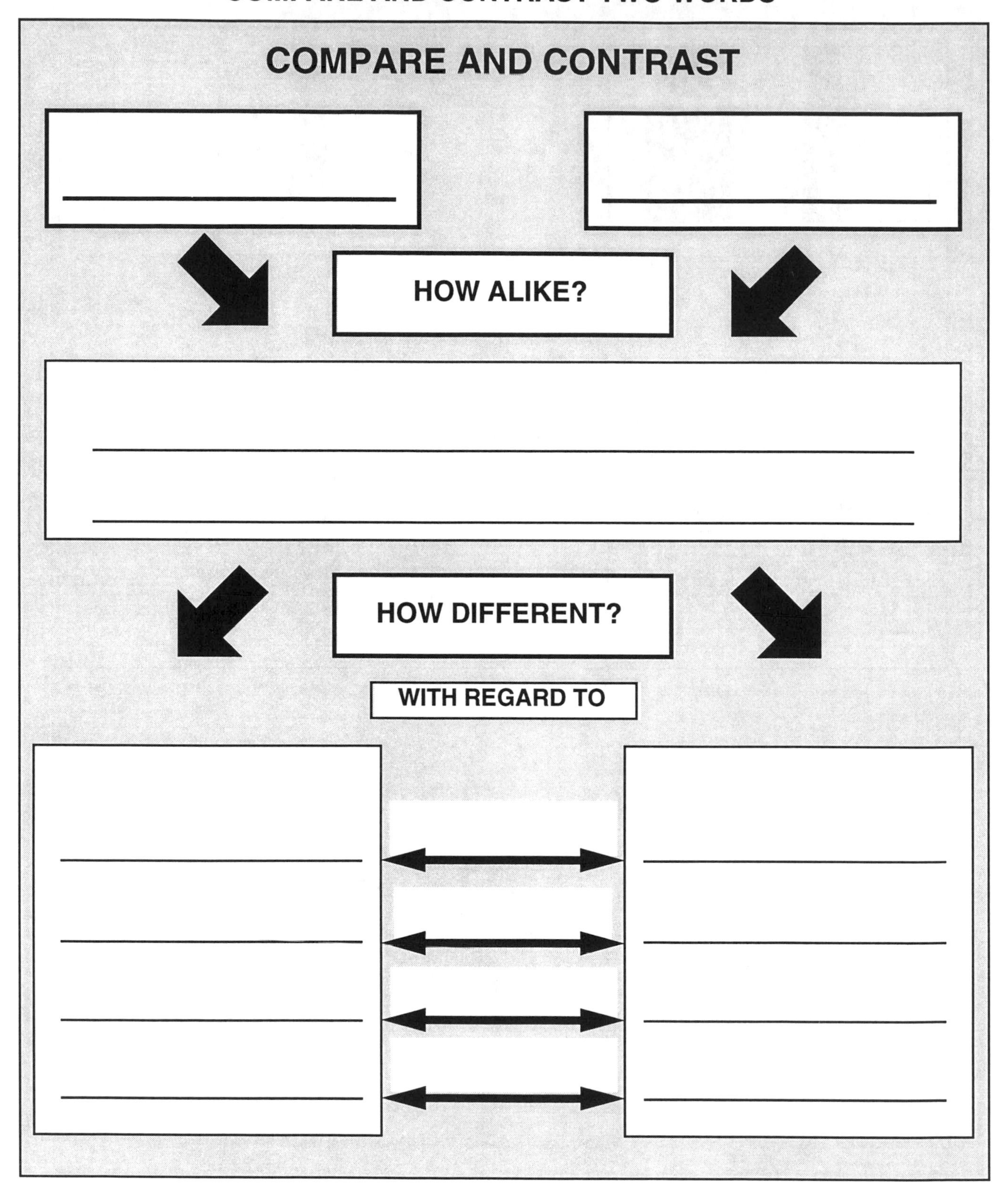

TRANSPARENCY MASTER 27

FINDING LOCATIONS

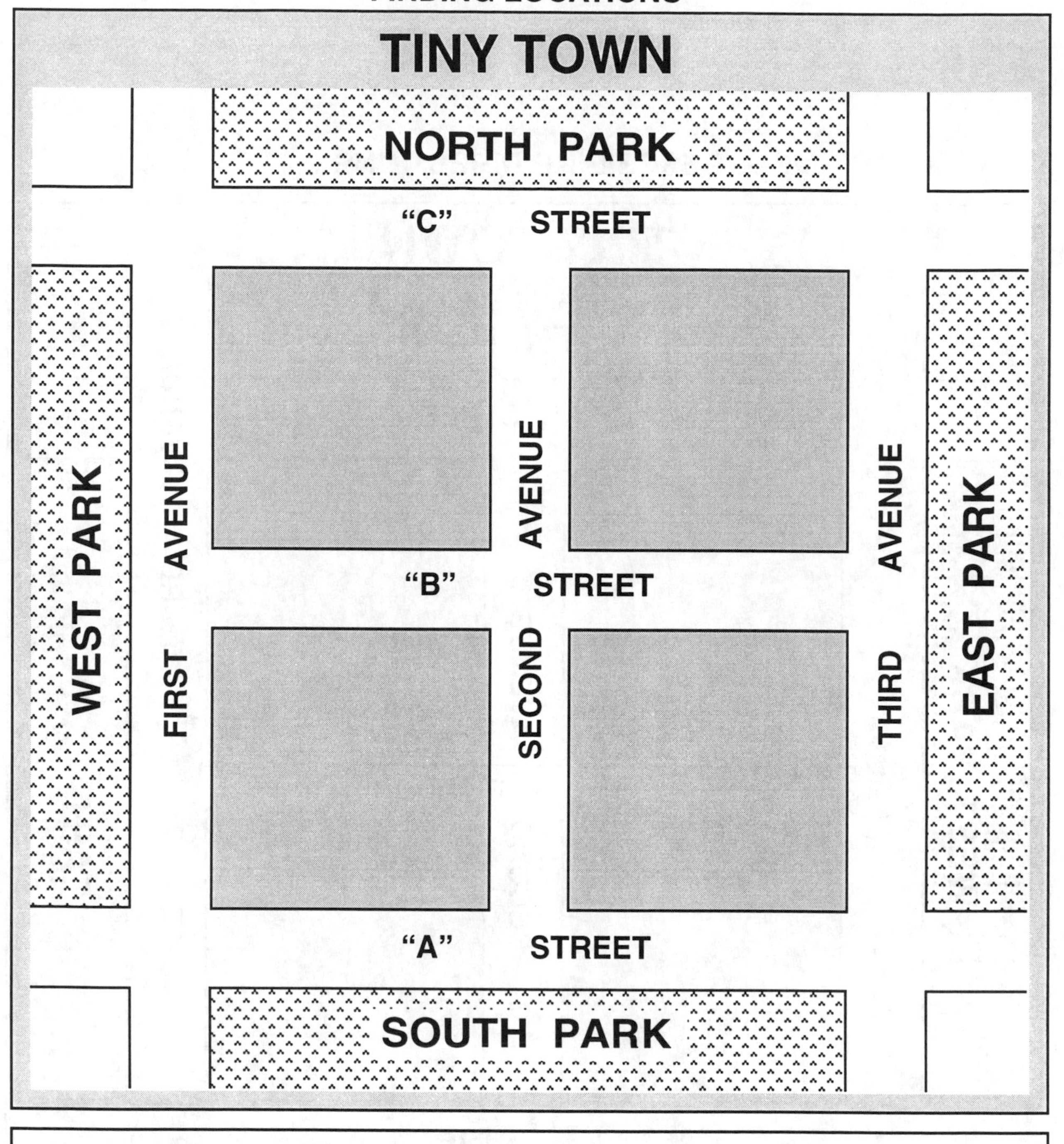

1. If you start at First Avenue and "A" Street and travel two blocks east, where will you be? ______________________
2. If you start at Third Avenue and "A" Street and travel one block north, where will you be? ______________________
3. If you start at Second Avenue and "B" Street and travel one block south, where will you be? ______________________

TRANSPARENCY MASTER 28

DESCRIBING DIRECTIONS

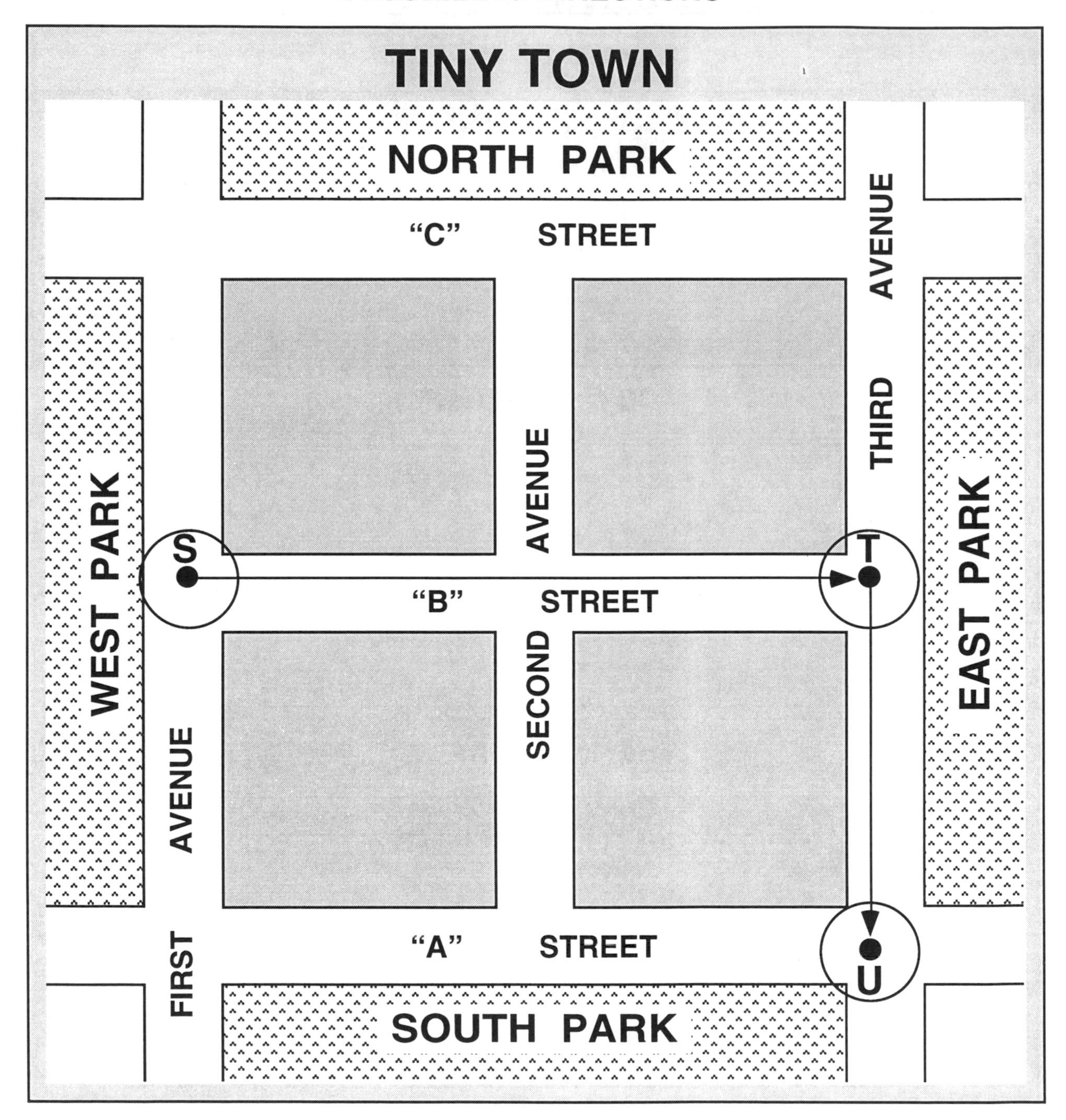

TRANSPARENCY MASTER 29

PART/WHOLE GRAPHIC ORGANIZER

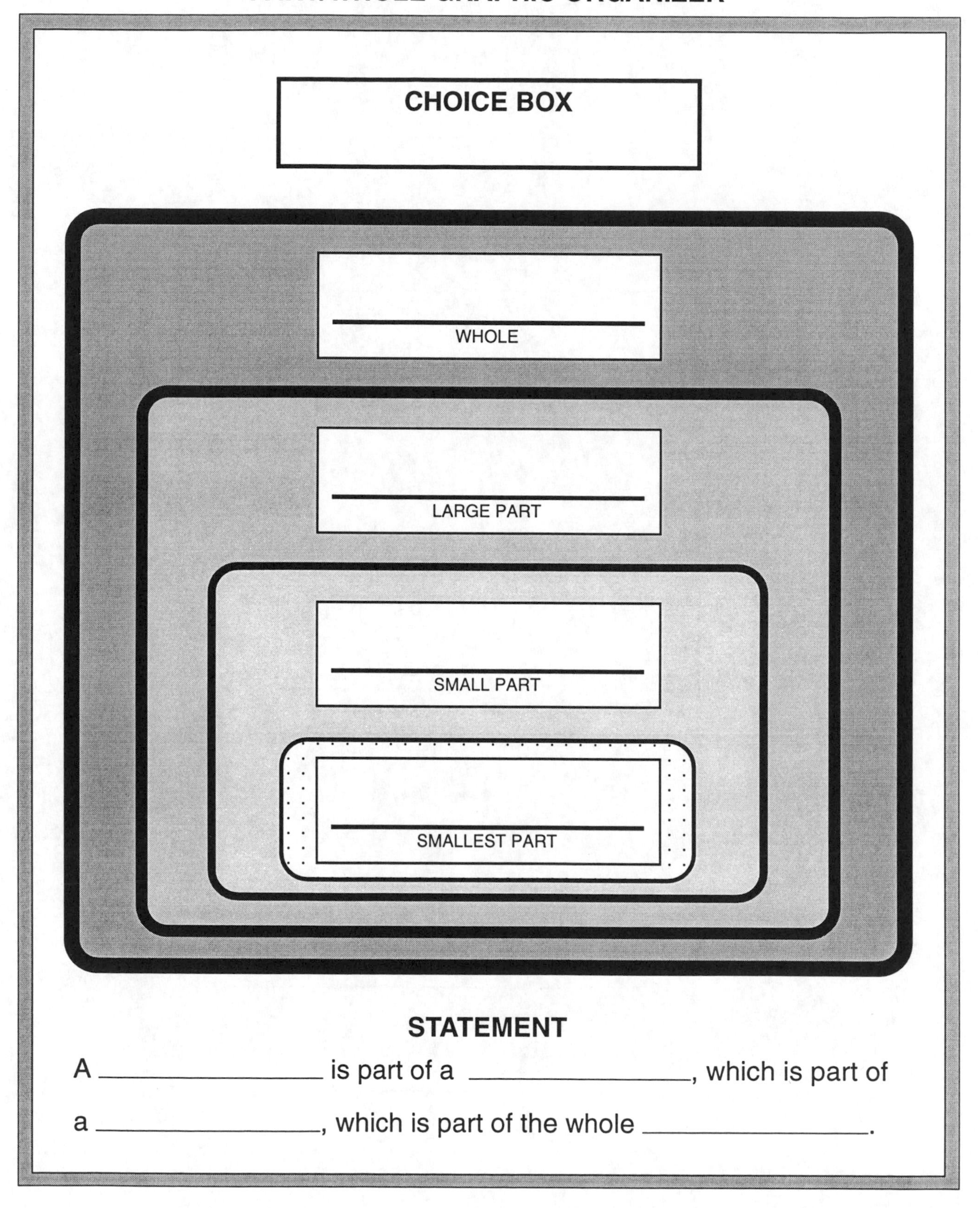

NOTES

NOTES